Conduction and breakdown
in dielectric liquids

Conduction and breakdown in dielectric liquids

Proceedings of the
5th international conference
organized by the Department of
Applied Physics of the
Delft University of Technology,
Noordwijkerhout, the Netherlands, 28-31 July 1975

Edited by
J. M. Goldschvartz
A. K. Niessen
W. Boone

Preface by
B. S. Blaisse

Springer Science+Business Media, B.V. 1975

Organizing committee
J. M. Goldschvartz (chairman)
Delft University of Technology

A. K. Niessen
Philips Research Laboratories, Eindhoven

W. Boone
N.V. Kema, Arnhem

Advising committee
I. Adamcsewski / Gdansk, Poland
B. S. Blaisse / Delft, the Netherlands
J. H. Calderwood / Salford, Great Britain
N. J. Félici / Grenoble, France
E. O. Forster / New Jersey, United States of America
Ch. Frei / Genève, Switzerland
T. J. Gallagher / Dublin, Ireland
Y. Inuishi / Osaka, Japan
T. J. Lewis / Bangor, Great Britain
F. Scaramuzzi / Rome, Italy

Originally published by Nijgh-Wolters-Noordhoff Universitaire Uitgevers B.V., Rotterdam in 1975

Softcover reprint of the hardcover 1st edition 1975

ISBN 978-90-298-0300-7 ISBN 978-94-010-1923-1 (eBook)
DOI 10.1007/978-94-010-1923-1

CONTENTS

EHD

PRE-BREAKDOWN

BREAKDOWN

MECHANISMS IN BREAKDOWN

CRYOGENIC LIQUIDS

MISCELANEOUS

PREFACE

The aim of the 5th international conference on conduction and breakdown of dielectric liquids is the same as that of the preceding conferences, namely to have a forum where all experts in this vast field can meet, can present their latest results and can exchange their ideas. It comes within this framework that the subjects of this conference cover the whole field of conduction and breakdown of polar and non polar liquids, including cryogenic liquids.

In comparison with many other international conferences this is a relatively small one. This fact has the great advantage that all participants can be lodged in the congress centre "De Leeuwenhorst" at Noordwijkerhout. The result is that everybody who wants to talk about a problem with a colleague can easily contact him and make an appointment.

When the organizing committee, whose members are J.M. Goldschvartz, chairman, A.K. Niessen and W.Boone began their task, the first idea was to organize the conference in the Aula complex of the Delft University of Technology, since the conference is sponsored by the Department of Applied Physics of the university. Then the participants had to be lodged not only in Delft but also in surrounding towns as the number of hotel beds in Delft is very limited. Thus, the decision to move into "De Leeuwenhorst" was practical and wise.

The subjects of the sessions of the conference are: mobility and conduction of charge carriers; high-field conduction; influence of impurities; injection of charge carriers; electrohydrodynamics; pre-breakdown; breakdown and mechanisms of breakdown; cryogenic liquids and of course also miscellaneous subjects with could not be covered by one of the above mentioned titles but are nevertheless important for the conference.

The contributed papers are printed in this Proceedings, which is given to all participants before the beginning of the conference. This was only possible because the organizing committee got through an enormous amount of work and the Delft University Press gave its full cooperation.

The quality and number of the papers ensures the success of the conference. Moreover, I hope the participants will come in touch with each other thus having a fruitful and enjoyable time.

B.S. Blaisse

ELECTRON TRANSPORT IN HIGH MOBILITY LIQUID HYDROCARBONS AND TETRAMETHYLSILANE

U. Sowada, G. Bakale[+], K. Yoshino[**] and W.F. Schmidt[*]

Introduction

Excess electron mobilities in liquid hydrocarbons exhibit a large spectrum of values ranging from 70 $cm^2V^{-1}s^{-1}$ for neopentane to 0.1 $cm^2V^{-1}s^{-1}$ for n-hexane at room temperature or from 400 $cm^2V^{-1}s^{-1}$ in liquid methane to $10^{-3}cm^2V^{-1}s^{-1}$ in liquid ethane at T = 111°K. In liquids with a low electron mobility the motion is thermally activated and at higher field strengths the mobility increases with field strength. The localized electron model has been applied to interpret these results [1,2]. Liquids exhibiting a large electron mobility ($>$ 10 $cm^2V^{-1}s^{-1}$) show a small temperature dependence of the mobility and a decrease of the mobility at higher field strengths. A model of quasifree electrons or band conduction seems to be applicable. Here we wish to discuss the relevant theoretical models and present some new data on tetramethylsilane.

Experimental Data

The mobility measurement technique, the electrical circuit and the parallel plate cells have been previously described [3]. At low field strengths the drift velocity v_D increases proportionally to the electric field strength F while above a certain critical field F_c the drift velocity increases approximately as $F^{0.5}$. In figure 1 the results obtained in methane, neopentane, and tetramethylsilane are displayed. For comparison also data of liquid argon, krypton, and xenon are shown.
The critical velocity ($v_c = \mu(o)\cdot F_c$) is comparable to the velocity of sound in the liquefied rare gases, which indicates the prevalence of elastic energy losses. In the molecular liquids the critical velocity is much greater than the velocity of sound. Here inelastic energy losses seem to be the determining factor.

*Hahn-Meitner-Institut für Kernforschung Berlin GmbH, Bereich Strahlenchemie, 1 Berlin 39, Germany
+Now: Case Western Reserve University, Cleveland/Ohio, USA
**On leave from Osaka University, Dpt.Electr. Engineering, Osaka, Japan

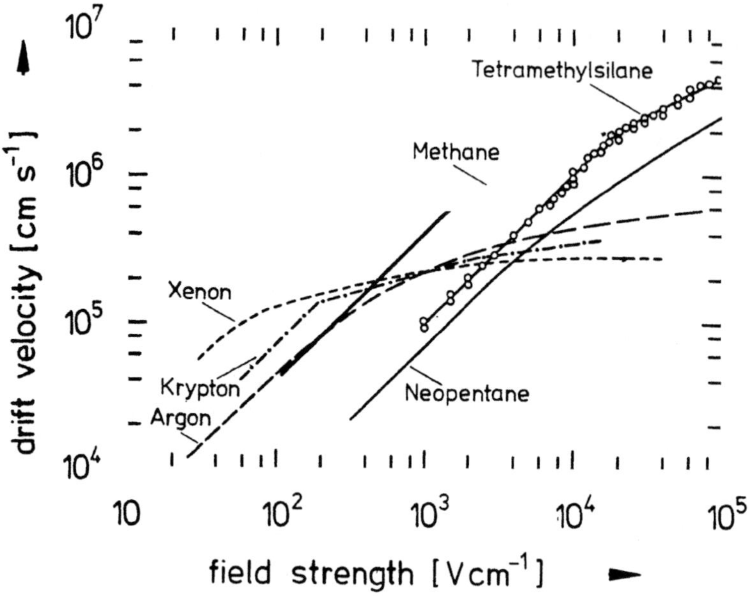

Fig. 1: Electron drift velocity as a function of the field strength. Argon: 85 K [4]; Xenon: 163 K [4]; Methane: 111 K [4]; Neopentane: 296 K [5]; Tetramethylsilane: 195 K, 296 K, this work. Krypton: 121 K,[4] and this work.

Models

In the classical description of electron motion through a system of scatterers the observed drift velocity is the result of acceleration by the electric field and energy losses due to elastic or inelastic collisions. The mobility is given by

$$\mu = \frac{e}{m} \cdot \tau \tag{1}$$

where τ is a time which depends on the electron energy and the scattering mechanism. At low field strengths the electrons will be in thermal equilibrium with the liquid and $\tau = \Lambda_1 / v_{th}$, where Λ_1 is the mean free path between collisions and v_{th} the mean thermal velocity. Taking into account the Maxwell velocity distribution gives for the mobility

$$\mu = \frac{2}{3} \left(\frac{2}{\pi \, m \, k_B T} \right)^{1/2} e \Lambda_1 \tag{2}$$

At higher field strengths the energy gain from the field will increase the mean electron energy above thermal energies. Equilibrium is obtained if

$$\left(\frac{\partial\langle\varepsilon\rangle}{\partial t}\right)_{\text{field}} + \left(\frac{\partial\langle\varepsilon\rangle}{\partial t}\right)_{\text{collisions}} = 0 \qquad (3)$$

A more detailed investigation on the motion of excess electrons in liquefied rare gases or polyatomic liquids was presented by Cohen and Lekner [6], and Davis et al. [7], respectively. They took into account the modification of the scattering process by the liquid structure.

The limit F_c of the low field region is related to energy loss by

$$f \approx \frac{1}{3}(e\, F_c\, \Lambda_o)^2 \,/\, (k_B T)^2\, S(0) \qquad (4)$$

Here f is the fractional energy loss per collision (for elastic scattering $f = \frac{2m}{M}$) and $S(0) = \frac{\Lambda_o}{\Lambda_1}$, the structure factor for thermal energies.

Λ_o is given by

$$\Lambda_o^{-1} = 4\,\pi\, a^2\, n \qquad (5)$$

with a the scattering length and n the number density of atoms or molecules. In principle, f is determined by eq. 4 if a is known. The scattering length a can be obtained from the value of V_o, the energy of the conduction state by the semiempirical method published by Fueki et al. [8]. These calculations have been done for the liquids of fig. 1 and the results are summarized in table 1.

Table 1

Experimental Data					Calculated Values					
Liquid	T[K]	μ^*	V_o^+	F_c^{**}	$\Lambda_1^{\S}$	$\Lambda_o^{\S}$	S(0)	f	f_{ela}	f/f_{ela}
Ar	85	400	-0.2	0.3	154	4.3	0.028	$3.7 \cdot 10^{-5}$	$2,7 \cdot 10^{-5}$	1,37
Kr	121	700	(-0.78)[1]	0.18	320	4.76	0.015	$1.5 \cdot 10^{-5}$	$1,3 \cdot 10^{-5}$	1,15
Xe	163	1900	-0.65	0.05	1000	3.6	0.0036	$0.15 \cdot 10^{-5}$	$0,8 \cdot 10^{-5}$	0,2
Me	111	400	-0.15	1.5	175	2.7	0.015	$40 \cdot 10^{-5}$	$7 \cdot 10^{-5}$	5,7
NP	296	70	-0.43	20	50	3.22	0.064	$330 \cdot 10^{-5}$	$1,5 \cdot 10^{-5}$	220
TMSi	296	100	-0.6	20	72	1.83	0.025	$274 \cdot 10^{-5}$	$1,2 \cdot 10^{-5}$	230

* $[cm^2 V^{-1} s^{-1}]$; + [eV]; ** $[kV\ cm^{-1}]$; § [Å]

V_o-values from W. Tauchert, this conference and [9]

1) estimated value by Raz and Jortner

Although these calculations have only approximate character, the different behavior of excess electrons in atomic and polyatomic liquids is clearly demonstrated. While in the liquefied rare gases elastic losses account for the observed mobility and the critical field F_c in polyatomic liquids the fractional energy loss f is much greater and transfer of energy to internal vibrations has to be considered. This would explain the similar behavior of neopentane and tetramethylsilane.

Acknowledgement: We thank "Deutsche Forschungsgemeinschaft" for financial support.

References

[1] W.F.Schmidt, G.Bakale, and U.Sowada, J.Chem.Phys. 61, 5275 (1974); [2] G.Bakale, U.Sowada, and W.F.Schmidt, 1974 Ann.Rep.Conf.Electr.Insul.Dielectr.Phenom.; [3] G.Bakale and W.F.Schmidt, Z.Naturforsch. 28a, 511 (1973); [4] L.S.Miller, S.Howe, and W.E.Spear, Phys.Rev. 166, 871 (1968); [5] G.Bakale and W.F.Schmidt, Chem.Phys.Lett. 22, 164 (1973); [6] M.H.Cohen and J.Lekner, Phys.Rev. 158, 305 (1967); [7] H.T.Davis, L.D.Schmidt, and R.M.Minday, Phys.Rev. A3, 1077 (1971); [8] K.Fueki, D.F.Feng and L.Kevan, Chem.Phys. Lett. 13, 413 (1972); [9] R.A.Holroyd and M.Allen, J.Chem. Phys. 54, 5014 (1971)

A MAXIMUM BOUND FOR ELECTRONIC MOBILITIES IN LIQUID HYDROCARBONS

M.R. Belmont*

Introduction

Calculations have been made [1→3] of the delocalised or quasi-free electron mobility in highly purified liquid hydrocarbons. For the evaluation of the distribution function and mobility the Cohen Lekner [4] analysis of plane wave scattering was used, the resulting expression for μ_o is

$$\mu_0 = \frac{2}{3}\left(\frac{2}{\pi m kT}\right)^{\frac{1}{2}} \frac{e}{n\, S(o)\sigma_1} \quad (1)$$

where n is the number density of the liquid molecules, k is Boltzmann's constant, T the absolute temperature, m the electron rest mass, e the electronic charge, σ the scattering cross section, S(o) is related to the pair correlation function of the liquid and brings in the allowed phonon excitations of the medium. It's value is given by $S(o)=nkTK_T$ where K_T is the isothermal compressibility of the liquid. Reasonable agreement was obtained with experimental results for the spherical isotropic molecules of neopentane [5, 6, 9] and methane [7]. However the technique failed when applied to the results for the linear molecule n-hexane [5,8]. A weakly temperature dependent drift mobility was predicted, having the value of 1.7×10^{-2} $m^2 v^{-1} sec^{-1}$ [1] or 1.5×10^{-2} $m^2 V^{-1} sec^{-1}$ [3] at 300°k. This was in strong contrast to the experimentally observed mobility μ [5,8] which has a magnitude of approximately 10^{-5} $m^2 v^{-1} sec^{-1}$ at 300°k and exhibits an exponential dependence on temperature as given by equation (2)

$$\mu = A \exp\left(\frac{-\phi}{kT}\right) \quad (2)$$

A is essentially temperature independent compared with the exponential term, and ϕ is the activation energy. It was noticed [1→3] that by empirically equating the calculated value for μ_o with A in equation (2) and using the experimental value for ϕ the correct value of μ could be obtained. The authors [1→3] claimed that this procedure should be valid for the other n-alkanes as well as n-hexane. Unfortunately activation energies are not available for other normal paraffins to ensure that this approach does not represent merely an isolated coincidental agreement.

Capture Cross Section

One feels that the agreement for n-hexane may be purely fortuitous for the following reasons. Firstly, most authors

*Glamorgan Polytechnic, Llantwitroad, Treforst, Pontypridd, Mid Glam

in this field assume that the form of the observed temperature dependence of strongly indicates that it is controlled by some thermaly activated release process involving localised levels. This may be a) quasi-free electron inter-trap motion via an energy band or b) hopping i.e. activated tunnelling between sites as discussed by Mott [10] , Davis and Mott [11] Miller and Abraham[12] and others. Both classes of process, predict the temperature dependence given by equation (2) but A would not be solely determined by μ_o in either case. Secondly if one assumes process a) which is favoured by many workers [1→3, 5→8, 13] then equating μ_o with A requires that the product $\nu\tau_o$ is unity. Here $1/\tau_o$ is the attempt to escape frequency and ν is the trapping frequency. $1/\tau_o$ has been reasonably assumed to correspond to phonon frequencies [3] i.e. $\tau_o \sim 10^{-13}$ sec. This then requires that $\nu \sim 10^{13}$ sec^{-1} . Such a high capture rate requires unreasonably large values of active capture cross section for the proposed neutral liquid traps. This is because the dipole moment required to quarantee localisation of an electron is surprisingly large [27]. The chance of a single molecule exhibiting such a moment as a result of transient thermal distortion destroying the cancellation of group moments or of several molecules simultaneously aquiring the required moment to produce the same overall effect is very small. Thus the trap density must be low hence δ would need to be very large. Finally it is also interesting to note that the distance to capture is equal to or less than the momentum mean free path values quoted for μ_o .
It seems from the above reasoning one must conclude that $\nu \ll 10^{13}$ sec^{-1}, with the result that any quasi-free electron mobility for n-hexane must be much lower than the calculated figure of approximately $1.5 \times 10^{-2} m^2.V^{-1}-sec^{-1}$ [1,3].

Maximum bound on μ_o

If one accepts for the moment that delocalised electrons can exist in n-alkanes then the following simple calculation sets an upper limit to μ_o of 1.4×10^{-3} m^2 v^{-1} sec^{-1} in n-hexane at 300°k. Schmidt and Allen [5] found that the drift mobility was independent of applied field strength $\mathcal{E}$ up to 8.3×10^6 $V.m^{-1}$ in n-hexane and 1.4×10^7 $V.m^{-1}$ in pentane [5] . Then unless some variation in the trap parameters is exactly cancelling a variation in μ , which seems improbable, one can assume μ_o is stress independent in the two alkanes up to these field strenghts. There will be some stress at which μ_o becomes stress dependent as the mean value of the drift component V_d of the electrons velocity becomes comparable with the thermal component V_c . The absence of stress dependence in the results of Schmidt and Allen means that V_d is small compared to V_c up to the maximum field strengths used. Bearing in mind their order of experimental error one can say with confidence that at 300°k

$$\frac{Vd}{Vc} \leqslant 0.1 \qquad (3)$$

Now by definition $V_d = \mu_0 \mathcal{E}$ and V_c can be obtained when inequality (3) holds from the expression

$$\frac{3}{2} kT = \frac{1}{2} m^* V_c^2 \qquad (4)$$

where m^* is the electron's effective mass. Equation (4) assumes the classical limit of the Fermi Distribution which is certainly true for the very dilute excess electron populations in the insulating liquids considered.
To establish a minimum value for m^* one first notes that the molecular overlap in liquid n-paraffins is weak. Thus the resulting energy bands will be narrow. The electron's wave vector thus varies slowly with energy and consequently m^* is large. Hence m^* is unlikely to be smaller than in a narrow band single crystal organic solid, so one can reasonable write

$$\frac{m}{m^*} \lesssim 1 \qquad (5)$$

near the bottom edge of any quasi free electron band in n-alkanes. In (5) m is the electronic rest mass.
Substituting for V_d and V_c into inequality (3) and using the maximum stress reached by Schmidt and Allen gives an upper bounds for μ_0 of 1.4×10^{-3} $m^2.v^{-1}.sec^{-1}$ in n-hexane and 8.2×10^{-4} $m^2.V^{-1}.sec^{-1}$., in n-pentane, both at 300°k.
The method used is obviously approximate as the in-equalities (2) and (5) are rather flexible. However as maximum allowable values have been used the upper bound on μ_0 is too high rather than too low and thus represent the very largest possible value of μ_0 which is one tenth the value calculated for n-hexane [1→3].

Alternative Model
With such a low maximum mobility value for supposedly highly delocalised electrons one is forced to re-examine the use of a quasi free scattered plane wave approach to the problem [1→3]. Exactly the same reappraisal was required for amorphous solids when workers in that field tried to fit conventional crystalline band models to their results. The situation there was resolved by the randomly perturbed periodic potential theories which lead to the current models [11,14→16] for electron energies in the amorphous phase. These same models will now be considered as alternative descriptions for conduction in n-alkanes. The presentation due to Davis and Mott [11] and Mott 15 will form the basis of the discussion, although this treatment is not wholey quantative it gives a good physical insight into the problem. In any case the spread in reported experimental data, particularly the activation energies [5, 8] in n-hexane is too large to justify detailed quantative treatment.
The basic model is of an increasingly more diffuse extension of the energy band beyond the conventional band edge for crystalline solids, as shown in figure 1a.

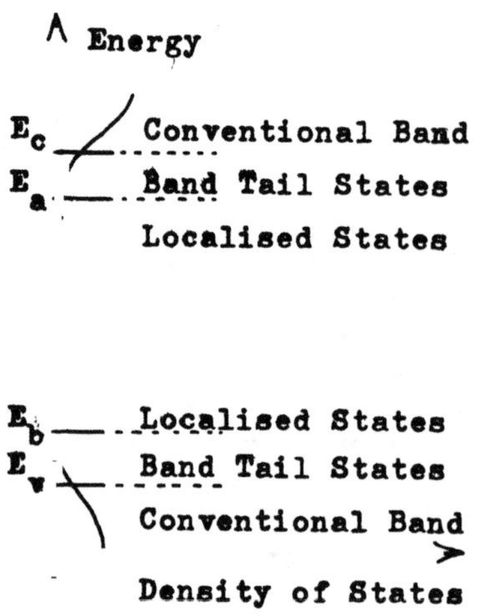

Fig. 1a

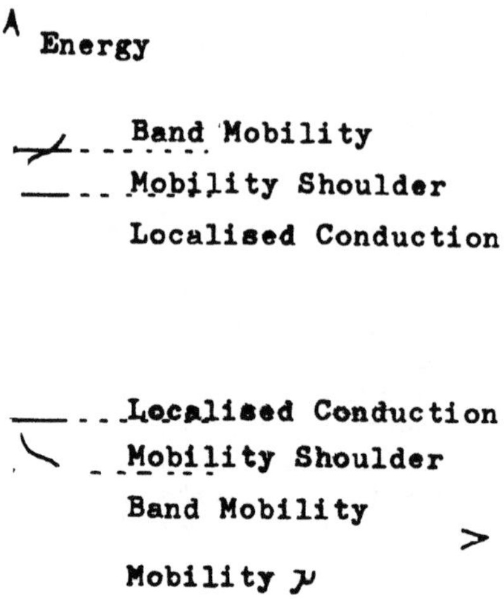

Fig. 1b

As the density of states falls away from the band edge so does the inter-site tunnelling probability. Thus Mott suggests that within the regions $E_c \rightarrow E_a$ and $E_v \rightarrow E_b$ the conduction is still delocalised in character, however the effective mass rises very rapidly as one moves out into the gap region. The consequences are the mobility shoulders Fig. 1b, where the delocalised mobility falls by orders of magnitude. Outside E_a and E_b delocalisation is supposedly no longer possible, charge motion is then via thermally activated inter-site tunnelling (activated hopping). For a wide band insulator the Fermi level will be far removed from the band edges and the diffuse or 'tail states' provided a very large density of carriers is not injected. Now even at very high field strengths few carriers gain energies above a few tenths of an electron volt. Thus the majority of charge transport will occur in the 'tail states' and not the band proper. Davis and Mott [11] suggested that for electrons in calcenogide glasses μ_o in the region $E_c \rightarrow E_a$ is of the order of 10^{-3} $m^2.V^{-1}.sec^{-1}$, at $300^{\circ}k$. It is interesting to note that this is the order of magnitude of predicted by inequality (3).
In terms of this model the experimental temperature dependence of μ would be explained in two ways:
a) At moderate and high temperature by a capture and release mechanism from localised sites lying between E_a and E_b with delocalised inter-site motion via the mobility shoulders, this is described by equation (6)

$$\mu = \mu_t \beta \exp\left(-E/_{kT}\right) \tag{6}$$

μ_t is the mobility within region $E_c \rightarrow E_a$, E the effective site depth below E_c and β is dependent on the capture parameters and the distribution in energy of localised states.
b) At low temperatures transport would be by thermally

activated tunnelling, without delocalised electron motion between sites. The mobility μ is then given by

$$\mu = R \exp(-W/_{kT}) \tag{7}$$

Due to the low tunnelling probability below E_a the inter-site transfer term R in equation (7) is very much less than $\mu_{t\beta}$. The hopping activation energy W, is often very low, typically 0,01 eV [17] for amorphous aluminium oxide, silicon oxides and silicon nitride. It has been suggested [15] that for hopping in liquids the continuously changing atomic or molecular configuration removes the requirement for an activation energy i.e. W=0. The order of magnitude of μ is then roughly that for ionic mobility in liquids.
The magnitude of the activation energy 0.18 ev→0.2 ev [5, 8] and the absolute value of μ in n-hexane tends to indicate equation (6) behaviour. Furthermore the change over to hopping at low temperatures often observed in amorphous solids does not seem to be present in liquid n-hexane for the activation energy remains constant down to the freezing point at 200°k.

Local Anistropy
Having suggested the proposed mechanism for mobility in linear alkanes one might ask why spherical molecular liquids like tetramethylsilane and methane do not exhibit similar behaviour instead of the highly delocalised conduction which experimental results point to [5, 7]. The dilemma is resolved by recognising that localised states and hence the band extensions or 'tail states' leading to them, only appear in a disordered system when the product of a potential fluctuation and the square of its effective radius are above a minimum value. Gubanov [20] shows that this minimum is lowered if there is any anistropy in the medium. For systems exhibiting any degree of chain or layer structure the density of localised levels may be many orders of magnitude larger than for wholey isotropic materials. This argument has been used to explain why 'tail states' are found in amorphous chain or layered semiconductors [21→23] while there is some doubt about their existance in isotropic amorphous silicon and germainium [24, 25]. Vapour pressure measurements 26 indicate that linear alkanes exhibit substantial rotational hinderance, the molecules packing locally along the chain axis. Thus an excess electron would 'see' a locally anistropic environment. The density of localised sites below E_c is thus much larger for linear alkanes than for the spherical molecular liquids. Hence the alkanes would be most likely to exhibit a trap controlled mobility, while the spherical molecules would exhibit quasi free electron transport. The low value for μ_0 in n-hexane predicted by inequality (3) would fit in with the very high effective mass values expected in the 'tail states' between E_c and E_a.
The anistropy argument also predicts the variation in observed in mixtures of n-hexane and neopentane, for as the concentration of hexane falls so does the degree of local anistropy.

References

1. K. Fueki, D.F. Ferg and L. Kevan, Chem.Phys.Lett. 13, 616, 1972
2. K. Fueki, Canad.J.Chem. 50, 3370, 1972
3. H.T. Davis, L.D. Schmidt and K.M. Minday, Chem.Phys. Lett. 13, 413, 1972
4. M.H. Cohen and J. Lekner, Phys.Rev. 158, 305 1967
   J. Lekner, Phys.Rev. 158, 130, 1967
5. W.F. Schmidt and A.O. Allen, J.Chem.Phys. 52, 4788, 1970
6. R.M. Minday, L.D. Schmidt and H.T. Davis, J.Phys.Chem. 76, 442, 1972
7. W.F. Schmidt and G. Bakale, J.Chem.Phys. 17, 617, 1972
8. R.M. Minday, L.D. Schmidt and H.T. Davis, J.Chem.Phys., 1971
9. J.P. Dodelet and G.R. Freeman, Canad.J.Chem. 50, 2667, 1972
10. N.F. Mott, J.Non-Cryst.Sol. 1, 17, 1967
11. E.A. Davis and N.F. Mott, Phil.Mag. 22, 903, 1970
12. A. Miller and E.Abrahams, Phys.Rev. 120, 745, 1960
13. Mc.Cubbin and Gurrey, J.Chem.Phys. 43, 983, 1965
14. M.H. Cohen, H. Fritzsche and S.R. Ovshinsky, Phys.Rev. Lett. 22, 1065, 1969
15. N.F. Mott, Phil.Mag. 24, 1, 1971
16. A.I. Gubanov, 'Quantum Electron Theory of Amorphous Conductors', Consultants Bureau N.Y., 1965
17. A.K. Jonscher, Thin Solid Films, 1, 213, 1967
18. P.G. Le Comber and W.E. Spear, Phys.Rev.Lett. 25, 509, 1970
19. E.L. Rossiter and G. Warfield, Dept. of Elect. Engg. Device, Phys.Lab., Princeton Univ.Tech.Rep. no. 10, 1967
20. A.I. Gubanov, Sov.Phys.-Semicord, 6, 1202, 1973
21. H.P.D. Lanyon, Phys.Rev. 130, 134, 1963
22. B.T. Kolomiets et.al., J.Non-Cryst.Sol. 4, 45, 1970
23. A.E. Owen and J.M. Robertson, J.Non-Cryst.Sol. 2, 40, 1970
24. T.M. Donovan et.al., Phys.Rev.Lett. 22, 1058, 1969
25. D.T. Pierce and W.E. Spicer, Phys.Rev.Lett. 27, 1217, 1971
26. L.H. Thomas, J.Chem.Soc. (A), 2609, 1968
27. M.R. Belmont, Thin Solid Films, 22, 281, 1974

ELECTRON DRIFT MOBILITY MEASUREMENTS IN LIQUID 3-METHYLPENTANE AND THE EFFECT OF IMPURITIES

I. Kalinowski, J. G. Rabe, and W. F. Schmidt*

Introduction

For the elucidation of the electron transport process in disordered systems measurements of the mobility of excess electrons in many systems have been performed. Several methods for the determination of the drift velocity of excess charges in an electric field are available [e.g. 1]. Bakale and Schmidt used a burst of high energy X-rays to produce a homogeneous distribution of charge carriers of either sign between the parallel plates of an ionization chamber [2].
Under the influence of an electric field the charge carriers drift towards their respective electrodes giving rise to a current which decreases in time. In liquid hydrocarbons a fast and a slow decay have been observed which were attributed to the drift of electrons and positive ions. Sometimes the purity of the liquid is not sufficient and electron attachment to impurities occurs. The portion of the current which is carried by the electrons does not decrease linearly with time but the apparent time constant of the decay may be still influenced by the strength of the electric field. It is the purpose of this paper to show how drift mobilities can be obtained in these cases.

Theoretical Considerations

After a short burst of high energy X-rays is applied to the liquid a uniform density $n_{\pm}$ of positive and negative charge carriers exists between the plates of the measurement cell. With an applied voltage V an ionization current i(t) is measured which is given by

$$i(t) = n_{\pm}(t) \quad e_o(\mu_- + \mu_+) \cdot \frac{V}{d} \cdot q \tag{1}$$

where e_o means the electronic charge, $\mu_{\pm}$ the mobility of the positive or negative carrier, respectively, d the electrode separation and q the electrode area. In the case of electronic conduction

$$\mu_{el} \gg \mu_+$$

so that eq. 1 reduces to

* Hahn-Meitner-Institut für Kernforschung Berlin GmbH, Bereich Strahlenchemie, 1 Berlin 39, Germany

$$i_{el}(t) = n_{el}(t)\, e_o\, \mu_{el}\, \frac{V}{d} q \qquad (2)$$

Electrons may be lost for the conduction process due to three different processes:

1. Excess electrons are neutralized on the anode
2. Volume recombination between electrons and positive ions occurs
3. Electrons are scavenged by impurities and the mobility of the negative ion is much smaller than the electron mobility.

In case 1 the current decreases linearly in time as

$$i_{el}(t) = i_{el}(o)\,(1 - \frac{\mu_{el} V}{d^2} \cdot t) \qquad (3)$$

The decrease of $n_{el}(t)$ due to the processes 2 and 3 is described by

$$\frac{d\, n_{el}}{dt} = - k_1\, n_{el}\, n_+ - k_2\, n_{el}\, n_s \qquad (4)$$

where k_1 and k_2 are the rate constants for recombination and attachment, respectively, n_s is the concentration of scavengers. The relative importance of the two processes can be estimated for the conditions of our experiments. It turned out that $k_1\, n_{el}\, n_+$ was always several orders of magnitude smaller than $k_2\, n_{el}\, n_s$ and, therefore, recombination was neglected. Equation 4 becomes

$$\frac{d\, n_{el}}{dt} = - k_2\, n_{el}\, n_s \qquad (5)$$

which has the solution

$$n_{el}(t) = n(o)_{el}\, e^{-k_2 n_s t} \qquad (6)$$

The electron current in the presence of scavenging is then given by

$$i_{el}(t) = i_{el}(o) e^{-k_2 n_s t} \,(1 - \frac{\mu_{el} V}{d^2}\, t) \qquad (7)$$

If the drift time $d^2/\mu_{el} V$ is comparable to $1/k_2\, n_s$ then the decay will not be linear but will still depend on V. Measurements at increasing voltages V give decay curves $i_{el}(t)$ with decreasing decay times. The constants $k_2 n_s$ and μ_{el} are determined by a curve fitting procedure. We choose a value of μ_{el} and plot

$\ln \frac{i_{el}(t)}{i_{el}(o)\,(1 - \frac{\mu_{el} V}{d^2}\, t)}$ versus t for the different voltages V.

With the correct μ_{el} a set of parallel straight lines is obtained, from which $k_2 n_s$ can be determined.

Experimental

The electrical circuit and the cell have been previously described[2]. Electrode distances varied between 0.1 and 0.05 cm. Purification of the 3-methylpentane was effected by chromatography through columns of activated silica gel, preirradiation under high vacuum with ^{60}Co-γ-rays and further degassing by trap to trap distillation in a high vacuum. The measurements were performed in the temperature range 258 K to 371 K and with electric field strengths up to 250 kV cm^{-1}.

Results and Discussion

The decay of the electronic component of the ionization current was recorded oscillographically and in figure 1 typical data obtained in liquid 3-methylpentane at 43°C are shown. The solid lines represent the dependence according to eq. 7 with a mobility of μ_{el} = 0.34 $cm^2V^{-1}s^{-1}$ and k_2n_s = 2.5 x 10^5 s^{-1}. The values for k_2 n_s varied between 10^5 and 6 x 10^5 for different preparations. Scavenging rate constants have been measured for several solutes in a number of solvents. Values up to 10^{14} l $mole^{-1}$ s^{-1} have been reported [e.g. 3]. Taking 2.5 x 10^{12} l $mole^{-1}s^{-1}$ (which may be realistic for our experiments) we estimate the order of magnitude of the residual impurity concentration to 10^{-7} mole/l or 6 x 10^{13} cm^{-3}. This figure should be compared to the

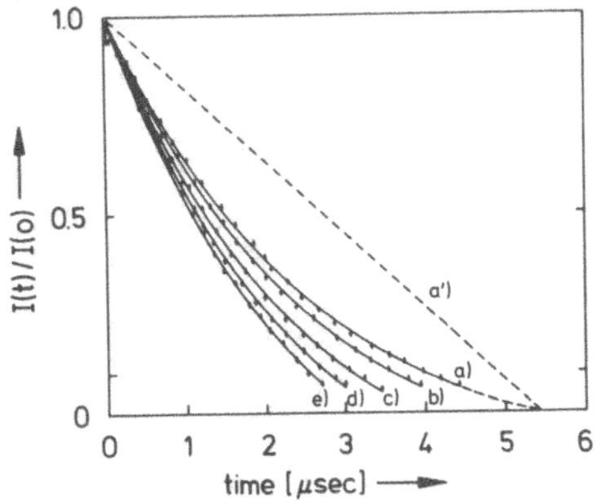

Fig. 1: Decay of the electron current in 3-methylpentane at t = 43°C
a) 57.1 kV cm^{-1}; b) 66.7 kV cm^{-1}; c) 76.2 kV cm^{-1}; d) 85.7 kV cm^{-1}; e) 95.2 kV cm^{-1};
a') the linear decay without attachment of curve a)

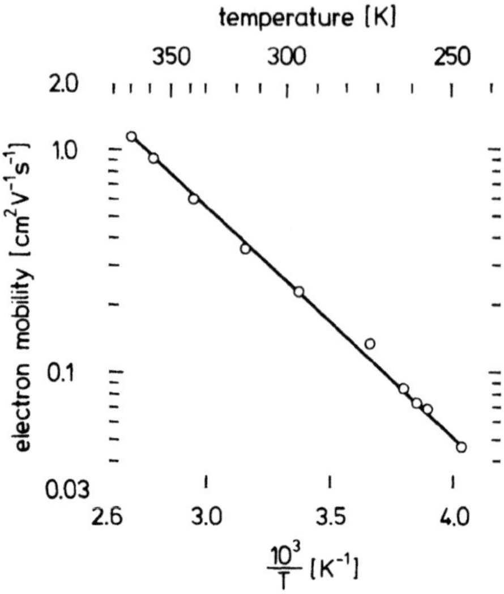

Fig. 2: Temperature dependence of the electron mobility in 3-methylpentane

number of 3-methyl pentane molecules 4 x 10^{21} cm^{-3}. A very small concentration of electron attaching impurity is sufficient to cause deviations from the linear current decay or in other words the liquid has to be purified to such an extent so that less than 1 scavenger molecules remains in 10^8 solvent molecules.

The temperature dependence of the mobility is shown as an Arrhenius plot in figure 2. The activation energy was determined to E_a = 0.2 eV. A trapping model seems to be applicable. Some evidence for field dependent mobilities above 200 kV cm^{-1} was obtained.

References. [1] A.Hummel and W.F.Schmidt, Radiat.Res.Rev. 5, 199 (1974); [2] G.Bakale and W.F.Schmidt, Z.Naturforsch. 28a, 511 (1973); [3] W.F.Schmidt, HMI-Report B-156 (1974).

HOLE CONDUCTIVITY IN LIQUID HYDROCARBONS

M. P. de Haas, J. M. Warman and A. Hummel*

Initial indications that the positive ion ("hole") formed on ionization of liquid saturated hydrocarbons by high energy radiation might have a mobility significantly larger than that expected for a molecular ion came from steady-state and optical pulse radiolysis studies of cyclohexane solutions. The formation of a high mobility hole in cyclohexane was recently confirmed by results obtained[1] using a microwave absorption technique to measure the radiation induced conductivity in this liquid on a nanosecond timescale. We report here a brief account of further studies of the radiation induced conductivity in liquid hydrocarbons.

The basis of the present method of studying ion formation in pulse irradiated liquid hydrocarbons is the attenuation of microwaves which results when propagation occurs in a conducting medium. Briefly, microwaves are incident upon a cell, containing the liquid of interest, which consists of a length of X band waveguide closed at one end by a glass transmission window and at the other by a short circuiting metal plate. The level of microwave power reflected from the cell is monitored as a function of time immediately following irradiation of the liquid with pulses of 3 MeV electrons from a Van de Graaff accelerator.

For absorptions less than 5% the fractional change in reflected power is directly proportional to the conductivity σ of the liquid and for the present conditions is given by

$$\Delta P/P_0 = 57\sigma \qquad (1)$$

for σ in units of $\Omega^{-1}m^{-1}$, at the frequency of maximum absorption (8.45 to 8.60 GHz depending upon the dielectric constant of the liquid).

The conductivity is given by

$$\sigma = Nue \qquad (2)$$

where N is the number density of ion pairs, u is the sum of the mobilities of the positive and negative ions and e is the electronic charge. If no decay of the ions occurs during the pulse then N is given by GDQd/100 where G is the yield of ion pairs per 100 eV absorbed, D is the absorbed dose in eV per gram per nanocoulomb (average value $1.35 \pm 0.15 \times 10^{15}$ in the present experiments), Q is the integrated beam current in nanocoulomb (measured by deflection of the beam onto a coaxial target connected to an electrometer) and d is the density of the medium in gram/cm^3. The product of the yield of ion pairs and the mobility (in units of $cm^2V^{-1}s^{-1}$) may then be obtained from

*Interuniversitair Reactor Instituut, Mekelweg 15, Delft, The Netherlands

Table I

Liquid	CH				MCH	IO
Temperature ^{o}C	22	41	61	83	22	22
G_{fi} $(100\ eV)^{-1}$	0.16[4]	0.20[4]	0.24[4]	0.29[4]	0.12[5]	0.36[6]
$u(-)^{a}$ $(cm^{2}V^{-1}s^{-1})$	0.22[2]	0.30[4]	0.40[4]	0.54[4]	0.068[7]	5.4[2]
u(-) (" ")	0.20	0.25	0.36	0.45	0.066	6.4
$u(+)x10^{3}$ (" ")	15	16	18	13	5.8	1.0
$\bar{u}(S^{\pm})x\ 10^{3}$ (" ")	0.8	0.8	1.0	1.0	1.1	1.1

(a) values of reference 4 normalized to 0.22 at 22°C.

$$Gu = 80 \quad \Delta P/P_{0}Qd \qquad (3)$$

Because of the large range of Gu values measured, the same pulse width could not be used throughout but was chosen, within the available range of 2.5 to 250 ns, to be as short as possible while still resulting in a readily measureable absorption signal with a lifetime considerably longer than the pulse width.

The liquids cyclohexane (CH), methylcyclohexane (MCH) and iso-octane (IO) were purified by distillation on a 90 theoretical plate column, deaerated by bubbling with dry N_2, evacuated on a vacuum line and allowed to stand over sodium-potassium alloy for several days before use.

From the fractional absorption of microwave power following pulse irradiation of pure CH, MCH and IO at 22°C values of Gu = $G[u(-) + u(+)]$, where u(-) and u(+) are the electron and hole mobilities, of 3.5×10^{-2}, 8.6×10^{-3} and $2.30\ cm^{2}V^{-1}s^{-1}(100eV)^{-1}$ respectively were obtained. If geminate ion pairs do not contribute significantly to the total yield of ions on the timescale of the present observations and, if the major charge carrier is the electron as indicated by the results which follow, then the above values of Gu should be equal to G_{fi} u(-) where G_{fi} is the yield of free ions. Values of G_{fi} and u(-) determined by other workers are listed in Table I. The values of G_{fi} u(-) of 3.5×10^{-2}, 8.2×10^{-3} and $1.95\ cm^{2}V^{-1}s^{-1}(100eV)^{-1}$ calculated from the literature data do in fact agree quite well with the Gu values obtained for the pure liquids.

The rate constants for reaction of excess electrons with SF_6 in the liquids studied have been found[2] to be greater than $10^{12}mol.dm^{-3}s^{-1}$ (the rate constant for MCH is presumed to be approximately equal to that found in n-hexane, a liquid with a similar electron mobility). In the presence of 10^{-2}M SF_6 the electron lifetime will therefore be reduced to less than 0.1 ns and, if the electron is the major charge carrier the end-of-pulse value of Gu, = $G[u(+) + u(S^{-})]$ where $u(S^{-})$ is the mobility of the molecular ion resulting from electron capture by SF_6, will be considerably lower than in the pure liquid. A decrease by a factor of approximately ten in Gu on addition of SF_6 to CH has been previously observed[1] and a similar decrease was found in the

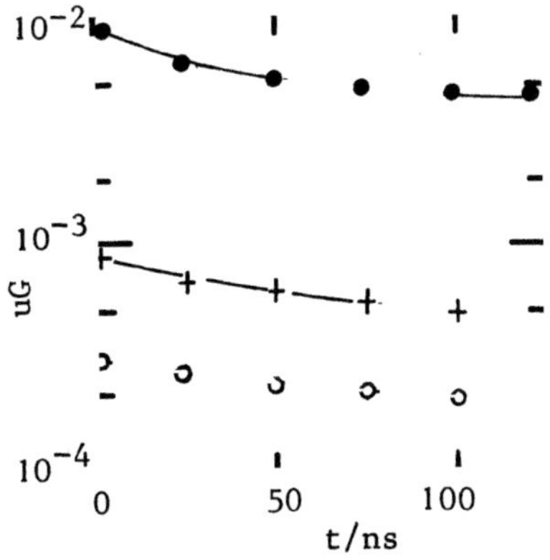

Figure 1. The product of the ion pair yield, G $(100eV)^{-1}$, and the sum of the ion mobilities, u $[cm^2V^{-1}s^{-1}]$, as a function of time following pulse irradiation of MCH containing: no additives, ●; 10^{-2} mol/dm^3 SF_6, +; and 10^{-2} mol/dm^3 SF_6 + 5 x 10^{-3} mol/dm^3 NH_3, ○.

present study for MCH, Figure 1. In IO an even more pronounced decrease in Gu, by a factor of 3 x 10^3, was observed on addition of SF_6. The electron is therefore confirmed to be the major charge carrier in the three liquids studied.

In CH it has been found[1] that a further substantial decrease in Gu occurs on addition of NH_3 to a solution containing SF_6. This has been explained by the reaction of holes with NH_3 to form positive ions, S^+, of a much lower mobility. The value of Gu in the presence of both NH_3 and SF_6 is then given by $G[u(S^+) + u(S^-)]$. A further reduction in Gu is also found to occur on addition of 5 x 10^{-3}M NH_3 to a solution of SF_6 in MCH, Figure 1. For IO however, Gu remains almost unchanged on addition of NH_3.

From the values of Gu determined in the pure liquid and in solutions containing SF_6 and $SF_6 + NH_3$, values of u(-), u(+) and $\overline{u}(S^{\pm}) = 0.5\,[u(S^+) + u(S^-)]$ may be derived if it is assumed, as a first approximation, that $G = G_{fi}$. The values obtained for CH over the temperature range 22 to 83°C and for MCH and IO at 22°C are listed in Table I.

The values of u(-) obtained in the present investigation are in reasonably good agreement with electron mobilities determined by other workers. While the values of $\overline{u}(S^{\pm})$ found are of the magnitude expected, they are possibly somewhat too high since it is known that for molecular ions a significant yield of geminate ion pairs may be present on a nanosecond timescale.

The hole mobilities in both CH and MCH are seen to be considerably larger than for molecular ions in the same liquid, whereas for IO u(+) is approximately equal to $\overline{u}(S^{\pm})$. These findings are qualitatively in agreement with the conclusions reached on the basis of the results of an optical pulse radiolysis study[3] of the kinetics of charge transfer to solute molecules in the present liquids.

In contrast with the more than twofold increase in u(-) in cyclohexane in going from 22 to 83°C, u(+) is seen to be almost independent of temperature.

A possible explanation of rapid hole transport in CH and MCH may be resonant charge transfer between positive ions and neighboring molecules. This would require the geometrical configurations of the ion and ground state molecule to be closely similar, a condition which is possibly not fulfilled for IO. Clearly more experimental data are required however, before a full understanding of the nature of hole transport in liquid saturated hydrocarbons is possible.

References

1) M. P. de Haas, J. M. Warman, P. P. Infelta and A. Hummel, Chem. Phys. Letters, in press.

2) A. O. Allen, T. E. Gangwer and R. A. Holroyd, J. Phys. Chem. 79, 25 (1975).

3) E. Zador, J. M. Warman, A. Hummel, Chem. Phys. Letters 23, 363 (1973).

4) J. P. Dodelet and G. R. Freeman, Can. J. Chem. 50, 2667 (1972).

5) J. H. Baxendale, C. Bell and P. Wardman, J. Chem. Soc., Farad. Trans. I, 69, 776 (1973).

6) W. F. Schmidt and A. O. Allen, J. Chem. Phys. 52, 2345 (1970).

7) A. O. Allen and R. A. Holroyd, J. Phys. Chem. 78, 796 (1974).

CARRIER MOBILITY IN LIQUID CRYSTALS

K. Yoshino*, K. Yamashiro* and Y. Inuishi*

Introduction

Liquid crystals is becomming more important as an element of the display device. However the detail of the mechanism of the electrical conduction which has close relation both with the light scattering and the memory effect are not well understood.
In this paper we present the mobility measurements in various liquid crystals at various phases (isotropic liquid, nematic, smectic and cholesteric states) as functions of the temperature, the applied electric and magnetic fields. Effect of surface condition of the electrodes will be also mentioned.

Experimental

Several liquid crystals [cholesteric type: cholesteryl nonanoate (CN), cholesteryl propionate (CPr), cholesteryl palmitate (CPa) etc.; smectic type: n-p-cyanobenzylidene-p-n-octylaniline (CBOA), ethyl p-methoxy benzylidene amino cinnamate (EMBAC) etc.; nematic type: p-azoxyanisole (PAA) etc.] were obtained from Fuji Dye Co. Ltd. and were purified by recrystalization for several times from appropriate solution. The samples were sandwiched between two nesa coated or gold evaporated glass electrodes and all measurements were done in argon (99.99%) atmosphere. Mobility was measured by the methods of transient SCLC, polarity inversion of the applied field or radiation induced condition. The transmission light of He-Ne laser was monitored by a photomultiplier 1P21 simultaneously. The magnetic field up to 15 kG was applied perpendicularly or parallel to the electric field if necessary.

Results and Discussion

1. Smectic and Nematic Liquid Crystals

Figure 1 shows the temperature dependence of the carrier mobility in CBOA. The carrier mobility decreases in the isotropic state with decreasing temperature and shows the abrupt decrease at the isotropic→nematic transition point. Such a step-wise change of the mobility at the phase transition was also observed in PAA[1] as shown in Fig.2 which shows the validity of Walden's rule between mobility and viscosity, indicating the ionic transport process. The mobility also decreases in the nematic state and shows step decrease at the nematic→smectic transition temperature by about one order. The small mobility value in the smectic state of the order of $10^{-6}\sim 10^{-7}$ $cm^2/Vsec$ is explained by the increased

*Faculty of Engineering, Osaka University, Japan

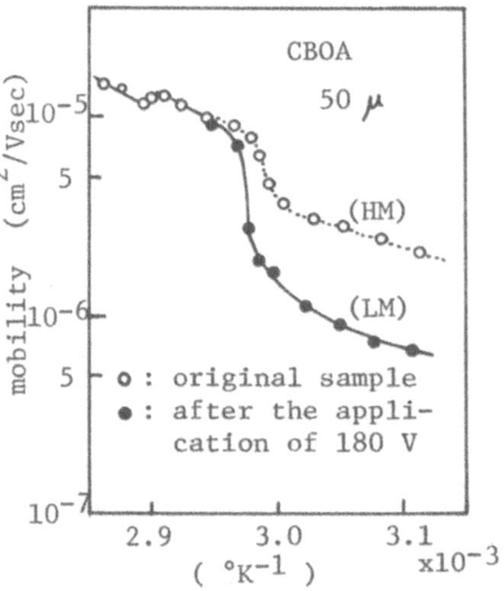

Fig.1 Temperature dependence of mobility in CBOA.

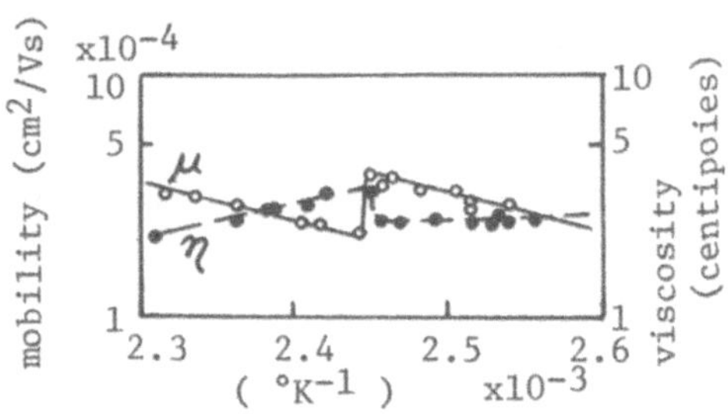

Fig.2 Temperature dependence of mobility.

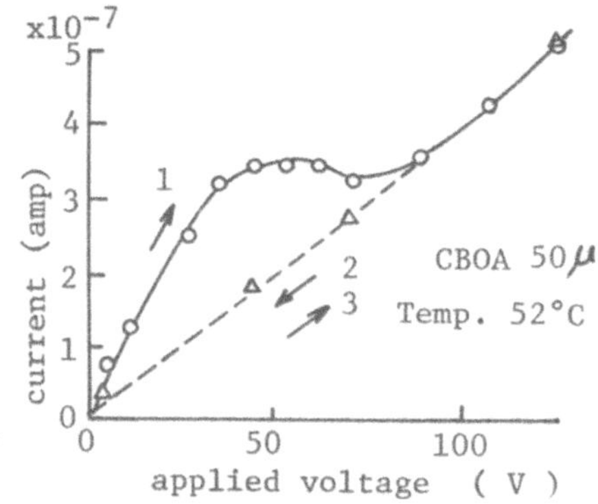

Fig.3 Voltage-current characteristics in CBOA.

viscosity in the smectic state, indicating also the ionic transport process. The mobility in the smectic state is strongly dependent on the applied voltage as shown in Fig.1. The sample on which the high voltage (90～180 V) has once applied in the smectic state shows the lower mobility values by about factor 3 compared with the original sample(dotted line). The activation energy of the mobility is smaller in the high mobility state (HM state) than the low mobility state (LM state). The optical properties of the HM state and the LM state are fairly different. In the HM state, the transmission of the light polarized parallel to the rubbed direction of the electrode plate is larger than the light polarized perpendicularly. However, such a polarization dependence of the transmission was not observed in the LM state. The HM state can be easily transformed into LM state by the application of high voltage. However, the transformation from the LM state to the HM state is quite difficult and the LM state is sustained for long time after the removal of the applied field, indicating the memory effect. The heating of the sample upto the nematic state is the only method to recover the initial HM state. The existence of the two states of the LM and HM states can explain satisfactorily the apparent negative resistance of the current voltage characteristics shown in Fig.3.

These results can be explained by the following speculation of the molecular alignment. The independence of the transmission on polarization indicates that all molecules aligned with their axes perpendicular to the glass surface under the LM state. On the contrary, in the HM state the molecule axis near the surface is forces to align parallel to the rubbed direction of the glass plate. In the middle of the sample cell, however, the molecule axis is not completely aligned parallel to the electrode surface, though it lies in the same plane with the direction of rubbing. Under these conditions, the sample is effectively uniform for the light polarized perpendicularly to the rubbed direction. For the light polarized to the rubbed direction, however, the reflective index is nonuniform in the middle part of the sample, resulting in the light scattering. As CBOA is p-type material, application of the sufficient high voltage induces the reorientation of the molecule axis from the perpendicular to the parallel alignment along the field direction, resulting in the change from HM to LM state. The observed larger dielectric constant in the LM state compared with HM state is consistent with this argument. Because of high viscosity in the smectic phase compared with the nematic state, reorientation requires fairly high voltage and the transformed state remains for a long time after the removal of the applied field. The observed larger activation energy of mobility in the LM state indicates the larger potential barrier to slip out from the smectic layer compared with that for the movement in the smectic layer. These results seem to support the assumption of the anisotropic conduction in the smectic made by E.F. Carr.[2] Magnetic field dependence of the mobility in the smectic state also supports the above discussion.

Effects of phase transition between various smectic states on the mobility was also studied. Field dependence of the mobility in PAA (nematic) was also studied. The carrier mobility decreases with increasing applied voltage and showed the minimum value of about $(2/3)\mu_0$ at about 5 V, where μ_0 is equal to the mobility value measured at lowest field. Then the mobility increases again leading into saturation. The decrease of mobility was explained by the transformation of the initial parallel molecular alignment to the electric field into the perpendicular alignment by the field effect, because PAA is n-type. Further increase of mobility at high field may be explained by the reorientation of the axis by the flow alignment and the hydrodynamic flow of liquid. These mobility behavior have correlation with the Williams domain formation and the strong DSM light scattering.

2. Cholesteric Liquid Crystals

Figure 4 shows the temperature dependence of the dark conductivity in CPa. In solid state, conductivity increases remarkably by raising temperature and shows a sharp peak at the solid liquid crystal transition point. Small step-wise change of the conductivity was also observed at the liquid crystal→isotropic liquid transition point. On the contrary, Shaw[3] observed nearly uniform increase of conductivity in all phases with increasing temperature and considered that the results indicate the same conduction mechanism in all phases. Shaw et al estimated the carrier mobility from the quadratic dependence of the current on the applied voltage by SCLC analysis to be $1\sim 8$ cm^2/Vsec. However, we could not observe such a quadratic dependence. Though there are several reports on the extremely large carrier mobility of electrons in dielectric liquids.[4] The life time of these free electron is

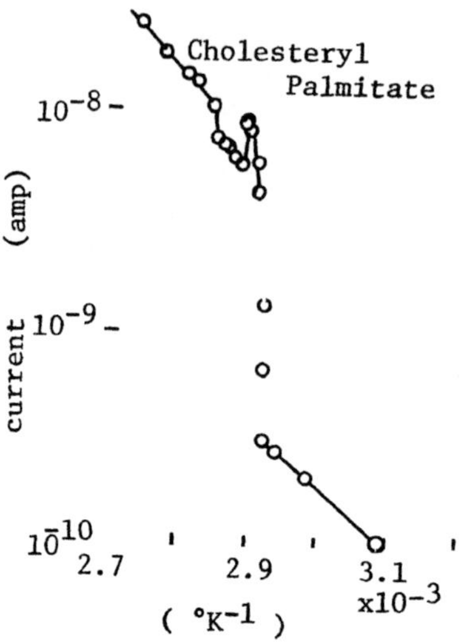

Fig.4 Temperature dependence of the electrical conductivity in cholesteryl palmitate.

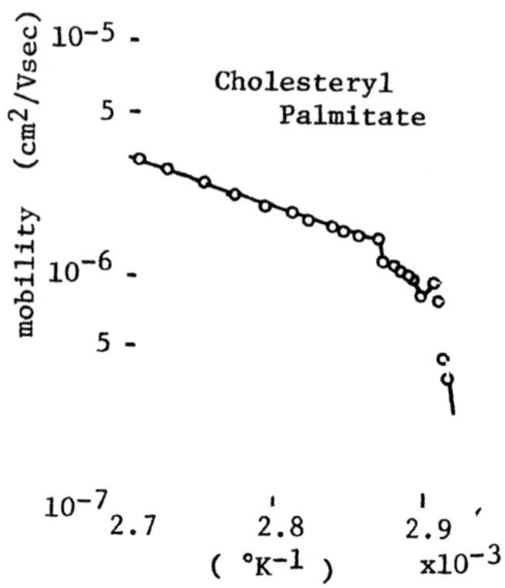

Fig.5 Temperature dependence of the carrier mobility in cholesteryl palmitate.

considered to be short. Accordingly the reported large electron mobility in all phases obtained by dc method seems to be unusual.

We also estimated the carrier mobility in the cholestric liquid crystals to the order of $10^{-5} \sim 10^{-7}$ cm^2/Vsec from polarity reversal method as shown in Fig.5 which shows step-wise change at the cholesteric→isotropic transition point. The step-wise change of viscosity at the phase transition[5] point was also reported, indicating the Waldens rule between the viscosity and our mobility. The walden's rule was also confirmed in other cholesteric liquid crystals such as CA etc. The Waldens rule between mobility and viscosity, and the magnitude of the mobility of the order of $10^{-5} \sim 10^{-7}$ cm^2/Vsec indicated that the ionic transport is dominant in the cholesteric liquid crystals investigated, being similar to the nematic and smectic liquid crystals.[6]

References

1) K. Yoshino, S. Hisamitsu and Y. Inuishi; J. Phys. Soc. Japan, 32 (1972) 867.
2) E.P. Carr; Phys Rev. Letters, 24 (1970) 807.
3) D.G. Shaw and J.W. Kauffman; Phys. Stat. Sol. (a) 4 (1971) 467, J. Chem. Phys. 54 (1971) 2424.
4) W.F. Schmidt and A.O. Allen; J. Chem. Phys. 50 (1969) 5037, ibid. 52 (1970) 4788, R.M. Minday, L.D. Schmidt and H.T. Davis; J. Chem. Phys. 50 (1969) 1473, P.T. Tewari and G.R. Freeman; J. Chem. Phys. 49 (1968) 4349.
5) P.S. Porter and J.F. Johnson; J. Appl. Phys 34 (1963) 55.
6) K. Yoshino, K. Yamashiro and Y. Inuishi; Japan J. Appl. Phys. 14 (1975) 216.

ELECTRICAL CONDUCTION IN NEMATIC LIQUID CRYSTALS

N. Félici*, B. Gosse*, J.P. Gosse*

Devices using nematic liquid crystals (LC) effects such as dynamic scattering (DSM) and storage require the passage of a DC to operate. Applying a voltage (∿20V) across a LC film (∿ 50µm) creates a strong turbulence forming scattering centres ; this turbulence is generally thought to be caused by ion injection as suggested by Félici (1), but Helfrich's theory (2) which explains the electrohydrodynamic instabilities under AC drive, has not been completely ruled out under DC drive. On the other hand, one the major concerns in LC device technology is the problem of operating life (3). Under DC exitation, a loss of performance occurs, the active scattering area of the cell diminishes and other LC characteristics as color, response time, transition temperature are altered. When carefully purified, these mildly polar compounds have a conductivity in the range 10^{-11} to 10^{-9} $(\Omega cm)^{-1}$ due to dissociated impurities. Their electrical conduction is very similar to that of other polar liquids and our electrochemical approach (4) is no doubt relevant.

Materials

We investigated various typical LCs which are relatively simple organic compounds : Schiff bases : X-⟨φ⟩-C(H) = N-⟨φ⟩-Y for example p-methoxybenzilidene p-butylaniline (MBBA) X = CH_3O, Y = C_4H_9 (5) ; tolanes : X-⟨φ⟩-C ≡ C-⟨φ⟩-Y (6) ; and biphenylnitriles : X-⟨φ⟩-⟨φ⟩-C ≡ N.

Experimental facts

The investigated LCs are chemically stable against hydrolysis, except MBBA which forms p-butylaniline (pBA) and p-anisaldehyde (pAde) according to the reaction : MBBA + H_2O $\rightleftharpoons$ pBA + pAde

As concerns oxidation and reduction, their redox potentials in acetonitrile on a platinum disk elektrode are given in table I, vs. the Ag/Ag^+ reference electrode.

Compound / Reaction	MBBA	formed by MBBA hydrolysis: pBA	pAde	Tolane	Biphenylnitrile: Alkyl	Alkoxy
$E_{1/2}$ oxid. (V)	1,1	0,5	-	1	1,6	1,3
$E_{1/2}$ red. (V)	-2,4	-	-2,3	- 2,8	- 2,4	- 2,4

* Laboratory for Electrostatics, Centre National de la Recherche Scientifique, B.P. 166, Centre de Tri, 38042 - GRENOBLE CEDEX, (France)

It is seen that the potential for reduction is quite similar for the various compounds ; at the opposite, the potential for oxidation is quite low for pBA and rather high for biphenylnitriles.

The first step of LC reduction is always $LC + e \rightleftarrows LC^{\cdot -}$; the oxidation of the anion radical $LC^{\cdot -}$ gives LC again. $LC^{\cdot -}$, which has been detected by EPR for all the investigated materials, is chemically reactive with protons, water and easily reduced compounds. Its life-time is about 10s for MBBA and pAde, and still longer for biphenylnitriles.

Oxidation of LC gives protons (with a faradic efficiency of 1) and neutral by-products more difficult to oxidize than the primitive LC, except in the case of pBA (from MBBA). We have met only once a different oxidation process : with heptoxybiphenylnitriles (HOBN) which form $HOBN^{\cdot +}$.

The chemical evolution of LCs under DC excitation was studied in air-tight cells (gap = 2mm, E = 25 $KVcm^{-1}$). The chromatograms of a MBBA sample which initially contains 300 ppm of water show that pBA and pAde are formed at the beginning of the experiment (fig.1). After about 300 hours, partial destruction of pBA has occured with faradic efficiency of 0.25 to 0.5, and the current density under constant voltage starts growing with time. After 36 days, the transition temperature has decreased from 46,8°C to 44,3°C, and the colour has changed from yellow to light brown. Later on, a partial disappearance of pAde is noticed together with an increasing formation of new compounds. By then, the sample has turned black and its transition temperature is low (36°C).

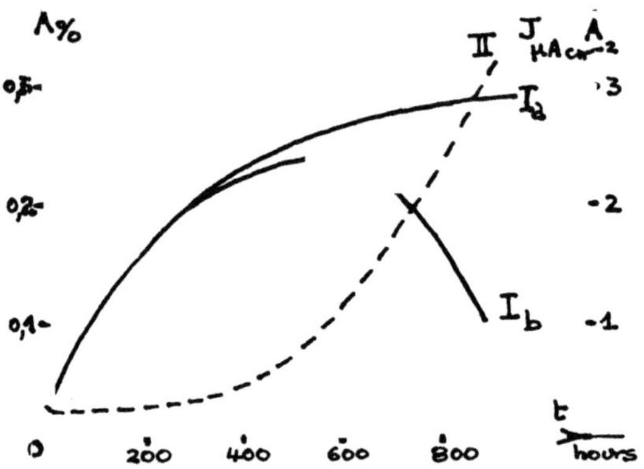

Fig.1 : Impurity percentages (I_a: pAde, I_b : pBA) in MBBA, 300 ppm H_2O, during a long application of a DC voltage

Tolanes and biphenylnitriles are altered more slowly than MBBA ; gas chromatography detects a few new compounds in small amounts, the current-time curves only show a slight increase in conductivity (fig.2) ; their transition temperature decreases by about 1°C after 900 hours, the corresponding decrease in MBBA is 5°C.

Discussion

Two of the basic problems in LC are : the operating life and the contribution to current of electrolytic dissociation.

An acceptable operating time for a LC device utilizing DSM is about 10^4 hours, with a current density of 1 $\mu A\ cm^{-2}$ (7) ; assuming an electrode area of 1 cm^2 and a gap of 10 μm, the amount of LC in the cell is 10^{-3} cm^3 or roughly 5.10^{-6} mole. The quantity of electricity passed through the cell is about 70 F/mole. Thus, the amount of material decomposed or altered by any mechanism whatsoever, must

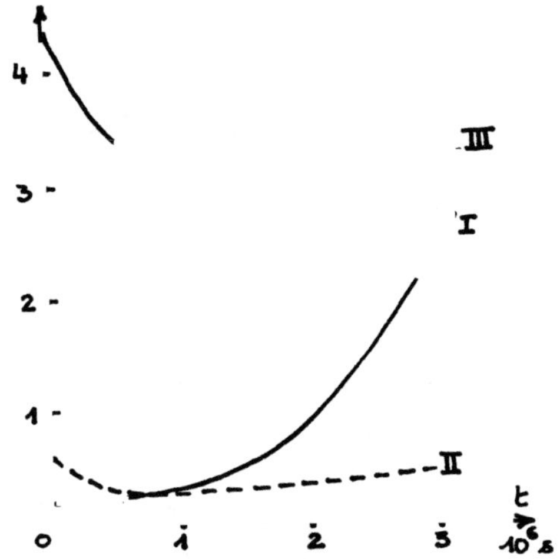

Fig.2 : Current density-time curves during a long application of a DC voltage V=5KV, l=2mm) to crystal liquids (I : MBBA, II : MP tolane, III : biphenylnitrile HOBN)

be much less than 10^{-2} mole/F. In particular, the contribution of electrolytic dissociation to current must be negligible. If not so, a corresponding amount of matter will be irreversibly decomposed. The same is true for mechanisms other than dissociation that entail irreversible, detrimental alteration : polymerization, creation of dissociated compounds for instance. We may conclude that only electrode reactions which may be called "cyclic processes", are acceptable. We can consider the two following examples : a) a cyclic process involving a LC anion radical and b) a cyclic process involving H_2 ;

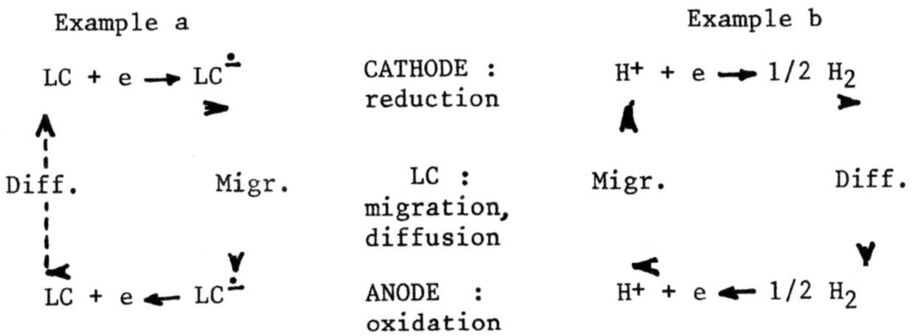

Cyclic process will insure a long operating life if the losses by side-reactions are low enough ; in the first case, it means that the anion radicals must have a life-time much longer than their transit time. In the second, diffusion of H_2 must maintain a sufficient H2 concentration near the anode to prevent competing processes causing oxidation of other reactants which do not give cyclic processes.

In our study,we met a very unfavourable case : MBBA. This LC is always partially hydrolysed into pBA and pAde. At the cathode MBBA and pAde are reduced to anion radicals but these are partially lost before they can reach the anode because they react with water or protons formed by oxidation, for instance :

$$MBBA^- + H^+ \longrightarrow MBBAH$$

$$MBBAH + H^+ + e \longrightarrow \text{by-products.}$$

At the anode pBA is oxidized and protons are injected. The ensuing processes are presumably similar to those following the oxidation of primary aromatic amines like aniline (9). In this case, one finds products denoting rapid follow-up reactions leading to dimeric and polymeric products which are strong bases in organic media. With aniline, for instance :

dimerization

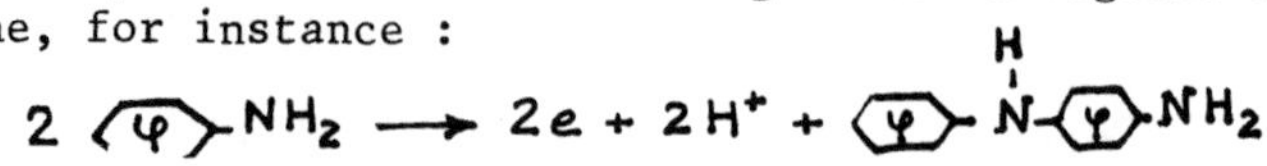

oxidation of the dimer

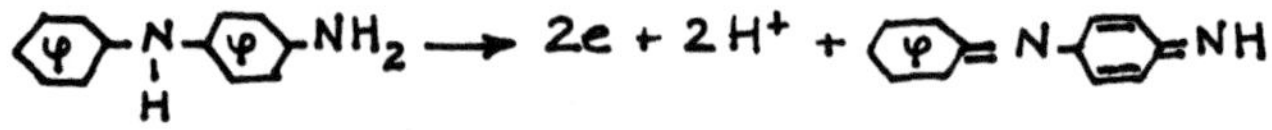

and a cascade of head to tail condensations and further electrooxidations giving members of the so-called "aniline black" series such as emeraldine

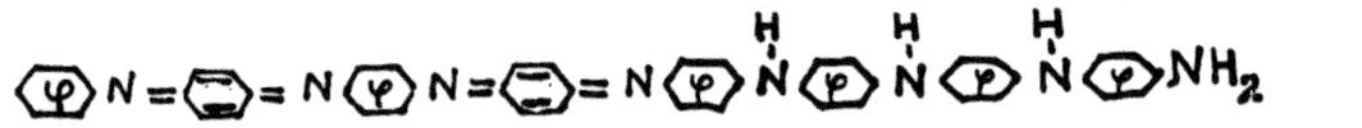

The preceding makes clear why the operating life of MBBA cannot exceed a few hundreds of hours.

A much favourable case is HOBN which is regenerated by mutual annihilation of the radicals or by their neutralization at the opposite electrode. Nevertheless, chemical reactions with radicals cannot be to completely avoided and they entail a slow destruction of the LC.

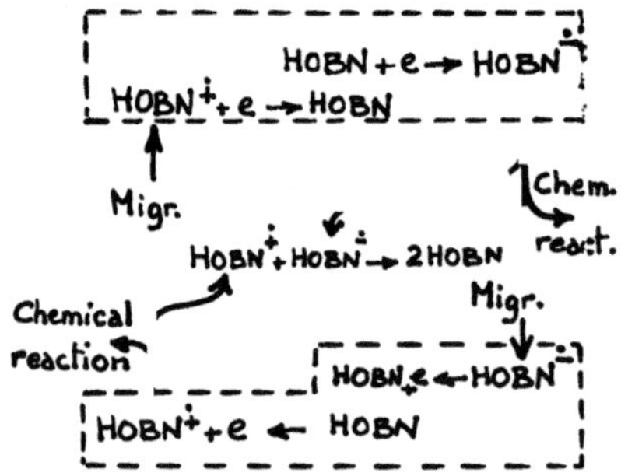

Closing remarks

It is clear that progress in DC operated LC devices can be achieved by synthetizing new LC molecules giving one or two stable radical ions. Another solution already tested is the deliberate introduction of dopants in LC, i.e. reactants giving also stable radicals.

The fact that the contribution of electrolytic dissociation to current is negligible in many cases (a prerequisite for a reasonable operating life), is very important for the physics of LCs. In particular, theories of DSM that imply electrolytic conduction are ruled out. This only applies, however, to the case when DC is used, for dissociable impurities are always present at the start.

1. N. Félici, R.G.E., 78, 717, 1969
2. W. Helfrich, J. Chem. Phys., 51, 4092, 1969
3. L.T. Creagh, I.E.E.E. Proceedings, 61, 814, 1973
4. B. Gosse, J.P. Gosse, N. Félici, J. Appl. Electrochem. to be published nov. 1975
5. A. Denat, B. Gosse, J.P. Gosse, J. Chim. Phys., 327, 1973
6. A. Denat, B. Gosse, Chem. Phys. Lett., 22, 91, 1973
7. A. Sussman, Appl. Phys. Lett., 21, 126, 1972
8. Y. Ohnishi, M. Ozutsumi, Appl. Phys. Lett., 24, 213, 1974
9. R.N. Adams, "Electrochemistry at solid electrodes", p. 327 M. Dekker Inc. New York 1969

CONDUCTION ADJACENT TO OIL-IMMERSED INSULATION SURFACES IN A NON-UNIFORM FIELD

R.E. James*

Introduction

It is often assumed that the introduction of an insulation surface between two oil immersed electrodes lowers the electrical strength of the gap. The mechanism by which such a reduction might occur is not fully understood and it is extremely difficult to determine quantative values for practical arrangements. The work described in the paper was part of a long term investigation (1) based on design situations arising in complex apparatus. Although the results presented here are only for d.c. measurements it is hoped they will contribute towards a better understanding of some aspects of the behaviour of surfaces when subjected to other forms of voltage. To investigate a range of initially non-uniform but symmetrical field distributions the concentric cylindrical electrode system shown in Fig. 1(a) was adopted. The arrangement allowed the sample to be made thin and a study of the surface characteristics to be carried out uninfluenced by effects produced by the bulk of adjacent solid insulation. The latter is present when a conventional spacer is tested between two electrodes as shown in Fig. 1(b). The configuration used in the present work achieves a good approximation of the required condition without the area tested becoming reduced to an unacceptable value; as arises with a single fibre, Fig. 1(c).

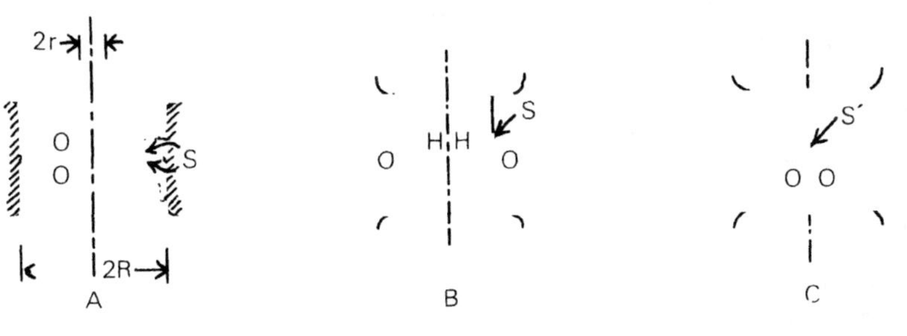

Fig. 1 Possible configurations for surface measurements
O = Oil
S = Surface(s)
H = Solid
S'= Surface (fibre)

Description of Apparatus

The glass test cell (1) was in the form of a cross with the h.v. entry at the bottom and a viewing window above, Fig. 2. One feature of the system was the capability of being able to change the punched sheet sample and h.v. electrode without exposure to atmosphere. The sample holes were pre-punched to allow tests with h.v. electrode diameters of 3.17, 6.35 and

*Department of Electric Power, University of New South Wales, P.O. Box 1, Kensington, N.S.W. 2033, Australia.

12.7 mm respectively and, also, to enable measurements to be made without a surface bridging the gap 10 groups of holes could be accommodated in one roll.
The l.v. electrode consisted of an inner ring divided horizontally and enclosed by a guard arrangement. The cellulosic dielectric materials were dried and oil impregnated in the test cell. The commercial quality transformer oil was processed in a separate all glass system and then fed to the cell through a filter.

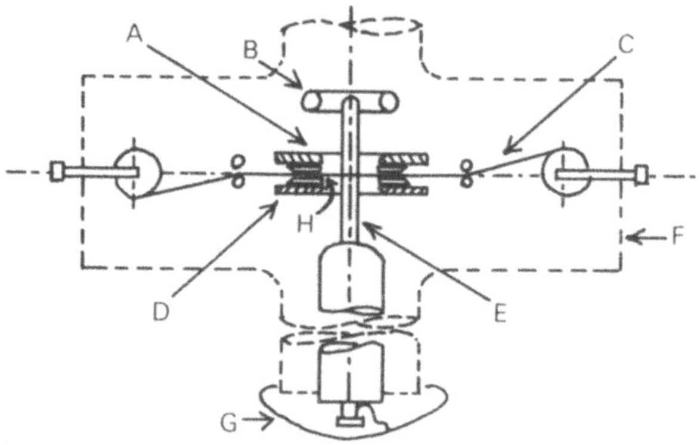

Fig. 2 Lay-out of test cell
A = Test gap
B = Stress ring
C = Sample sheet 50 mm wide-pre-punched
D = L.V. electrodes
I.D. = 25.4 mm (inner insulated from guards)
E = Concentric h.v. electrodes (3). Can be withdrawn axially to allow relocation of sample in test gap
F = Glass sections
G = H.V.
H = Inner

Gap designation

H.V. electrode diameter	Type of gap
3.17 mm	3/25
6.35 mm	6/25
12.70 mm	12/25

The oil moisture content was 10 ppm., the resistivity between 10^{15} -10^{16} ohm-metre and air content 1-2%. The oil and dielectric would be expected to be in equilibrium in respect of moisture content (2). The positive high voltage was applied from a 150kV highly stabilised power supply having a ripple of better than 0.02%. A fast acting h.v. trigatron circuit was included for diversion of current if an unexpected oil flashover occurred, or breakdown tests were being carried out. The conduction current through the test gap was detected by measurement of the voltage across an impedance connected between the inner l.v. electrodes and earth. Before tests the accuracy and screening of the low voltage circuits were checked.

Results

It is not possible to reproduce the detailed results but the general trends were similar to those shown. In Fig. 3 are given short term variations of current with and without a surface in position. The curves were deduced from a number of differend tests: the values are referred to the final current for the particular condition.

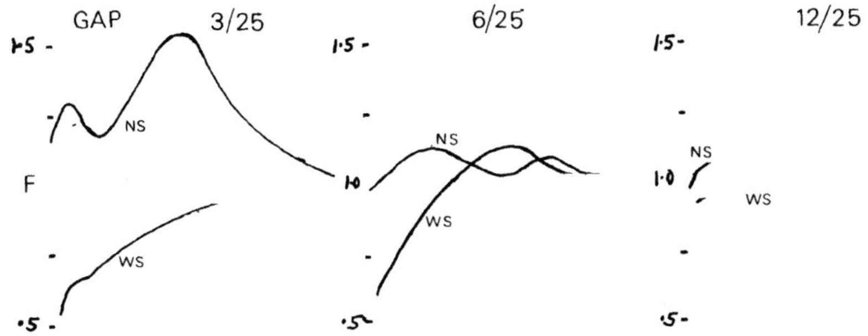

Maximum stress 195 kV/cm   150 kV/cm   150 kV/cm
Average stress 58.5 kV/cm 68.2 kV/cm 120.5 kv/cm
F = Final current
NS = no surface
WS = with surface

Fig. 3 Estimate of current variations at 65 kV
- 10 minute period following increase in voltage
- stress values ignore space charge effects

For ease in presentation and to relate to other work, the current values at the end of a period for a given voltage were plotted on log-log paper, Fig. 4. The values are the average of readings from several test series.

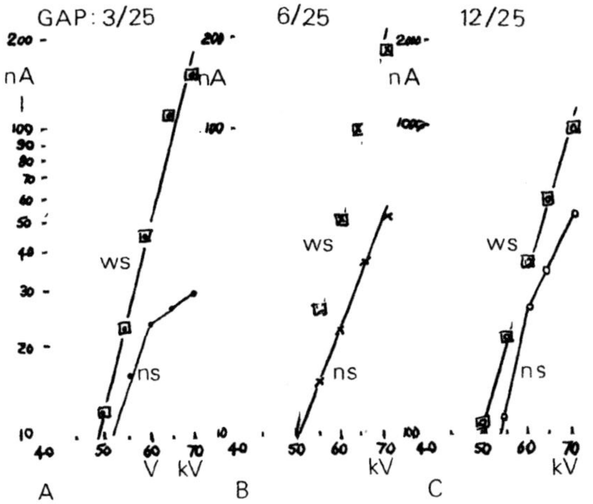

WS = with surface
NS = no surface

Fig. 4 Variation of final value of current

The ratios of current magnitudes with and without a bridging surface are plotted in Fig. 5.

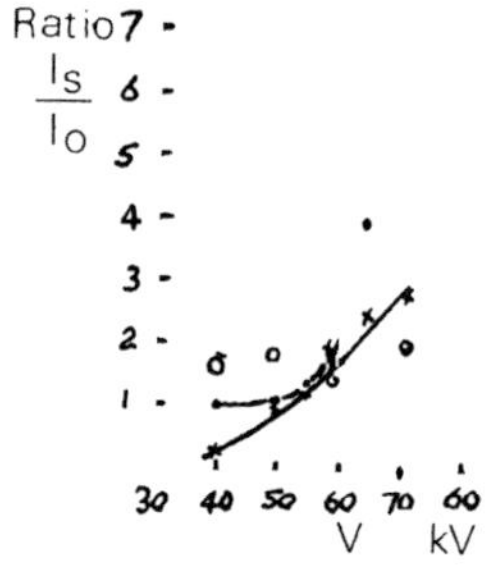

Fig. 5 Ratio of current magnitudes
I_s = with surface
I_o = no surface

Discussion

The techniques developed allow measurements of current changes adjacent to a thin dielectric at different positions along the surface. Several materials can be investigated in succession without reprocessing of the oil or sheet dielectric. In addition three gap configurations may be studied so assisting in the establishment of hypotheses involving the physical structures. Results to date have shown some interesting effects: the most important of these appears to be the tendency for a surface to stabilise conditions in cases of erratic oil behaviour. This was especially true for electrode systems 3/25 and 12/25 where sudden changes in the log-log plots occurred at about 60 kV with oil only. To relate these changes to possible EHD effects (3, 4) an attempt was made to find a relationship between current and voltage indicative of a critical condition for each gap.

Due to the nature of the configuration and the inherend difficulties involved, the derivation of a precise relationship may be impossible. However an empirical equation of the form

$$\frac{I}{V} \propto \frac{1}{(R-r)^2} \quad \frac{1}{\log \frac{R}{r}}$$

seems to give a fair estmate of the relative I/V values of the expected point of instability over the gap range $2 \leq \frac{R}{r} \leq 8$. The relationship indicates that the 6/25 gap should have become unstable at 76 kV. Such an incipient condition may have been present as a number of flashovers in the oil were recorded at about this voltage: the current magnitudes were not being measured at the time. The voltage values were higher than those noted for the other two gaps.

Field plots show that the fringing in the 3/25 gap might be expected to produce more oil movement than in the others. Oil flow could have resulted in the carrying of particles (or ions) away from the measurement area although remaining within the gap itself. Such activity could explain the sudden change of the oil line in Fig. 4(a): this discontinuity is not apparent with the surface present. It is postulated that the introduction of a dielectric sheet tends to stabilise conditions, Fig. 3, possibly producing vortices. Parts of the latter may follow the insulation surface and give an increase in carrier concentration in the adjacent oil: thus increasing the current at the inner l.v. electrodes. Similar reasoning may be applied to the other gaps but the order of the effect is smaller, probably because of the different stress distributions and reduced spacings. The various factors are closely inter-related and no simple quantative assessment can be made of the EHD and other effects. However, the arguments presented at the last D.C. Conference and elsewhere (5) help in deducing qualitative explanations of some of the phenomena observed. It is hoped to assess the effects in more detail during further experimental and theorectical studies.

Conclusions

The work reported seems to support the view that a thin surface in a tangential field does not lower the electrical strenght: this agress with a.c. breakdown tests (1) where the modal value increased slightly in the two gaps (3/25 and 12/25) tested. Increases in conduction current could be due either to modification of existing oil movement or to creation of motion which alters the charge distribution near the surface. The material of the latter seems to have minimal effect assuming no contamination is present: see also reference 6.

Acknowledgements

The author wishes to acknowledge the facilities provided by the Electrical Engineering Department of Portsmouth Polytechnic, (U.K.) and the encouragement given by colleagues during that period.

References

1. James, R.E.: Ph.D. Thesis, London University, 1973
2. Franklin, E.B.: Electrical Times, 1965
3. Atten, P. and Moreau, R.: CR Acad. Sc. Paris t. 270, 1970
4. Hewish, T.R. and Brignell, J.E.: J. Phys.D: APP.Phys. vol. 5., 1972
5. Felici, N.J.: Direct Current and Power Electronics vol. 2, 1971
6. Cherney, E.A. and Cross, J.D.: IEEE Trans. EI 9, 1974

THE EFFECT OF SOLID POLYMERIC MATERIALS ON THE IONIC CONDUCTIVITY OF LIQUIDS UNDER ELECTRIC STRESS

R. Tobazéon* and E. Gartner*

Introduction . It has recently been shown that sheets of polymers are able to produce (under A.C. or D.C. voltages) a genuine deionization of polar liquids in contact with them (1)(2)(3). This effect - called "absorption effect" - consists in a field accelerated penetration into the polymeric material of the liquid and its ions, producing modifications of the original characteristics of the solid and the liquid according to their physico-chemical properties. This effect contrasts with the Garton effect (4) where carriers are supposed to be blocked at liquid/solid interfaces, and free to return to the liquid when the field reverses ; the Garton's analysis has been shown to apply fairly well to mineral solids (mica, ceramics) and paper capacitors impregnated with slightly conductive liquids ; surprisingly, papers of any density exhibit but a poor absorption effect. In electrical engineering, premature failures of buried or immersed polyethylene (PE) cables have been often attributed to treeing ("treeing" describes dentritic-type patterns observed in cables aged under stress) ; electrochemical trees are caused by ingress of water (5)("water-treeing") The influence on PE of ion content and nature in water has been studied (6)(7). This phenomenon of liquid penetration is not restricted to the water/PE association, and our purpose is to show its generality and to study the physico-chemical factors which are involved.

The field-induced absorption phenomenon . The deionization of a strongly polar liquid such as nitrobenzene (NB) through the action of polypropylene (PP) sheets has been strikingly demonstrated in the following way: a given alternating voltage is applied to a liquid layer between polymer foils covering metallic electrodes ; tanδ and the capacitance of the liquid/solid association are time-dependent; their evolution testifies to a progressive ionic purification of the liquid ; this has also been ascertained by field measurements into the liquid using the Kerr technique, and by the continuous decrease in concentration (evaluated by photometry) of a coloured electrolyte voluntarily added to the liquid prior to voltage application. Ions of both signs are absorbed in equal quantities into the polymer foils ; the liquid oozes through the films, sometimes destroying the vacuum-deposited metallization in contact with the electrodes. The liquid conductivity decreases more rapidly when voltage and frequency are raised ; its final value varies with the liquid gap and initial ionic concentrations. After removal of voltage, ions are released, but liquid conductivity increases only slowly. Deionization occurs also when a single unmetallized foil is introduced between the bare metallic electrodes (dividing the cell in two compartments) or when one electrode is covered with a polymer sheet ; nevertheless the final resistivity of the liquid is

* Laboratory for Electrostatics, Centre National de la Recherche Scientifique, B.P. 166, Centre de Tri
38042 - GRENOBLE CEDEX, France

poorer than above (certainly because bare electrodes inject ionic carriers into the liquid). The deionization process has been applied with success to a recycled circulating liquid.
The great generality of the phenomenon : is demonstrated by using several types of polymer, of liquid and different kinds of added electrolyte.
1) Deionization of various liquids with the same polymer. The following results have been obtained with PP foils (10 or 12 microns in thickness) enclosing a liquid filled space of 1.5 to 2 mm in width ; the cell was subjected to a 50 Hz constant voltage (3 to 8kV).
- Strongly polar liquids. Among the non-ionized liquids, we have mainly studied nitrobenzene distilled under vacuum ; its resistivity rises from $2x10^8\Omega$.cm to $7x10^{10}\Omega$.cm in about 20 hours. In similar conditions, the resistivity of propylene carbonate (PC) rises from $10^7\Omega$.cm to $2x10^9\Omega$.cm, and comparable values are obtained for mixtures of these liquids with polychlorobiphenyls (PCBs) or dibutylphtalate (DBP). When the NB is recirculated in a circuit containing much more liquid than in the cell to which the AC voltage is applied, the upper limit of the resistivity is $10^{10}\Omega$.cm. The liquid resistivity has been brought, thanks to an auxiliary D.C. electrodialysis cell, up to $7x10^{11}\Omega$.cm ; when the electrodialysis is stopped, it is observed that the A.C. voltage applied to the cell which contains PP foils is not able to maintain such a high value : the resistivity falls to 5 to $7x10^{10}\Omega$.cm in a few minutes.
We have also tried some auto-ionized solvents difficult to deionize by conventional electrodialysis ; the pure grade liquids were distilled (no special chemical treatment was made, so that the initial purity was not exceptional). In a few hours, the increases of resistivity were : for ethanol, from 10^7 to 10^9 Ω.cm ; acetonitrile : $2.5x10^6$ to $6x10^8$ Ω.cm ; n-ethylacetamide : 10^7 to $2x10^8\Omega$.cm. With glycerol (initial resistivity : $2x10^7\Omega$.cm) and water (initial resistivity $5x10^5\Omega$.cm), the increase of resistivity was small and difficult to appreciate because the dielectric properties of the polymer change considerably (cf further §)during voltage application.
- Weakly polar liquids. Slightly conductive samples of these liquids gave in similar conditions than above the following results : chlorobenzene : $8x10^9$ to $5x10^{11}\Omega$.cm - 1460 Aroclor : $4x10^{10}$ to $2x10^{11}\Omega$.cm. The kinetics of deionization is very slow when the liquid resistivity exceeds, say $10^{11}\Omega$.cm. On the above tested samples, the impurities were unknown ; appropriate salts in known concentration have been dissolved in several weakly polar liquids, and strong increases of resistivities were observed : chlorobenzene : from 10^8 to $4x10^{10}\Omega$.cm - 1460 Aroclor : 10^7 to $1.5x10^{10}\Omega$.cm ; 1500 Aroclor : $3x10^8$ to $5x10^9\Omega$.cm. DBP : $3x10^7$ to $2x10^9\Omega$.cm. A nematic liquid crystal, the methoxybenzilidenebutylaniline (MBBA) was poorly deionized by PP,especially in the nematic phase,even with an added electrolyte.
- Non polar liquids. In these liquids, electrolytes are poorly and slowly dissociated. With an used transformer oil, the initial resistivity ($10^{12}\Omega$.cm) was scarcely increased although the voltage was applied during several tens of hours ; tests over much longer periods are going on. With benzene, we have also observed a slow evolution of the resistivity with conductive samples ($10^9\Omega$.cm).

2) Deionization with different polymers. Studies have been carried out on sheets of comparable thicknesses ($\sim 10\mu$m) of PP,PE, polyethyleneterephtalate (PET), polytetrafluorethylene (PTFE).
The efficiency of these polymers for the deionization of NB can be

classified : PP > PE > PET > PTFE (with PET and PTFE the final resistivity does not reach $10^{10}\Omega$.cm). The deionization process has been shown to occur with other polymer/liquid associations, but has not been studied extensively.
The crystallinity of the polymer plays an important role : with more crystalline PP or PET (stretched), the deionization starts less easily and is less efficient.

3) Influence of the ion concentration and nature. Tetrabutylammonium picrate has been dissolved at 10^{-5} Mole/l concentration into NB, PC and acetonitrile. (in each of these liquids this salt is fully dissociated). In a few minutes, the salt is eliminated from the liquid bulk, the conductivity of the liquids being divided by 10^3 or over ; in 10 hours, the resistivities reach $10^9\Omega$.cm (acetonitrile) and $5\times10^9\Omega$.cm (NB and PC). Beyond 10 yours, the decrease in conductivity is slow for PC, and faster for NB, while a slight increase is observed with acetonitrile.
For higher concentrations ($>10^{-4}$ Mole/l), the applied voltage has to be reduced, otherwise breakdown of the polymer occurs quickly ; moreover it seems (for 10µm foils of PP and liquid-filled gaps of 2mm) that 10^{-3}-10^{-4} Mole/l is an upper limit for the electrolyte concentration, above which deionization starts slowly and is inefficient (It mus be noted that the salt concentrated into the liquid entering the polymer may precipitate ; sometimes salt crystals have been collected at the polymer/liquid interface). With lower concentrations, the deionization has been repeated with the same PP foils on renewed samples of conductive NB : this experiment shows that the first ionic epuration is the slowest.
Several kinds of salts have been tried in different solvents (organic salts of ammonium generally, and mineral ones like lithium perchlorate and sodium chloride). No major differences were observed in the kinetics of deionization and on the final value of resistivity (under AC voltage).
The study of the kinetics of release of ions (after voltage removal) in the 3 liquids (acetonitrile, NB,PC) containing initially 10^{-5} Mole/l of a picrate salt shows that the liquid conductivity increases progressively, but the initial value is not quite restored after weeks (a small amount of salt may be definitely retained into the PP) ; the return to the initial resistivity is faster for acetonitrile than for NB and PC.

4) Nature of the losses and influence of the frequency. At the start of the deionization (50 Hz AC) the losses are mainly linear in nature ; when the liquid resistivity exceeds, say $10^{10}\Omega$.cm, the percentage of non-linear losses (the losses are analyzed with the aid of a comparison bridge and an oscilloscopic detector) increases. The maximum of the non-linear current I_{NL} is produced (for millimeter liquid gaps and voltages V of several hundred volts or more) after the voltage reversal (i.e. I_{NL} and V have the same sign), and not before this reversal as expected from Garton's theory. This could be attributed to a delayed extraction of ions from the PP, but from the measured transit times it is clear that the charge transport is ensured by an electrohydrodynamic mechanism, the convective current following the conduction one. When the distance is reduced to microns or less (as in PP capacitors), the voltage applied to the liquid does not exceed volts, and the peak of I_{NL} can be observed before the voltage reversal.

It has been observed that the kinetics of deionization is varied when the frequency is changed ; the phenomenon is very slow in D.C. and its efficiency is limited. The speed and efficiency are increased by increasing the frequency (it has been shown that deionization occurs between 20 Hz and 2000 Hz) ; at the present time, because of heat production which varies when the frequency is varied (and consequently causes modifications of the materials), we possess but qualitative results about the kinetics of deionization, and the role of the different parameters :

Consequence : modifications of the properties of the polymer. The most striking fact is that liquids enter polymer foils (sometimes crossing 100µm sheets) : profound modifications of the foils are observed (colouration, cracks) and weight increase up to 10 % have been measured. The electrical properties are altered (the measurements were done with mercury electrodes). The permittivity ε is increased : for instance, PP sheets after absorption of conductive liquids such as water, NEA, PC exhibit an increase of about 10 % ; with NB, acetonitrile the increase is 2 or 3 times greater. $\tan\delta$ rises from 2×10^{-4} to 10^{-2} (or more) in the range 10^{2}–10^{3} Hz. The losses are generally linear, but some swollen polymers exhibit marked non-linear behaviour. The D.C. currents were increased by orders of magnitude ; A.C. breakdown voltages were much reduced.

Conclusion . The penetration of the liquid and its ions into polymers is enhanced by the electric field ; consequently the liquid resistivity is increased. The final resistivity level depends on the physico-chemical properties of the liquid and of the polymer : the crystallinity, the swelling ability, the initial ionic density in the liquid are important factors affecting the behaviour of the solid/liquid association. The absorption occurs with different kinds of electrolytes ; its kinetics is fast with fully dissociated salts; otherwise, it is limited by the dissociation rate. Polymers swelled by conductive liquids undergo profound modifications of their original properties.

1. R. Tobazéon, Doctoral Thesis, Grenoble 1973
2. E. Gartner and R. Tobazéon, Proc. of the 4th Intern. Conf. on Conduction and Breakdown in Dielectric Liquids, Dublin,Jul.1972
3. R. Tobazéon and E. Gartner, Conf. on Elec. Insul. and Dielectr. Phenom. Nat. Acad. Sci., Downingtown,Oct. 1974
4. C.G. Garton, Proc. Inst. Elect. Engrs., 88, 1941, 103
5. 1972 Annual Report, Conf. on Elec. Insul. and Dielectr. Phenom., Nat. Acad. Sci., 1973
6. D.W. Auckland and R. Cooper, Conf. on Elec. Insul. and Dielectr. Phenom., Nat. Acad. Sci., Downingtown,Oct. 1974
7. C. Katz, N. Srinivas and B.S. Bernstein, ibid.

LIGHT EMISSION STUDIES IN THE INTERPRETATION OF HIGH-FIELD CONDUCTION IN DIELECTRIC LIQUIDS

P.B. McGrath and J.K. Nelson*

Introduction

Many studies have been made on conduction and breakdown in liquid dielectrics providing a wide range of data, yielding sometimes conflicting interpretations for apparently similar conditions.

Experiments involved with the use of uniform field electrodes (1) have shown conduction current and light emission to be electronic in origin for stresses up to breakdown level under direct voltage conditions. More recently the use of a highly non-uniform electrode configuration has shown that for n-hexane pre-breakdown conduction current may exceed 1μA and be accompanied by the appearance of light emission and gas evolution within the vicinity of the point electrode (2). For step voltage application pre-breakdown phenomena have been photographed for both uniform (3) and non-uniform (4) fields and differing views expressed as to the origin of the disturbance.

Interpretations of processes leading to electrical breakdown may generally be divided between electronic and cavitation mechanisms. It is apparent that experimental conditions are important in determining the nature of electrical characteristics and in order to reconcile the two viewpoints it is suggested that both mechanisms can be shown to operate, and a transition effected by temperature or stess variation.

Experiment

Transformer oil of Peruvian origin, saturated and in equilibrium with dried air at atmospheric pressure was filtered and admitted to a fully enclosed light-tight test cell. The cell was equipped with a viewing port and a facility to enable calibrated narrow band optical filters to be interposed between the test gap and a photomultiplier. A barium titanate transducer 1cm in diameter, was lightly suspended in the test liquid such that it was mechanically isolated from the test cell body.

Direct voltage was applied via a decoupling circuit to the point of a point/sphere electrode system. The stainless steel sphere was 5mm in diameter and the tungsten point electrolytically etched to an initial tip radius of approximately 1μm, although this increased with stressing time.

Many tests were performed and, whereas small variations occurred due to point erosion, the results presented depict typical and reproducible characteristics.

*Department of Electrical and Electronic Engineering, Queen Mary College (University of London), Mile End Road, LONDON E1 4NS

Integrated conduction current, light emission and acoustic vibrations were measured simultaneously using vibrating capacitor electrometers: the circuits being carefully decoupled from the supply and each other.

For elevated temperature measurements the whole test cell was slowly heated to allow dissolved gas to remain in equilibrium with the oil sample. However, in an attempt to maintain accuracy of results, the ultrasonic transducer was not used under these conditions as its behaviour was an unknown function of temperature.

Discussion of results

Fig. 1 shows characteristics for conduction current, light and sound emission at room temperature for a point/sphere electrode geometry. It is evident that region 2 represents a transitional state between two conduction regimes. The width of this current transition region was found to be dependent on the condition of the point electrode and the voltage required for transition in light emission increased with temperature. Below the lower knee of the transition, depicted by region 1, it is suggested that an electronic mechanism is responsible for the light emission and the absence of acoustical vibrations further supports the viewpoint that the applied stress is below that necessary for cavitation. This is similar to the characteristic observed for a sphere/sphere electrode system, Fig. 2, where no sound is detectable before total gap breakdown. It is probable that this light emission originates from downward transitions of π electrons associated with polycyclic molecules present in the oil, being excited into higher energy states as a result of the applied field (5).

For the non-uniform electrode configuration, increasing the applied voltage above 10 kV presents a considerable increase in conduction current and light emission, and marks the onset of acoustical disturbance, as shown by region 3. Viewing the test gap directly

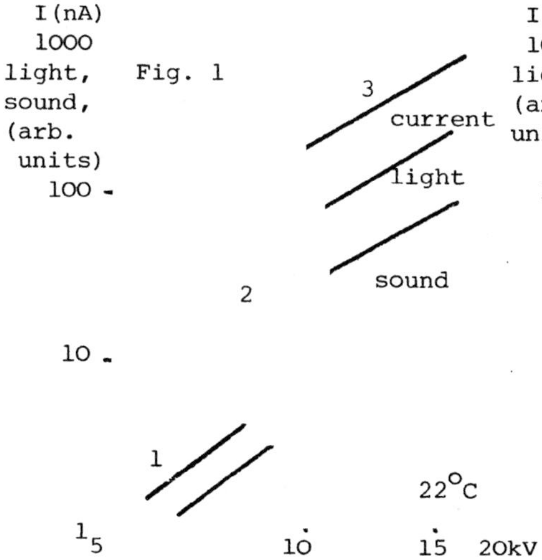

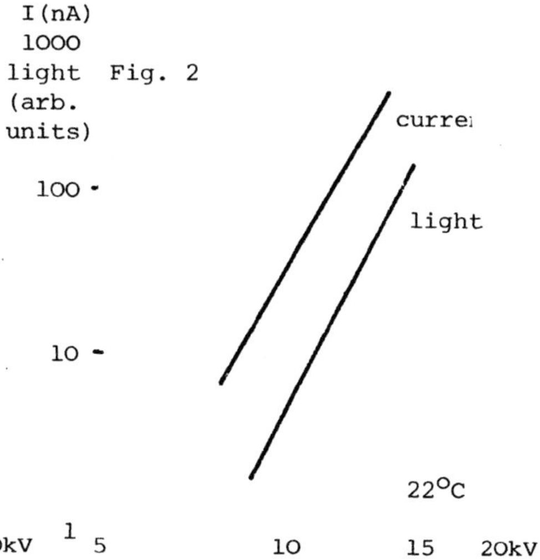

it was possible to observe localised light emission at the tip of the point electrode, and, on using a laser light scattering technique, a spray of bubbles could be seen to emanate from this region as described by Singh et al (2). Furthermore the two processes occurred simultaneously even at elevated temperatures.

Fig. 3 indicates that whereas inception voltage for electroluminescence is relatively insensitive to temperature variation, that for higher level light emission, as depicted by region 3 of Fig. 1, shows a marked increase with temperature.

As cavitation and high level light emission have been seen to occur simultaneously, Fig. 3 indicates that it is more difficult to provide the necessary conditions for cavitation as temperature is increased. This result is in direct conflict with the view that cavitation is primarily dependent on surface tension as this would imply that cavitation initiation occurs more readily with reduction in surface tension.

Selecting applied voltage as a fixed parameter the behaviour of conduction current and light emission may be related to temperature as shown by Fig. 4. For the low level conduction regime depicted by region 1 of Fig. 1 light emission is insensitive to temperature variation but conduction current increases substantially over the range 20 to 80°C. On increasing the applied voltage such that conduction occurs in the post transition region, current can be seen to be independent of temperature. The light emission which accompanies this level of conduction current ceases at elevated temperatures when the selected fixed voltage becomes insufficient to maintain the light emission, supporting light inception dependence on temperature as shown by Fig. 3. This result leads to the conclusion that for transformer oil conduction current is insensitive to both high level light emission and the cavitation phenomenon, although very small depressions in current were sometimes observed during gas evolution.

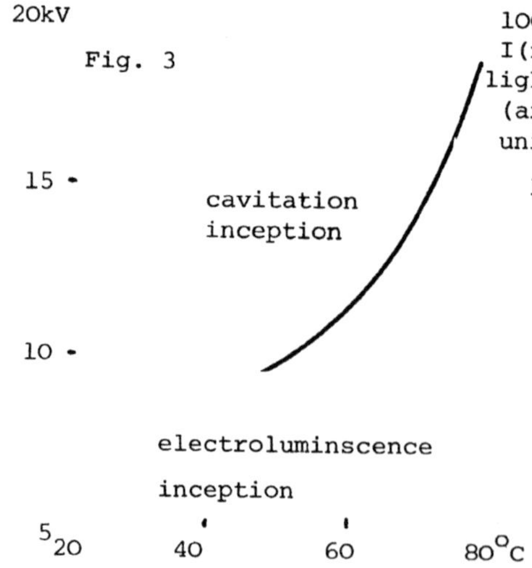

Fig. 3

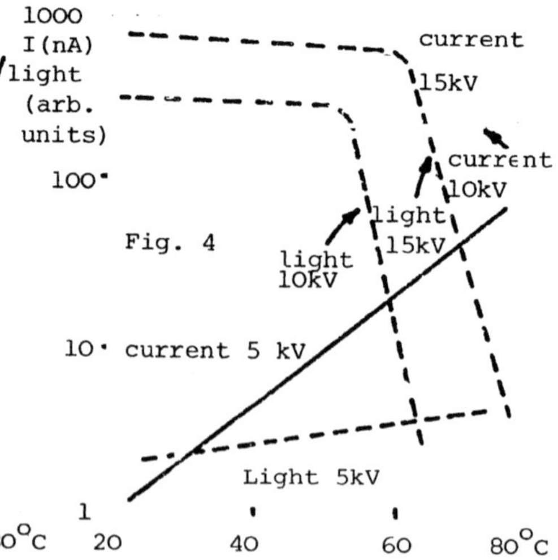

Fig. 4

Spectral analysis of the light emitted from stressed transformer oil indicates that when cavitation occurs the spectrum differs from that observed when electroluminescence is the only emission mechanism. This shift in spectrum is shown by Fig. 5 and it can be seen that this may be accomplished by a change of temperature for constant applied voltage. Comparison of Figs 3 and 4 indicates that the spectral curve for elevated temperature is associated with electroluminescence whereas at room temperature the higher emission characteristic of the cavitation mechanism becomes dominant.

Conclusions

Allan and Hizal (6) presented an arguement for pre-breakdown disturbance formation in terms of an elevated temperature liquid plasma region, produced by an electron collision process. It is difficult to reconcile this viewpoint in the light of the results presented here, where a gaseous mechanism has been shown to operate under controlled conditions of temperature and stress. Furthermore, the results depicted by Figs 3 and 4 provide conflicting evidence for liquid lift and bubble stability predictions based essentially on surface tension and its attendant temperature dependence.

Recently Shammas et al (7) presented results for hexachlorodiphenyl showing a decrease in corona inception voltage with temperature over the range 20 to 60°C. However, for the present work, results clearly provide an opposing temperature dependence for transformer oil. But the findings of Rashwan and Kao, using hexane and benzene, support the characteristic of integrated light emission found for transformer oil, namely a reduction of intensity with temperature increase. The differing pressure dependence of light intensity for benzene and hexane shown by Rashwan and Kao (8) may be explained by considering the emission mechanism for the two liquids. Benzene readily produces light by electroluminescence, and will thus show little pressure dependence, but hexane, containing no

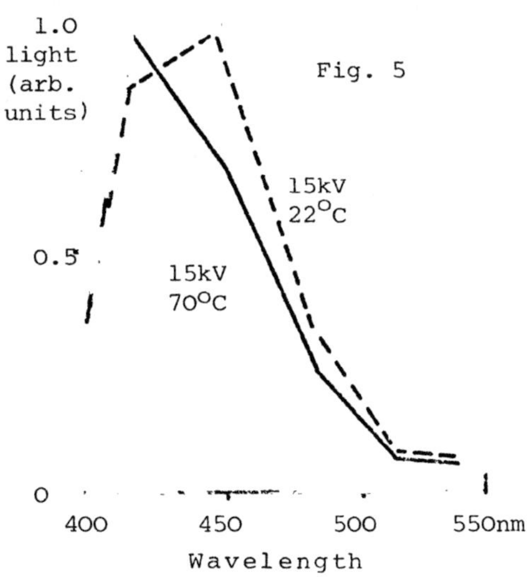

Fig. 5

Fig. 2. sphere/sphere electrodes in transformer oil, 0.75 mm gap.

All other figures, negative point/sphere electrodes in transformer oil, 3 mm gap.

aromatic compounds, will produce light only whilst cavitation occurs, and so will exhibit a reduction of light emission with increase in hydrostatic pressure.
It is suggested that as conduction current and light emission have been shown to be critically dependent on the experimental conditions it is necessary to identify the underlying mechanism before interpreting or comparing liquid dielectric measurements.

References

1. Darveniza, M. and Tropper, H., Electrical properties of organic insulating liquids containing fluorescent solutes, Proc. Phys. Soc. 1961, 78 pp. 854-868.

2. Singh, B., Chadband, W.G., Smith, C.W. and Calderwood, J.H., Pre-breakdown processes in electrically stressed insulating liquids. J. Phys. D: Appl. Phys. 1972 5 pp. 1457-1464.

3. Thomas, W.R.L., An ultra high speed laser schlieren technique for studying electrical breakdown in dielectric liquids, NRC-NAS Conf. on Elect. Ins. and Diel. Phen., October 1973, pp. 130-136.

4. Chadband, W.G. and Wright, G.T., A pre-breakdown phenomenon in the liquid dielectric hexane. Brit. J. Appl. Phys., 1965 16 pp. 305-313.

5. Malik, A.K., Electroluminescence and conduction in mineral oils subjected to time varying electric stresses, Ph.D. Thesis, London University, 1970.

6. Allan, R.N. and Hizal, E.M., Pre-breakdown phenomena in transformer oil subjected to non-uniform fields, Proc. IEE 1974, 121, 3, pp. 227-231.

7. Shammas, N., Smith, C.W. and Calderwood, J.H., Coronas in insulating liquids, J. Phys. D: Appl. Phys 1974, 7, pp. 2587 -2592.

8. Rashwan, M.M. and Kao, K.C., The effects of hydrostatic pressure and temperature on the electroluminescence in dielectric liquids, Conf. on Elect. Ins. and Diel. Phen. October 1972, pp. 60-66.

FREQUENCY DEPENDENCE OF STRESS-INDUCED CAVITATION IN FLUOROCARBON LIQUIDS

J.K. Nelson and I.F.M. Hashad*

Introduction

The excellent electrical and thermal properties of fluorocarbon liquids has prompted a study of the processes leading to breakdown in low boiling point perfluoro n-hexane in non-uniform fields. Interest in this fully fluorinated liquid stems from the possibility of its use in the nucleate boiling mode. Operation in this mode is attractive for certain industrial applications for which high heat flux densities are encountered. However, the gas phase introduced as a means of providing the enhanced convective heat transfer will limit the electric strength of the liquid. Nevertheless a substantial reduction in overall size may be achieved for certain items of high voltage plant.

One of the penalties of employing such a liquid as an insulant is that the possibility of stress-induced cavitation (1) is increased even at room temperature.

Experimental

All the measurements reported were carried out with alternating voltages derived from a specially constructed H.V. transformer having an extended frequency range (40-1500Hz). This transformer was driven from an oscillator through a 600W amplifier and low-pass filter. A thyristor crowbar circuit was employed at the transformer input terminals to limit electrode damage at breakdown. The voltage was applied to a sealed test cell which could be evacuated prior to the admission of the liquid from an integral glass multi-stage filtration and degassing system. The cell was incorporated into a lens Schlieren system for viewing the point/plane gap, but could be modified to accept an EMI 6256S photomultiplier for direct measurement of pre-breakdown light emission. The point electrode was an etched tungsten needle of about 1μm tip radius. Conventional measurements of the conduction current pulses were made (the a.c. component being rejected with a high-pass filter) together with recordings of acoustic disturbances in the liquid detected by a ceramic transducer having a resonant frequency of 400 kHz.

Results

As the voltage is raised at any given frequency, a point is reached when the liquid conducts by a series of characteristic pulses.

*The authors are with the Department of Electrical and Electronic Engineering, Queen Mary College (University of London), Mile End Road, LONDON E1 4NS.

These pulses are reproduced in the oscillogram of Fig. 1. together with the output of the ultrasonic transducer mounted in the cell. It is clear from many such recordings that, when allowance has been made for the sonic transit time, the acoustic output is correlated with the current. Furthermore the emission of pre-breakdown light also accompanies this disturbance as is evident from Fig. 1.

Fig. 2 indicates the variation of the onset voltage for the disturbance in a 1.9 mm point/plane gap as a function of supply frequency. The marked minimum occurs at about 270 Hz. The figure also depicts the variation of breakdown voltage with frequency. Although the absolute strength values may have been modified by the inevitable tip damage, the form of the curve closely resembles that of the onset voltage.

The events taking place at various frequencies are portrayed in the sketches of Fig. 3 made from large format Schlieren photographs. Despite the fact that as the frequency is increased the voltage has to be reduced to avoid breakdown (see Fig. 2), the photographs indicate the presence of frequency dependent cavitation. At 80 Hz a region of finely divided bubbles is evident near the point electrode similar to those found for hexane under pulse conditions (2). As the frequency is raised the region extends into the gap and a number of larger bubbles are precipitated in the adjacent liquid. These trends continue until at 270 Hz (Fig. 3(c)) the gap is almost completely filled with cavities.

Results of polarity tests indicate that, unlike some previous findings (3), the onset of this pre-breakdown event occurs first when the point is positive. This is probably the result of the inhibiting effects of negative ion space charge for negative point conditions.

Discussion

Stress-induced cavitation occurs at the point electrode but it is thought that particulate contamination plays a vital role in the nucleation of cavities. Malik (4) has demonstrated that the amplitude of particulate motion decays with frequency and in transformer oil particles of the order of 1μm diameter are virtually frozen at 400 Hz. This behaviour may explain the minima of Fig. 2 since there will be an optimum frequency for which the 'loosening up' of the fluorcarbon liquid in the vicinity of the point will be a maximum. This will increase the probability of stress-induced cavitation by the electrostatic forces (5). The lack of symmetry observed in Fig. 2 is to be expected on account of the particle size distribution in the liquid.

The type of cavitation produced in the high field region is characteristically different from that occurring in the adjacent regions. The propagation of shock waves from the point may precipitate cavitation at nucleation sites. This takes the form

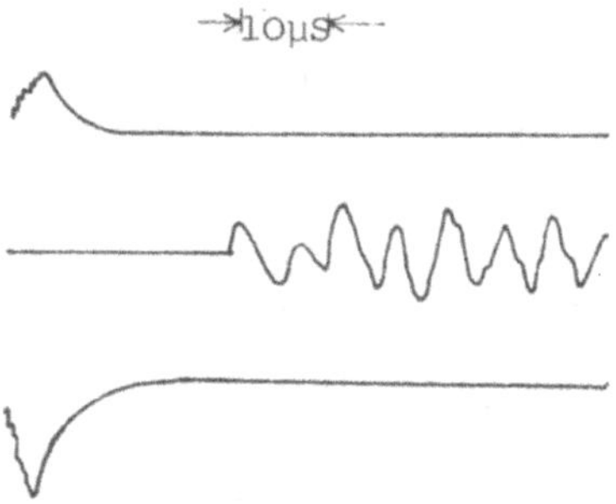

Fig. 1 (above) Oscillograms showing relationship between current pulses (upper trace) and light output (lower trace) together with the ultrasonic output.
Voltage: 4.2kV, 50 Hz;
Gap: 1.5 mm, point positive.

Fig. 2 (opposite) Inception voltage for pre-breakdown disturbance (i) and breakdown voltage (ii) as a function of frequency. Point/plane 1.9 mm gap.

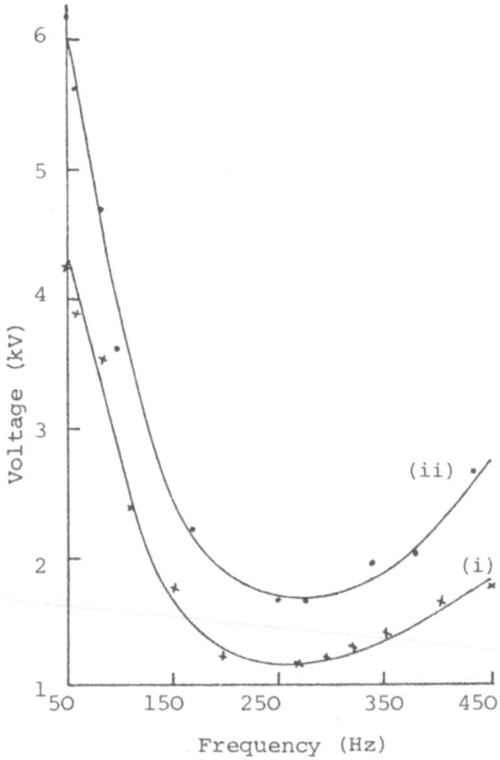

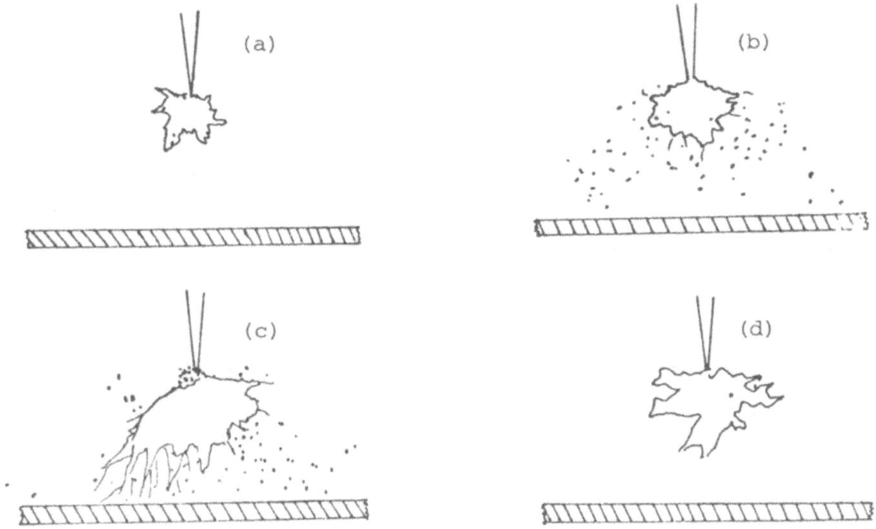

Fig. 3 Reproduced Schlieren images showing cavitation in perfluoro n-hexane. Gap: 2.34 mm. (a) 80 Hz, 2.8 kV (b) 180 Hz, 1.85 kV (c) 270 Hz, 1.85 kV (d) 400 Hz, 2.1 kV.

of small microbubbles in the high stress region since the pressure due to the electric field imposes a small radius constraint. Outside the high field region bubbles may grow to a much larger diameter.

The production of shock waves (see Fig. 1) both at and near (2) the point raises the question of the effects of cavity collapse. The small microbubbles will be inherently unstable and subject to rapid growth and collapse. In the lower field regions, the bubbles, both having large radii and being subject to lower intensity shock waves, can remain stable for several milliseconds. Shammas et al (6) have shown that pre-breakdown light emission in n-hexane is confined to the region round the point. This raises the question of the mechanism of light emission since there must be some doubt that the free path of electrons is sufficient to support a classical gas phase discharge. However, the energy density associated with the rapid collapse of such cavities may be sufficient to raise the local temperature to a level permitting thermal ionisation and the consequent emission (7). Although little quantitative evidence exists to verify the hypothesis of sonoluminescence in stressed liquids, it must be recognised that the shock waves detected will be associated with collapse rather than nucleation and the emission of light is intimately connected with acoustic disturbances.

Acknowledgements

The authors are indebted to the Marconi Company for the provision of samples and IFMH for the financial support of the Arab Republic of Egypt.

References

1. B. Singh, C.W. Smith and J.H. Calderwood, 4th International Conference on Conduction and Breakdown in Dielectric Liquids, p. 202, (1972) Dublin.

2. P.B. McGrath and J.K. Nelson, IEE Conf. on Dielectric Materials, Measurements and Applications, Cambridge (1975)

3. B. Singh, W.G. Chadband, C.W. Smith and J.H. Calderwood, J. Phys, D.: Appl. Phys., p. 1457 (1972).

4. A.K. Malik, Ph.D. Thesis, London University (1970).

5. Z. Krasucki, Proc. Roy. Soc. A, 294, p. 393 (1966).

6. N. Shammas, C.W. Smith and J.H. Calderwood, J. Phys. D: Appl. Phys. 7, p. 2587 (1974).

7. B.E. Noltingk and E.A. Neppiras, Proc. Phys. Soc. 638, p. 674 (1950).

HIGH-FIELD CONDUCTION PULSES IN N-HEXANE

C.J. Buffam and J.E. Brignell*

Introduction

In a previous conference paper[1] it was shown that it is possible to monitor the conveyance of charge by naturally occuring carriers. Pulse-height-analysis yields information about the amount of charge carried, but if the actual current or charge waveform can be captured it is possible to establish the mobility of the charge, the uniformity of the field through which it passes, its effective radius and perhaps whether charge multiplication occurs. The results previously reported relied either on random capture or the sacrifice of information to achieve triggering. Neither method is convenient, and they both fail at high speeds which characterise the most interesting physical phenomena.

Since that work was performed, an instrument has appeared which is designed to solve precisely the problem of capturing fast random waveforms. It is the Transient Recorder, a digital device which offers high sampling rates and an adequate number of samples per waveform. It operates in a mode of continual cyclic data storage which is arrested by the delayed trigger signal. Thus pre-trigger information is captured, which means that the entire waveform can be stored, whatever feature of it is actually used for trigger purposes.

An almost equally important characteristic of the device is its digital nature, which makes it particularly adaptable to direct connection to an on-line computer with a great increase in its capabilities. In this paper we describe how such a combination can be applied to the study of infra-breakdown phenomena in liquid dielectrics, and discuss some preliminary results.

Apparatus

The choice of techniques employed in this work is mainly conditioned by the fact that the amount of charge to be measured can be very small (< 1 fC) yet the transit times to be measured can be very short (< 1 μS); so the question of noise and interference is dominant. The test-cell[2] is of cylindrical construction in stainless steel with internal PTFE insulation and co-axial connections, and it is housed with the principal measurement equipment in a screened room. However, the computer-controlled power supply[3], the Transient Recorder and the computer all constitute interference sources and therefore must remain outside the screened enclosure. Thus interconnections through the screen pose a major practical problem.

* Department of Electrical and Electronic Engineering, The City University, St. John Street, London EC1V 4PB, England.

The signal is acquired through a charge sensitive amplifier connected to the low voltage electrode of the test-cell. The electrodes are spherical, as this is the only way of achieving sufficiently low breakdown probabilities in the high-stress region (the probability is proportional to the stressed area[4]). The signal from the amplifier is fed via a second amplifier to the Transient Recorder. This device, under computer control can be used to trap transient waveforms of duration from 0.1 μS to 6 hours which is more than sufficent for the task in hand. A great advantage of connection to the computer is that it becomes possible to apply digital correction for the inevitable deficiencies in the characteristics of the input channel[5]. For example, it is never possible to use a true charge-sensitive amplifier, as this would be a perfect current integrator, which would eventually bias itself off with unipolar conduction pulses, so a discharge resistor is included as a deliberate imperfection. In the computer, a simple digital filter of appropriate design can completely compensate for this deficiency and restore the charge waveform to its true shape. The use of charge- rather than current- sensitivity reduces noise on the waveform, which is a particular advantage in a digital channel of relatively low capacity (8 bits).

It is essential in this sort of measurement to be able to treat relatively large numbers of transients and here advantage can be taken of computer backing store. We have used a disc store of capacity 48 M bits, which allows 2000 complete waveforms to be stored in digital form. These may be later displayed on a VDU, plotted on an incremental plotter or used as the basis of direct digital processing, e.g. pattern recognition.

As far as the liquid sample for this work is concerned, the most important impurity is the solid particle. The test-cell is provided with a local filtering (.1 μm pore) and recycling system which allows the gap to be flushed; and the liquid is prepared in a separate purification system which also contains a filter of .1 μm. In our experience, however, the only way of achieving sufficiently low particle numbers to be able to distinguish their separate contributions and possible other effects is to keep the cell sealed and keep using it with fresh samples[6]. In this way the particle population of the cell slowly declines but will to some extent re-establish itself each time the electrodes are changed. Because of these considerations much of the early work on this system reported here is concerned with solid particles and their nature as charge carriers.

Measurements

Fig. 1 shows an example from the many thousands of waveforms captured as part of the preliminary study of solid particle motion. The time scale has been compressed more than usual to show the effects of the imperfect charge sensitive amplifier response, but the region of interest is the rising ramp which represents the transit of a single particle. Almost always this represents remarkably uniform motion. Note the period of quietude before the transient, which is a unique feature of this method of recording.

From such waveforms one may make a number of deductions about the behaviour of the particle. Directly, the height and shape of the ramp yield the charge and velocity of the particle. Indirectly, assuming that Stoke's law applies, we have the particle radius. Hence a graph of the form shown in fig. 2 can be drawn. This shows particle charge plotted against particle radius for four different values of applied stress. To indicate the order of scatter, experimental points are shown for one case, The slope of this line is a measure of the potential to which the particle has become charged. Previous experimental work has produced relatively low values (16 volts for artificial particles[7] in Freon 113, and 12 volts for natural ones in n-hexane[6]). It is interesting to note that the slope of line A in fig. 2 corresponds to 11 volts.

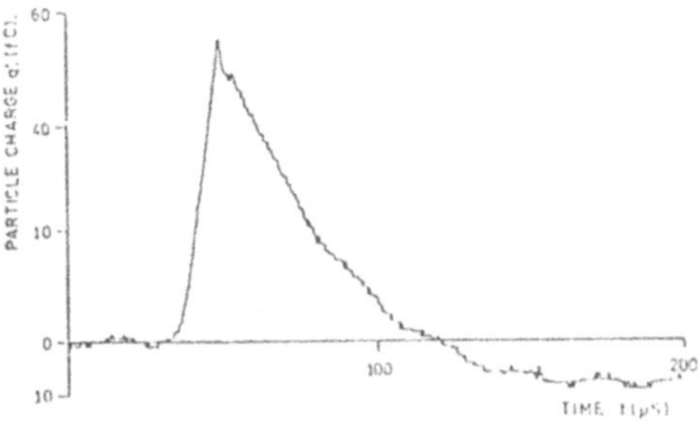

FIG. 1 CHARGE TRANSIENT (Uncorrected)

If the assumptions are valid, the particle velocity ought to be proportional to the applied stress. These quantities are plotted in fig. 3. The regression line shown corresponds to a particle mobility of 0.45×10^{-6} $m^2.V^{-1}.S^{-1}$. The value previously obtained was 0.37×10^{-6} $m^2.V^{-1}.S^{-1}$. Considering the above-mentioned defects of the previous method this agreement is remarkable, especially for a phenomenon which might be expected to exhibit wide variation.

Discussion

Current work is concerned with an attempt to investigate the suggested occurrence of much larger pulses not directly connected with particle motion. At the time of writing, the particle population of the test-cell was too high for this to be easily achieved. What has been done so far has strikingly confirmed the rather tentative conclusions drawn from former results.

Naturally occurring particles are extremely inefficient in their role as charge conveyors. Their dwell time is very much longer than their transit time. It appears that because of the insulating nature of the particle and/or layers on the electrode surface, changing takes place relatively slowly

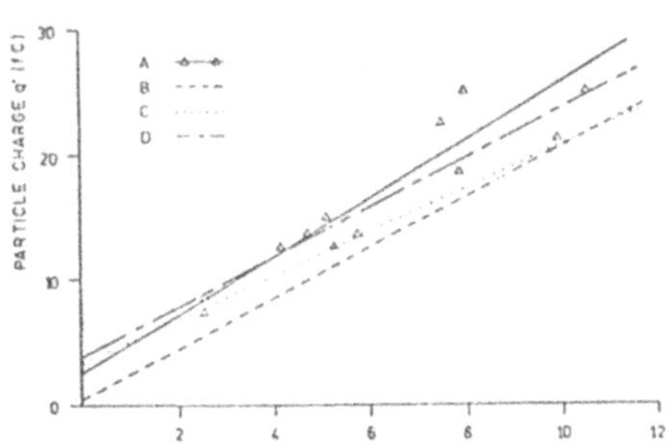

FIG 2 Particle charge v. particle radius for stresses of:-
A, 22.7 $MV.m^{-1}$; B, 30.3 $MV.m^{-1}$;
C, 37.8 $MV.m^{-1}$; D, 45.5 $MV.m^{-1}$.
The electrode gap was 132 µM.
Points are plotted for stress A.

and ceases as soon as the electrostatic force due to the applied field is sufficient to overcome the forces which bind the particle to the electrode. When the particle population is very high the behaviour can be much more complicated, e.g. the formation of chains, but the impulsive current is never significant in comparison with the total conduction current.

With the waveforms stored in the computer, it is relatively easy to apply automatic techniques of pattern recognition and to measure and plot the charge transport parameters without the necessity for human effort. Thus one may work with large numbers of results and obtain accurate average estimates.

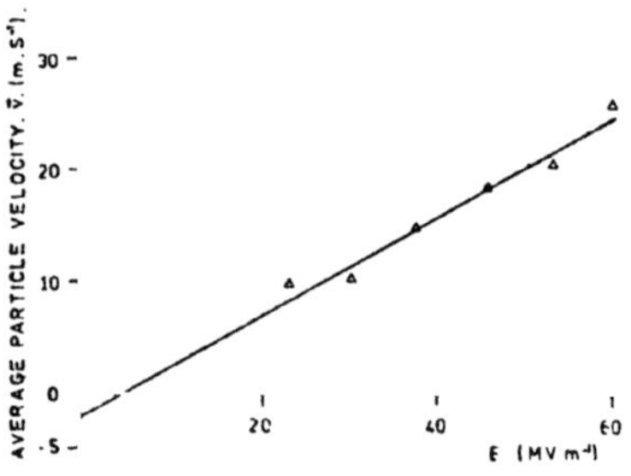

FIG 3 Average particle velocity v. Electric Stress for an electrode gap of 132 μM.

Acknowledgments

We wish to thank Dr. E.O. Forster for suggesting the Transient Recorder as a means of improving the measurement technique. The equipment and maintenance of one of us (CJB) were provided by the U.K. Science Research Council.

References

1. Rhodes, G.M., Brignell, J.E., 1972, Proc., 4th Int. Conf. 'Conduction and Breakdown in Dielectric Liquid' Typografia Hibernia.

2. Rhodes, G.M., 1971, Ph.D. Thesis, The City University.

3. Mears, B.R., Brignell, J.E., This conference.

4. Brignell, J.E., Metzmecher, K.D., 1971, J. Phys. D., 4, 253.

5. Brignell, J.E., Rhodes, G.M., 'Laboratory On-Line Computing', Intertext 1975.

6. Rhodes, G.M., Brignell, J.E., 1971, J. Phys. D., 4, L47.

7. Felsenthal, P., Vonnegut, B., 1967, Brit. J. Appl. Phys., 18, 1801-6.

ELECTRICAL CONDUCTANCE AND BREAKDOWN IN LIQUID HYDROCARBONS

W. R. L. Thomas and E. O. Forster*

Introduction

The transport of electrical energy through liquid hydrocarbons and the events leading to breakdown have been investigated extensively. Studies of the conduction mechanism usually involved measurements under dc conditions because sufficiently sensitive ac techniques have become available only in recent years with the advent of bridges capable of resolving resistances as high as 10^{15} ohm.(1) Research in the breakdown process has dealt mainly with either impulse conditions, which simulated lightening or switching surges, or ac stresses which were allowed to rise at a preselected rate until the material failed.
During the last two decades new, more sophisticated techniques have become available which permitted the determination of charge carrier mobilities in carefully purified hydrocarbon fluids(2) and of processes leading to breakdown.(3) With all this new information being generated, it appeared desirable to assess the state of knowledge in this area. This paper represents a critical review of the published literature followed by an attempt to unify the experimental findings into a more or less quantitative theory of conduction and breakdown.

Conduction Measurements (Including Mobilities)

Most of the early work was concerned with the interpretation of current-voltage measurements which were performed at various temperatures, with different electrode materials, fluids of different molecular weight and impurity content, etc. These studies have led to the determination of activation energies for conduction and they have shown that the presence of small amounts of oxygen could affect seriously the level of conduction currents. But these studies failed to identify unambiguously the nature of the charge carriers. Krasucki(4) for example postulated that a large number of results could be explained by the presence of submicron size conducting particles which transferred charge between the electrodes. Others believed the charge carriers to be ions or electrons. Many of these current-voltage measurements were complicated by the occurrence of more or less significant current pulses which appeared to be of rather low frequency.
To supplement the amount of information available from current-voltage measurements regarding charge carriers, methods for determining carrier mobilities became of interest. Some investigators employed photoinjection, while others produced carriers by γ-ray irradiation. The mobilities so measured were always of the order of 10^{-4}V cm^2/sec. The observed dependency of mobility on current was eventually interpreted to indicate that the charge motion measured was due to bulk liquid motion.

*Exxon Research and Engineering Company, P. O. Box 45, Linden, New Jersey 07036

In recent years fast carriers have been observed in carefully purified and degassed liquid hydrocarbons and the experimental results have been explained in terms of free electrons which are trapped occasionally in molecular cages but which otherwise can move freely through the liquid. While the results discussed so far were all obtained with rather low electric fields, the following section deals with measurements carried out near or at electrical breakdown, i.e. in the presence of extremely high fields.

Breakdown Measurements

The voltage at which electrical breakdown occurs, V_b, has been measured for a number of liquid hydrocarbons under a variety of conditions. Since electrical breakdown is a destructive process the repetitive use of either the same electrodes and/or the same liquid to obtain reproducible results is highly questionable. Recent experiments with impulse voltages have indicated that large pulses of high current but low duration occur at or near electrical breakdown.
The problem with the use of conventional current measuring devices is the inherent limitations in the response time of the instruments. This realization has lead to the use of optical methods such as the Schlieren technique. The time resolution in these experiments was limited by the lack of adequate illumination sources to about one microsecond. An improvement was made with the use of a mechanical streak camera. Another approach was used by several investigators who adopted more viscous fluids to slow down the breakdown process. Their results suggested that bubble formation occurred at very high electrical fields, a phenomenon first postulated theoretically by Sharbough and Watson. The bubble was observed to become elongated in the direction of the field. If conduction took place in the cavity it would become unstable and assume tree like figures. Recent pulse laser Schlieren pictures taken with low viscosity fluids such as n-hexane have confirmed these observations.[(3)]
Studies have also been made of fluorescent dielectric liquids under high electrical stresses. These observations indicated that electrons could gain significant energy in the liquid. Light emission occurring shortly before breakdown has been observed with the help of an image intensifier and converter camera. Using the Kerr effect and a two objective Schlieren system, Felici and his students at Grenoble studied the changes occurring in nitrobenzene as high electrical fields were applied.
Summarizing then the results reviewed here it is quite obvious that conductivity measurements alone do not give all the information required to predict modes of breakdown. No sufficiently detailed information regarding charge carriers is obtained. Measurements of large current pulses might provide useful information particularly regarding electrode surface effects if they could be carried out at high enough frequencies, i.e. in the 100 MHz range or above. Mobility measurements are also limited in the amount of useful information they provide. The discovery of "free" electrons in a number of ultra pure liquid hydrocarbons is of great importance, but does not explain the role of impuri-

ties in electrical conduction processes. In the next section an attempt will be made to develop a theoretical picture that explains equally well low and high field conduction as well as breakdown behavior of liquid hydrocarbons.

A Model for Conduction and Breakdown

The proposed scheme visualizes as the first step the creation of charge carriers at the electrode-liquid interface since the hydrocarbons under consideration are intrinsic insulators. It is quite reasonable to assume that electrons will be injected from the cathode surface into the liquid. At certain asperities on the electrode surface electron emission will be favored under the influence of the applied field. However this electron injection will not occur at constant rate for the following reasons. First it is field dependent and the electron cloud leaving the metal will reduce the field in the immediate vicinity of the emitting point. After a certain small time (10^{-9} - 10^{-7} sec) the electron cloud will move away from the surface, thereby allowing the local field at the asperity surface to increase, resulting in more electron emission. However, a steady space charge limited current will eventually be reached. Space charge relaxation in the metal electrode will occur in less than 10^{-12} sec, so that the electrode surface can be considered at a constant potential. Current fluctuations could also occur if the emission changed the electron emission characteristics of the asperity. The fate of these electrons will depend on the nature and purity of the liquid, the magnitude of the applied field, as well as the prevailing experimental conditions. For example in ultrapure liquid aliphatic hydrocarbons and working with carefully degassed surfaces, free electrons could probably travel by moving in and out of molecular cages. In other pure liquid hydrocarbons or in less pure systems the electrons could form ion-radicals in combining with O_2 or other electron acceptors. These molecular cages or ion-radicals will be the charge carriers which give rise to the observed currents. As the applied field increases, the rate of electron injection will increase as well as the drift velocity of the carriers, and the net result will be an increase in current which will not be necessarily proportional to that of the field. This behavior is visualized to be in large part due to the events occurring on the opposite electrode where charge carriers are supposed to be discharged. While it is easy to picture electrons arriving in form of molecular cages to find their way to the metal surface, it is harder to conceive of the loss of electrons from rather stable ion-radicals. These latter charge carriers appear to form space charge in front of the electrode and this space charge will modify the field ahead. This increased field could well facilitate the movement of the electron into the metal. The latter could well occur intermittantly leading to current pulses. It is also important to realize that more asperities on the cathode can become emission sites as the field is increased. Concurrently more discharge sites may become available at the anode
As the applied field is increased further, it will reach a critical value. Under these conditions the rate of electron emission

from the asperity is so rapid that liquid will be literally lifted from the surface and a region of low density will form in the vicinity of the asperity. These postulates are supported by the results obtained with the high speed laser illuminated schlieren system.[3] An example of these results is shown in Figure 1 which indicates the formation of density gradients caused by the heating of the liquid due to charge injection from the cathode. As this condition is reached, a new regime develops causing growth of the low density region into a cavity in which discharges can occur. In this region emitted electrons will have a sufficient mean free path to accelerate under the influence of the electric field and thereby gain enough energy to cause excitation and/or ionization of solvent molecules upon impact. Figure 2 illustrates this situation. The density gradients seen in Figure 1 have been replaced with dark regions which reflect the laser light indicating the existence of plasma like conditions.

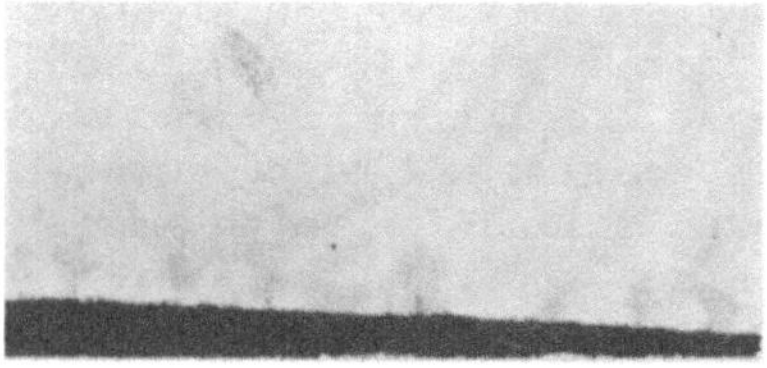

Fig. 1 Density gradients at cathode. Polished brass electrodes, separation 0.4 cm n-hexane, about 1.5 μsec before breakdown.

Fig. 2 Plasma formation at cathode; same conditions as in Fig. 1. About 1 μsec before breakdown.

These localized discharges release further energy to sustain cavity growth. At the same time, the ions they produce cause the formation of a plasma within the cavity and an instability of its surface due to the accumulation of charges. The latter leads to the formation of spikes which grow particularly in the direction of the field.[3] Under certain circumstances, one of these will eventually come close enough to the opposite electrode to induce the localization of considerable positive charges on the surface of the nearest asperities. The attraction of the opposite charges then facilitates breakdown of the last liquid barrier and the formation of a discharge path between the electrodes. Alternately, electron impact ionization may occur in the liquid because of the extremely high electric field at the tip of the spike. This will lead to a self-propagating streamer rapidly crossing between the electrodes, as observed in high pressure gas discharges. Either of these mechanisms results in a sudden passage of a large current leading to a total collapse of the field and extensive release of energy. As a consequence of these processes, shock waves are generated and the recombination of ions leads to excited molecules which give up light upon returning to the ground state. Figure 3 illustrates these features in detail. Although the model described above adequately accounts for much published data on breakdown strengths in highly purified liquids, it should be recognized that in less controlled situations the formation of unstable cavities may not be the only mechanism by which a very high localized field can be created in the liquid, leading to streamer breakdown. Under actual service conditions, industrial insulating liquids may contain long thin oriented particles. Also, unpolished electrodes may possess long thin asperities. Either of these two conditions may provide a sufficiently high localized field for streamer initiation, and subsequent breakdown. Figure 4 shows an example of this process initiated by a small particle suspended in toluene.

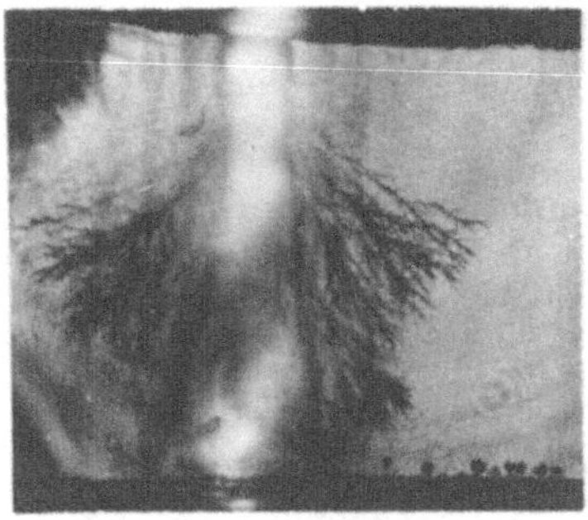

Fig. 3 The consequences of electrical breakdown; same conditions as in Fig. 1. About 0.2 μsec after breakdown.

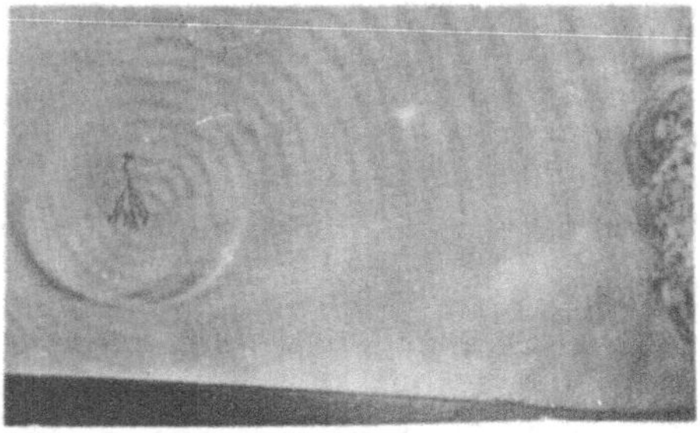

Fig. 4 Streamer initiation from a suspended particle in toluene; same conditions as in Fig. 1. About 0.2 μsec after breakdown.

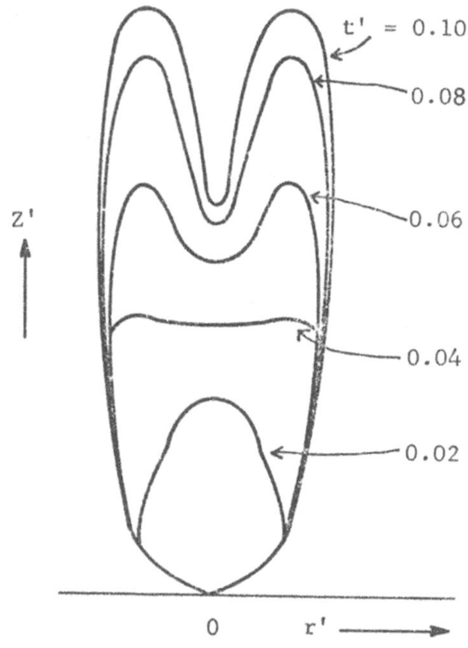

Fig. 5 Contours of Constant Temperature rise as calculated from mathematical model for case of charge trapping. Z' is direction towards anode, r' is parallel to cathode surface, t' is time in μsec elapsed from beginning of el. field application.

Some of the ideas presented in the preceding paragraphs have been used by Thomas[5] to develop a computer program which describes mathematically the process of charge emission in the presence of space charge distortion of the electrical field. This program performs a complete two dimensional solution of the charge motion in liquid insulators. The calculations demonstrate that only small temperature increases occur in the liquid prior to cavitation and that the regions of increased temperature have complex irregular shapes owing to the instability of the surface of a space charge cloud. Figure 5 represents the results of such calculations. The contours drawn indicate the onset of the irregular shapes. Both these predictions have been verified by high speed laser schlieren photography,[3] thus lending further credence to the validity of a crucial part of the proposed model. The charge carriers in the cases studied by Thomas[5] were identified as electrons even in relatively impure n-hexane.

Summary

The model described in the preceding section appears to be capable of predicting quite well the observations reported in the literature regarding conduction and breakdown phenomena in liquid hydrocarbons. In its simplicity it is quite attractive and it might be even useful in the interpretation of similar phenomena in solids particularly plastics and elastomers. It certainly deserves further attention and attempts will be made in the near future to put it onto a more quantitative basis.

References

1. E. O. Forster, Proc. 4th Inter. Conf. on Cond. and Breakdown in Liquid Dielectrics, Dublin, Ireland, 148 (1972).
2. See for example, W. F. Schmidt and A. O. Allen, J. Chem. Phys. 52, 4788 (1970).
3. W. R. L. Thomas, 1973 Annual Rept. Conf. on Electrical Insulation and Dielectric Phenomena, NAS, Washington, D.C. (1974).
4. Z. Krasucki, Phenomenes de Conduction dans les Liquids, Isolants, Grenoble, 1968, p. 311.
5. W. R. L. Thomas, 1974 Annual Report Conf. Electrial Insulation and Dielectric Phenomena, NAS, Washington, D.C. (1975).

ON THE THEORY OF FILAMENTARY SINGLE INJECTION AND HIGH-FIELD ELECTRIC CONDUCTION AND BREAKDOWN IN DIELECTRIC LIQUIDS

K. C. Kao*

The presence of unavoidable impurities and charge-carrier injection from electrodes play a major role in high-field conduction and breakdown in liquids [1,2]. It should be emphasized, however, that it could lead to an incorrect conclusion if the field emission equation of Fowler and Nordheim and the thermionic emission equation of Schottky, which were originally derived for electron emission from a clean metal to vacuum, were used directly without modification to correlate with the experimental results on liquids. In liquids the effects of space charge [3] and electrode surface condition [4] can not be ignored. It is therefore most likely that at fields higher than the threshold field for carrier injection from the electrodes the electric conduction becomes filamentary because the electrode surface is not microscopically identical from domain to domain in asperity and surface condition, and thus there may be one or more microregions at which the potential barrier has a profile more favourable for carrier injection than that at other regions. Furthermore, the liquid itself is never microscopically homogeneous, particularly under high fields after the onset of electrohydrodynamic motion. Thus, for a given applied voltage across two parallel plane electrodes (with the edge effect neglected) the field E is not uniform longitudinally due to the space charge effect and the current density J is not uniform radially due to the filamentary current flow. The current filaments formed in chlorobenzene have been reported by Goodwin [5] and in anthracene by Kao et al [6]; and the idea is further supported by the fact that the electric breakdown channel is always filamentary no matter whether it is in a liquid, in a solid or in a gas, and that a breakdown produces only a tiny spot of damage on each of the electrode surfaces.

It is proposed that several filaments of currents I_j may simultaneously occur in a liquid gap. Thus the total current can be written as

$$I_T = \sum_j I_j \simeq HI = H \int_0^{r_d} J_z(r)\ 2\pi\ rdr \qquad (1)$$

where J_z is the current density in one filament of current I confined mainly in a filament of radius r_d, and H is a parameter which may depend on the field and electrode condition. In this analysis we consider only one filament along the z-axis with the plane at $z = 0$ as the cathode and that at $z = d$ as the anode in the cylindrical coordinates, and make the following assumptions: (a) the longitudinal component of the diffusion current can be ignored because of the large longitudinal component of the electric field, and the radial component of the drift current can be ignored because of the small radial component of the electrical field; (b)

*Materials Research Laboratory, Department of Electrical Engineering, University of Manitoba, Winnipeg, R3T 2N2, Canada

the mobilities of negative and positive charge carriers, μ_n and μ_p, are constant and independent of electric field; (c) the charge carriers are mainly produced by electron injection from the cathode and ionization of the molecules of impurities and of the liquid itself by extraneous radiation such as cosmic rays into positive ion-electron pairs with the generation rate G, and an electron may spend some time as a free electron and some time as a negative ion due to its attachment to a neutral molecule; and (d) the concentration of negative charge carriers n (mainly generated by injection from the cathode) is much larger than that of positive charge carriers p (mainly generated by extraneous radiation), and they tend to recombine to form neutral molecules again with the recombination coefficient α. The electric conduction in one filament is therefore governed by the following current flow equation

$$\vec{J}(r) = \vec{J}_z(r) + \vec{J}_r(r) = q\mu_n nE(z)\,\vec{i}_z + q\left[\mu_n nE(r) - D_n\frac{dn}{dr}\right]\vec{i}_r \qquad (2)$$

the continuity equation

$$\partial p/\partial t = G - \alpha pn \qquad (3)$$

$$\partial n/\partial t = G - \alpha pn - \vec{\nabla}\cdot\vec{J}(r)/q \qquad (4)$$

and the Poisson equation

$$\vec{\nabla}\cdot\vec{E} = \rho/\varepsilon = -q\,(p - n)/\varepsilon \simeq qn/\varepsilon \qquad (5)$$

since it is assumed that $n \gg p$; where $\vec{i}_z$ and $\vec{i}_r$ are, respectively, the unit vectors in the z and r directions, and other symbols have the usual meanings. In the steady state and on the basis of the fact that E in the r direction is negligibly small but dE/dr is not, and that physical reality requires n to be finite for all values of r and $dn/dr \to 0$ when $r \to 0$; then from Eqs. (2) - (5) we obtain

$$d^2n/dr^2 = q\mu_n n^2/\varepsilon D_n \qquad (6)$$

By expressing the current density at r = 0 as $J_{zo} = q\mu_n n_o E = q\mu_n \bar{n}_o E_{av}$, and using the boundary conditions: $E \to E_c$ (the electric field at the cathode) when $z \to 0$ and the applied voltage $V = \int_0^d E\,dz = E_{av}d$; the solution of Eq. (5) leads to

$$J_{zo} = \frac{9}{8}\varepsilon\mu_n \frac{E_{av}^2}{d} M = \frac{9}{8}\varepsilon\mu_n \frac{V^2}{d^3} M \qquad (7)$$

where

$$M = -(4Y^2 - 3)/6 + \left\{\left[-(4Y^2 - 3)/6\right]^2 - 16Y^3(Y - 1)/27\right\}^{\frac{1}{2}} \qquad (8)$$

$$Y = E_c/E_{av} \qquad (9)$$

$\bar{n}_o$ and E_{av} are, respectively, the average value of n_o (the value of n at r = 0) and that of E along the z direction. By solving Eq. (6) and expressing $J_z(r) = q\mu_n \bar{n}E_{av}$, where $\bar{n}$ is the average value of n along the z direction; then we obtain

$$J_z(r) = J_{zo}\left[1 + (3M\mu_n V/16D_n)^{\frac{1}{2}}\, r/d\right]^{-2} \qquad (10)$$

and

$$I = \int_0^{r_d} J_z(r)\,2\pi r\,dr = J_{zo}(2\pi d^2)(16D_n/3M\mu_n V)\left\{\ln\left[1 + (3M\mu_n V/16D_n)^{\frac{1}{2}}\, r/d\right] + \left[1 + (3M\mu_n V/16D_n)^{\frac{1}{2}} r/d\right]^{-1} - 1\right\} \qquad (11)$$

M as a function of Y and $J_z(r)/J_{zo}$ as a function of r are shown in

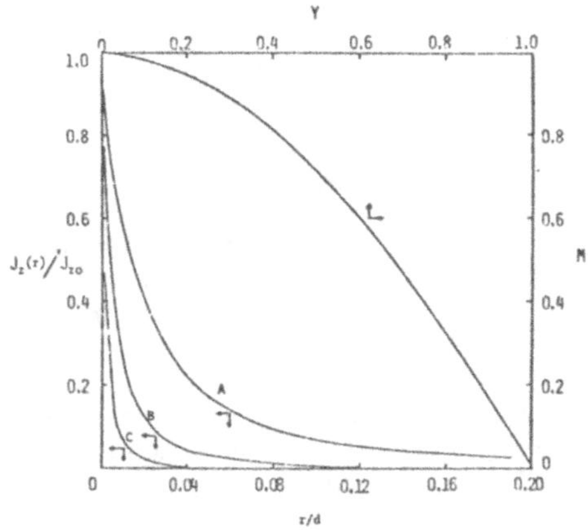

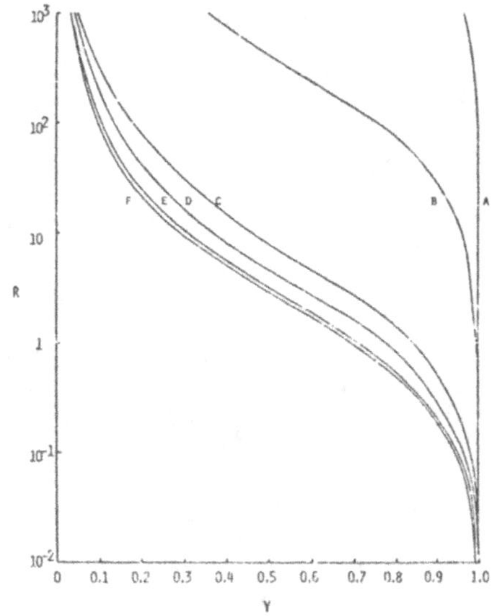

Fig.1 M as a function of Y and $J_z(r)J_{zo}$ as a function of r/d. A: M=16/300, d=0,02 cm and $E_{av}=10^5$ V/cm; B: M=16/300, d=0,02 cm and $E_{av}=10^6$ V/cm; or M=16/30, d=0,02 cm and $E_{av}=10^5$ V/cm; C: M=16/30, d=0,02 cm and $E_{av}=10^6$ V/cm.

Fig.2 Y as a function of R. A: S=10; B: S=5; C: S=1; D: S=0,5; E: S=0,1 and F: S=0.

Fig. 1. By assuming that the electron injection is mainly due to tunneling through a potential barrier at the cathode following the equation [2]

$$J_c = aE_c^2 \exp\,(-b/E_c) \tag{12}$$

The current density at the cathode J_c can be considered to be equal to J_{zo} since the current density in the z direction is continuous. Thus, substitution of Eq.(9) into Eq.(12) and then into Eq. (7) gives

$$Y = \frac{8}{27\,[Re^{-S} + 2/3]^2 + 4} + \left\{ \left(\frac{8}{27[Re^{-S} + 2/3]^2 + 4} \right)^2 + \frac{27Re^{-S}}{27[Re^{-S} + 2/3]^2 + 4} \right\}^{\frac{1}{2}} \tag{13}$$

where $R = 8ad/9\varepsilon\mu_n$ and $S = b/YE_{av}$ (14)

The plot of Y as a function of R for various values of S is shown in Fig. 2. It can be seen that E_c/E_{av} or the space charge effect depends on both d and V.

It is also proposed that at the field near the breakdown strength there is only one filament, in which the current density is much higher than that in other filaments. It is this filament to lead to breakdown when its current density reaches such a value that the Joule heating produced in it is sufficient to raise the temperature of the liquid to its boiling point temperature. Supposing that the breakdown strength is measured with rectangular pulses of duration τ, and assuming that μ_n follows the relation $\mu_n = \mu_o \exp(-W/kT)$, and that the thermal conduction term can be neglected; then the solution of the energy balance equation

$$C_p g \frac{dT}{dt} = J_{zo} E_{av} \quad (15)$$

gives the critical field to cause electric breakdown. This critical field is

$$(E_{av})_b = \left\{ \frac{8}{9} \frac{C_p gkd}{W\epsilon\mu_o M\tau} \left[T_o^2 \exp(W/kT_o) - T_b^2 \exp(W/kT_b) \right] \right\}^{1/3} \quad (16)$$

where k is the Boltzmann constant; T_o and T_b are, respectively, the initial and boiling point temperature of the liquid within the filament; and C_p and g are, respectively, the specific heat and the density of the liquid.

For a given gap length and at a constant temperature we can assume for most cases that the term $(3M\mu_n V/16D_n)^{\frac{1}{2}}(r_d/d)$ is much smaller than unity. Then from Eqs. (1) and (11) the total current can be approximated to

$$I_T = AH\,E_{av}^2 + BH\,E_{av}^{5/2} \quad (17)$$

where A and B are constants. In most cases A >> B and therefore I_T is approximately proportional to E_{av}^2 if H is assumed to be independent of the field. However, Eq. (17) can be made to fit most available experimental results on $I_T - E_{av}$ characteristics. By choosing the values of R between 2.1×10^{-2} and 1.6×10^{-1} corresponding to the range of d between 0.0055 cm and 0.042 cm to find the values of Y from Fig. 2 and then to find the corresponding values of M from Fig. 1; we have computed I_T as a function of E_{av} for n-hexane using Eq. (17) for three values of d, and the results are in good agreement with the experimental ones of Goodwin and Macfadyen [7] as shown in Fig. 3. It should be noted that the current density determined by dividing the total current by the electrode area may completely mislead the microscopic physical picture, and that the gap length dependence of I_T can be interpreted in terms of electron injection from the cathode and space charge effect in current filaments without involving the electron multiplication process. In a liquid not only ionic impurities but also neutral impurities will be drawn into the high-field region in the liquid gap from its surroundings by the electrophoretic and dielectrophoretic forces, and these impurities will form scavenging traps which indirectly reduce the average mobility of charge carriers [8, 9]. As time is required for these impurities to be drawn to the liquid gap, the impurity concentration in the liquid gap would be much smaller and the average mobility much larger under pulse conditions than under d.c. voltage conditions. Thus, we can choose $\mu_n = 0.1\ cm^2V^{-1}sec^{-1}$ for pulse conditions (instead of $10^{-3} cm^2V^{-1}sec^{-1}$ for d.c. voltage conditions) for n-hexane at room temperature. However, by choosing the values of R between 3×10^{-2} and 12×10^{-2} corresponding to the range of d between 6.35×10^{-3} cm and 25.4×10^{-3} cm, M/d and hence I_T become independent of d within this range of gap lengths. This explains the pulse conduction results of Watson and Sharbaugh [10]. The tem-

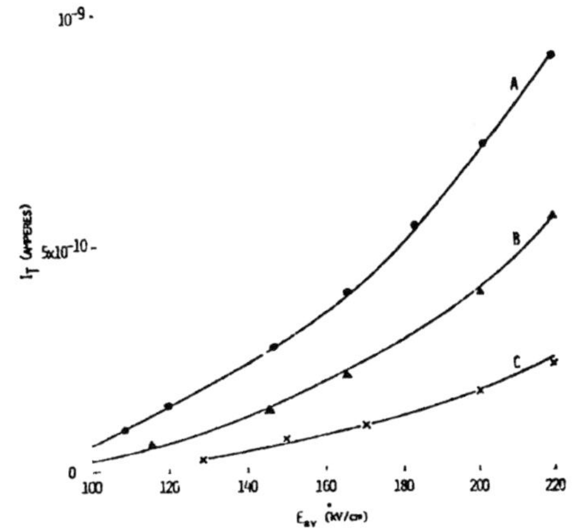

Fig. 3 Total current as a function of average electric field strength in n-hexane for three gap lengths.
A: d=0,042 cm; B: d=0,020 cm and C: d=0,0055 cm.
Theoretical curves;
● ▲ × Experimental results with phosphor bronze electrodes (after Goodwin and Macfadyen).

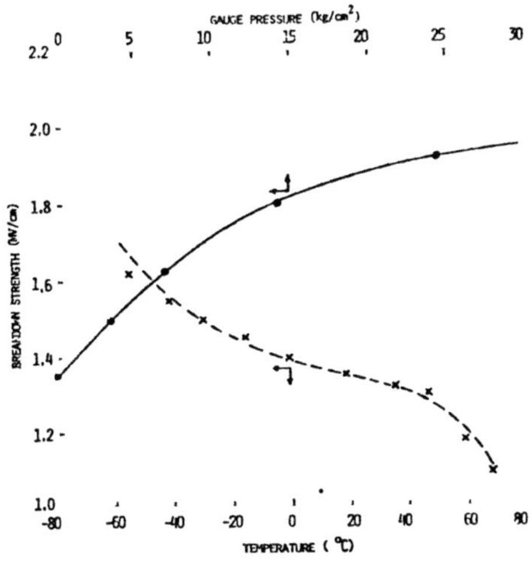

Fig. 4 Breakdown strength of n-hexane as functions of pressure and temperature for pulse duration of 4.5 sec and gap length of 0,02 cm.
and – – – –Theoretical curves; ● × Experimental results with stainless steel electrodes (after Kao and Higham).

perature dependence of I_T can also be explained through the temperature dependence of μ_n and D_n in Eq. (11).

The pressure dependence of $(E_{av})_b$ is mainly due to the pressure dependence of T_b and the temperature dependence of $(E_{av})_b$ is mainly due to the change of T_o. Using Eq. (16) and the physical parameters for n-hexane, we have computed the breakdown strength as a function of pressure and temperature, and the theoretical results are in good agreement with the experimental ones of Kao and Higham [11] as shown in Fig. 4. The above theory can also explain why for the breakdown strengths measured with linearly rising voltages the pressure dependence phenomenon in n-hexane disappears if the rate of rise of the applied average fields are higher than 4 MV per cm per sec as reported by Kao and McMath [12]. It can also explain the molecular structure dependence of breakdown strength and time lag.

ACKNOWLEDGMENT The author wishes to thank the National Research Council of Canada (under Grant No. A-3339) for supporting this work.

REFERENCES

1. K. C. Kao, pp. 44-52 in 1965 Annu. Rep. Conf. Elec. Insul., Publ. 1356, Nat. Acad. Sci., Washington, D.C.,U.S.A. (1966).
2. I. Adamczewski, "Ionization, Conductivity and Breakdown in Dielectric Liquids", London, Taylor and Francis (1969).

3. Z. Croitoru, "Progress in Dielectrics" 6, 104 (1965).
4. W. B. Green, J. Appl. Phys. 27, 921 (1956).
5. D. W. Goodwin, Proc. Phys. Soc. B69, 61 (1956).
6. W. Hwang and K. C. Kao, J. Chem. Phys. 60, 3845 (1974).
7. D. W. Goodwin and P. K. Macfadyen, Proc. Phys. Soc. B66,85(1953).
8. W. F. Schmidt and A. O. Allen, J. Chem. Phys. 52, 4788 (1970).
9. R. M. Minday, L. D. Schmidt and H. T. Davis, J. Chem. Phys. 54, 3112 (1971).
10. P. K. Watson and A. H. Sharbaugh, J. Electrochem. Soc. 107, 516 (1960).
11. K. C. Kao and J. B. Higham, J. Electrochem. Soc. 108,552(1961).
12. K. C. Kao and J. P. C. McMath, IEEE Trans. Elec. Insul. E15, 64 (1970).

TRANSIENT ELECTRIC FIELD AND SPACE CHARGE BEHAVIOR FOR DRIFT DOMINATED BIPOLAR CONDUCTION IN DIELECTRIC LIQUIDS

M. Zahn*

Introduction

When applied voltages are large so that diffusion phenomena is negligible, drift dominated conduction models have been used to describe charge transport in dielectrics and semiconductors.[1] Using an unipolar ion-mobility conduction law, recent work has solved for the bulk and terminal transient behavior for any initial and boundary conditions and for any type of applied excitation. In this paper we wish to extend this work by allowing for two charged carriers of opposite signs with charge densities q_+ and q_- and respective mobilities μ_+ and μ_-.

Field Equations

In the drift dominated limit, charge conservation for each carrier and Gauss's law for one dimensional variations with the coordinate x results in the relations

$$\frac{\partial J_\pm}{\partial x} + \frac{\partial q_\pm}{\partial t} = 0 \ , \ J_\pm = \pm q_\pm \mu_\pm E \ ; \ \varepsilon \frac{\partial E}{\partial x} = q_+ + q_- \tag{1}$$

DC Steady State

The dc steady state distributions can be found by setting all time derivatives to zero so that J_+ and J_- are constants,

$$E = \left[\frac{2x}{\varepsilon}\left(\frac{J_+}{\mu_+} - \frac{J_-}{\mu_-}\right) + E_o^2\right]^{1/2} \ ; \ q = q_+ + q_- = \frac{1}{E}\left(\frac{J_+}{\mu_+} - \frac{J_-}{\mu_-}\right) \tag{2}$$

where E_o is the steady state electric field at x = 0 which must be supplied as a boundary condition. The distributions of (2) are similar in form to that obtained for unipolar conduction where $J_- = 0$.[2] Note that despite the bipolar conduction, the resulting charge density is of one polarity throughout space depending on the relative values of the current to mobility ratio of each carrier. Recent measurements[3] using the Kerr electro-optic effect in nitrobenzene showed that under a dc voltage a bulk space charge distribution of one sign formed, while under an ac voltage a bipolar charge distribution could develop.

Transient Solution

The hyperbolic partial differential equations of (1) can be solved using the method of characteristics, where trajectories in the x-t plane are found upon which position and time are related so that the equations are converted to ordinary differential equations which can be easily integrated. Using (1) and

*Department of Electrical Engineering, University of Florida, Gainesville, Florida 32611

the chain rule for total derivatives we obtain the system equations in standard form as

$$\begin{bmatrix} 1 & \mu_+E & 0 & 0 \\ 0 & 0 & 1 & -\mu_-E \\ dt & dx & 0 & 0 \\ 0 & 0 & dt & dx \end{bmatrix} \begin{bmatrix} \partial q_+/\partial t \\ \partial q_+/\partial x \\ \partial q_-/\partial t \\ \partial q_-/\partial x \end{bmatrix} = \begin{bmatrix} -q_+\mu_+(q_+ + q_-)/\varepsilon \\ q_-\mu_-(q_+ + q_-)/\varepsilon \\ dq_+ \\ dq_- \end{bmatrix} \tag{3}$$

The first set of characteristic equations which also describe the particle trajectories for each carrier are obtained by finding those relations which set the determinant of coefficients on the left-hand side of (3) to be zero. To find the solutions along these trajectories we substitute the column vector on the right-hand side of (3) into any column in the left-hand matrix and again set the determinant of the resulting matrix to zero. The solutions are then

$$\frac{dq_+}{dt} = \frac{-q_+\mu_+}{\varepsilon}(q_+ + q_-) \; ; \; \frac{dE}{dt} = \frac{J(t)}{\varepsilon} + \frac{q_-E}{\varepsilon}(\mu_+ + \mu_-) \text{ on } \frac{dx}{dt} = \mu_+E$$
$$\frac{dq_-}{dt} = \frac{q_-\mu_-}{\varepsilon}(q_+ + q_-) \; ; \; \frac{dE}{dt} = \frac{J(t)}{\varepsilon} - \frac{q_+E}{\varepsilon}(\mu_+ + \mu_-) \text{ on } \frac{dx}{dt} = -\mu_-E \tag{4}$$

where J(t) is the total current per unit electrode area (with spacing d) given by the sum of conduction and displacement currents obtained by integrating over space the sum of the two charge conservation equations in (1),

$$J_+ + J_- + \varepsilon\frac{\partial E}{\partial t} = J(t) = \frac{1}{d}\left[\varepsilon\frac{dv}{dt} + \int_0^d (q_+\mu_+ - q_-\mu_-)E\,dx\right] \tag{5}$$

Note that if the system is excited by a current source, J(t) is known, while for the usual case of a voltage excitation v(t), J(t) must be calculated using the right-hand side of (5). The equations of (4) are in the standard form for numerical integration schemes based on the Runge-Kutta method. All that needs to be supplied are initial and boundary conditions.

A Simple Case Study

To illustrate the general approach with closed form solutions we consider the simple case where one carrier has zero mobility ($\mu_+ = 0$). Physically, we imagine some mechanism such as radiation ionizing a region creating an equal number of positive and negative carriers uniformly distributed in space. We assume that the lower electrode at x = 0 is at a positive voltage V with respect to the upper electrode at x = d. If the positive carriers are much larger than the negative carriers, their mobility is much less. In this limit the positive charge carriers do not move ($\frac{dx}{dt} = 0$) and at every position their density, q_+, is constant in time. Because of the assumption of uniform ionization, this density is also constant in space.

It is convenient to introduce the normalizations

$$\tilde{q} = q/q_+ \; ; \; \tilde{x} = x/d \; ; \; \tilde{V} = \varepsilon V/q_+d^2 \; ; \; \tilde{E} = \varepsilon E/q_+d \; ; \; \tilde{t} = q_+\mu_-t/\varepsilon \tag{6}$$

Then from (4) the solution for negative carriers is

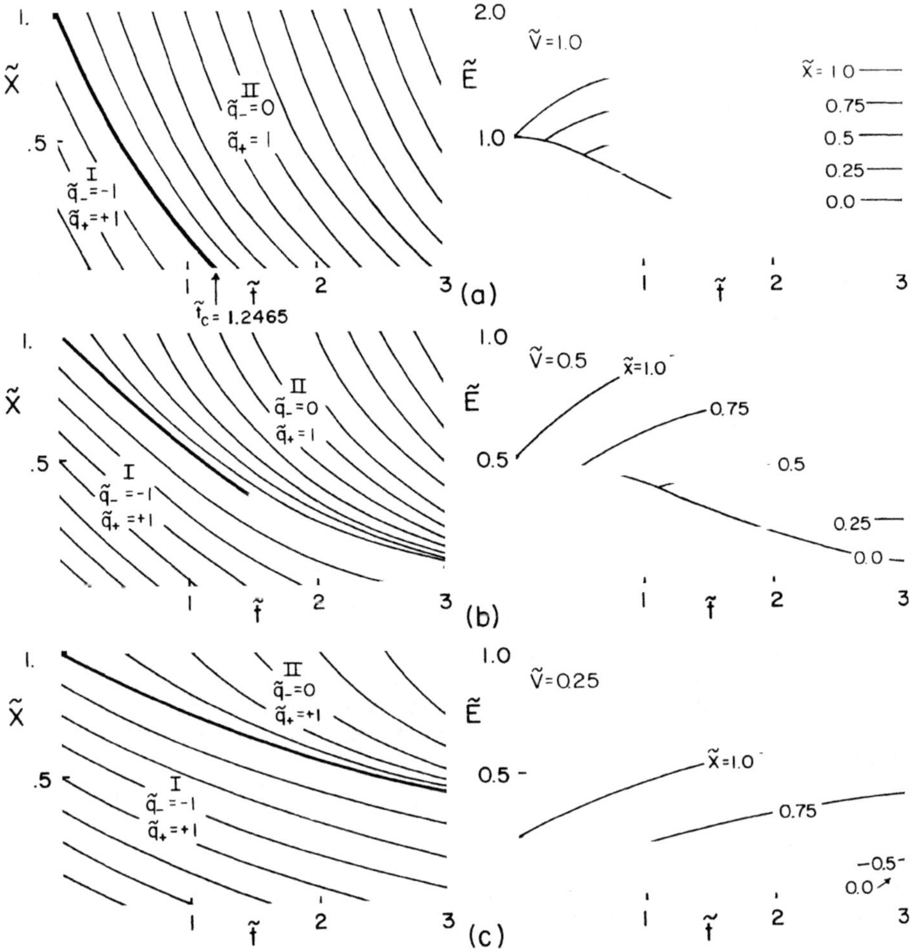

Fig. 1 Characteristic trajectories and time dependence of the electric field at various positions with $\mu_+ = 0$. (a) $\tilde{V} = 1.0$, (b) $\tilde{V} = 0.5$, (c) $\tilde{V} = 0.25$.

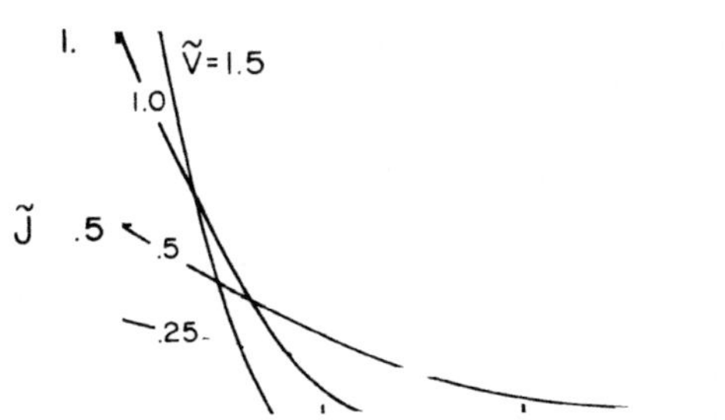

Fig. 2 Terminal current per unit electrode area versus time for various values of voltage.

$$\tilde{q}_- = \tilde{q}_o[-\tilde{q}_o + (\tilde{q}_o + 1)\exp(\tilde{t}_o - \tilde{t})]^{-1} \qquad (7)$$

where $\tilde{q}_o$ is the negative charge density at the start of a characteristic trajectory at time $\tilde{t}_o$.

The characteristic trajectories are separated into two regions by the demarcation curve $\tilde{x}_d$ emanating from the position $\tilde{x} = 1$ at $\tilde{t} = 0$. Above this curve, denoted as Region II, $\tilde{q}_o = 0$ so that $\tilde{q}_- = 0$ everywhere in Region II as no additional charges are generated. The initial condition transient of Region I has $\tilde{q}_o = -1$ so that $\tilde{q}_- = -1$ everywhere in Region I as we assume that at $\tilde{t} = 0$ the system is initially neutral. Because the average electric field value must be $\tilde{V}$, we obtain the solutions for the electric field and trajectories in each region as

Region I	Region II
$\tilde{x}(\tilde{t}) = \tilde{x}_o - \sqrt{2\tilde{V}}\tanh\sqrt{\tilde{V}/2}\,\tilde{t}$	$\tilde{x} = \tilde{x}_d(\tilde{t}) + (1 - \tilde{x}_d(\tilde{t}_o))\exp(\tilde{t}_o - \tilde{t})$
$\tilde{E} = \tilde{V} - .5 + \tilde{x}_d(\tilde{t}) - .5[\tilde{x}_d(\tilde{t})]^2$	$\tilde{E} = \tilde{x} + \tilde{V} - .5 - .5[\tilde{x}_d(\tilde{t})]^2$

$$\tilde{x}_d = \begin{cases} 1 - \sqrt{2\tilde{V}}\tanh\sqrt{\tilde{V}/2}\,\tilde{t} & 0 \le \tilde{t} \le \tilde{t}_c \\ 0 & \tilde{t} \ge \tilde{t}_c \end{cases} \qquad (8)$$

The parameter $\tilde{x}_o$ is the starting position at $\tilde{t} = 0$ for characteristic curves in Region I, and $\tilde{t}_o$ is the starting time at $\tilde{x} = 1$ for characteristic curves in Region II. Fig. 1 shows the results for voltages $\tilde{V} = 1.0$, $\tilde{V} = 0.5$, and $\tilde{V} = 0.25$. For $\tilde{V} > .5$, Region I only extends up to the finite time $\tilde{t}_c$ when $\tilde{x}_d = 0$. After $\tilde{t}_c$ the system is in the steady state with the electric field having a linear variation with position due to the uniform positive charge distribution. For $\tilde{V} = .5$, $\tilde{t}_c \to \infty$ while if $\tilde{V} < .5$, $\tilde{x}_d$ never reaches zero but rather approaches the steady state value $1 - \sqrt{2\tilde{V}}$. The negative charge is never completely swept out of the system.

For these cases with constant applied voltage, the terminal current per unit electrode area

$$\tilde{J} = \tilde{V} - .5[\tilde{E}^2(1) - \tilde{E}^2(0)] \;;\; \tilde{J} = \varepsilon J/\mu_- q_+^2 d \qquad (9)$$

is plotted in Fig. 2 versus time for various values of voltage.

References

1. M. A. Lampert and P. Mark, Current Injection In Solids, New York: Academic, 1970.
2. M. Zahn, C. F. Tsang, and S. C. Pao, "Transient Electric Field and Space-Charge Behavior For Unipolar Ion Conduction", J. Appl. Phys. Vol. 45, No. 6, pp. 2432-2440, June, 1974.
3. E. C. Cassidy, R. E. Hebner, Jr., M. Zahn, and R. J. Sojka, "Kerr-Effect Studies of an Insulating Liquid Under Varied High Voltage Conditions", IEEE Trans. Elect. Insul. Vol. EI-9, No. 2, pp. 43-56, June, 1974.
4. W. F. Ames, Nonlinear Partial Differential Equations In Engineering, New York: Academic, 1970, pp. 416-422.

SIMULATION OF CHARGE TRANSPORT FOLLOWING UNIPOLAR INJECTION

J. Evison and J.E. Brignell*

Introduction

In the recent literature on slower carriers in dielectric liquids, various hypotheses have emerged in explanation of certain curious features of charge transit wave forms, and incidentally this is remarkably paralleled in the literature of solid dielectrics[1]. Since the dielectric sample even in its simplest form is a complex non-linear system, it is often not easy to assess the consequences of such hypotheses.

Furthermore, in the study of the actual process of charge injection into a dielectric, the phenomenon can only be observed through the non-linear filter that is the dielectric, i.e. the current measured at any time is due to the total emission over a previous interval equal to the transit time and is influenced by the non-linear interaction between charges in drift motion.

The latter problem is a common one of needing to obtain the inversion of a non-linear operator to cancel its effect, and while much progress has been made in the application of such mathematical techniques as functional analysis to this problem, the analytical possibilities for solution are still somewhat limited.

At present the best approach to these problems appears to be one of digital simulation, and we describe below one approach to the development of such a technique. Ultimately it may prove to use such a simulation on-line to enable inaccessible quantities such as the injected charge to be measured by continuous correction of a model. Certainly it is convenient to have a model directly available in the experimental laboratory and accessible for manual adjustment with immediate visual display of results. For these reasons the model has been developed as a program for a laboratory on-line computer, but its usefulness suggests that it ought to be realized in a more transportable form (e.g. FORTRAN) for a wider availability among experimenters.

It is only possible to give a small indication of the wide application of such a technique in a conference paper, and a more detailed account[1] is necessary to achieve a fuller appreciation.

Brief description of the model

The model is one-dimensional and based on a quantisation of the dielectric thickness, d, into a number of planar cells. Each cell i, q_i, is considered to have a uniform charge, q_i and a stress level, E_i. A cycle of simulation commences with the

* Department of Electrical and Electronic Engineering, The City University, St. John Street, London, EC1V 4PB, England.

present charge distribution in the dielectric (i.e. q_i for each of the cells) and the stress distribution, E_i, is then calculated from the discrete form of Poisson's equation.

Each cell is so dealt with in turn, and then the distances moved by the front and rear edges of the cell charge are computed from the time increment of simulation, T, the carrier mobility and the various E_i involved. These distances are used to redistribute the charge accordingly over several cells and hence to calculate the change of charge induced in the electrodes. When this process has been completed for all the cells in turn, a new charge distribution has been established and the net current can be calculated[2] by dividing the induced charge increment by T. At the start of a new cycle further charge can be added to simulate continuous injection.

The accuracy of such a model depends on the fineness of quantisation of the gap, d. In the program the number cells was made variable, and for numbers greater than 100 the error in comparison with known solutions[3] was less than 1%.

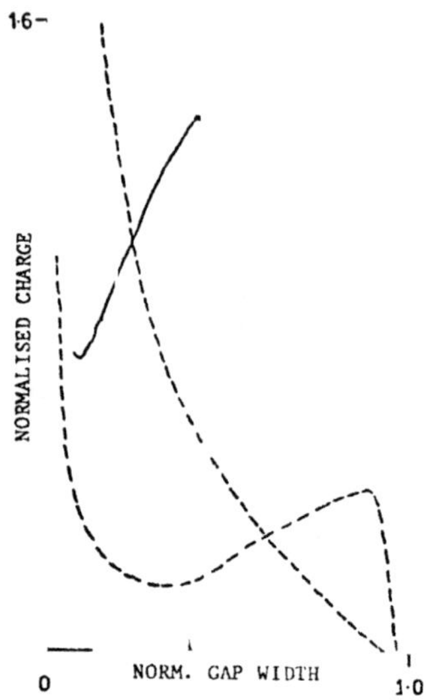

Fig. 1
Simulation of S.C.L. injection (with trapping)
____ trapped and free charge distributions at normalised time of 0.4
---- normalised time of 0.787

This basic model has been further developed to incorporate two separate carrier types each of variable mobility with the facility for declaring a probability for either carrier to be converted (e.g. trapped) into the other. A full description of this model and tests of its accuracy are given elsewhere[1]. An illustration of the power of the model and a demonstration of its accuracy is given by the case of space charge limited (SCL) injection with trapping. In order to do this, one carrier mobility is set to zero and trapping (with a lifetime of 0.5) is allowed in one direction only. The resulting trapped and free charge distributions are given in Fig. 1 for normalised[3] times of .4 and .787. The curves compare well with corresponding ones given by Many and Rakavy[3].

Applications

Several authors have devised digital solutions for the analytical expressions which apply to charge transport in a dielectric (e.g. refs. 4, 5, 6). Each method, however, is

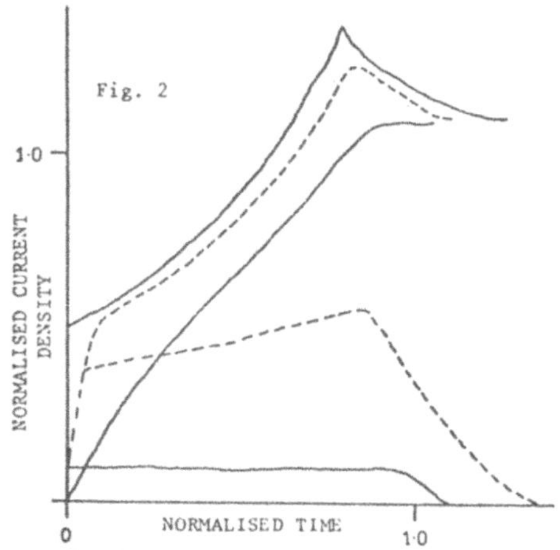

Space charge perturbed transients

applicable only to a restricted set of conditions. The power of this iterative method lies in its adaptability. The model copes straight forwardly with space charge perturbed (SCP) conditions, and the option of transfer between two types of carrier permits the emulation of a wide range of possible transport phenomena in solid and liquid dielectrics. One example is the case of a fast carrier in a liquid being trapped to form a slow ionic carrier. Furthermore, the ability to add or remove charge anywhere in the dielectric allows the model to deal with bulk as well as surface carrier generation.

Two illustrative examples have been chosen from the great deal of information that has been provided by the model. The first, shown in fig. 2, is of SCP transients following pulse and step injection. The step results, which appear not to have been previously published, are notable for showing that appreciable distortion, from the perturbation-free ramp, only occurs for currents close to SCL values. Such results are an aid to the experimenter who wishes to distinguish between these effects and, say, liquid motion effects.

The second application is a use of the model in an adaptive mode. Instead of working from a known *charge* injection to produce the resulting *current*, the model can be used in an automated trial-and-error mode to obtain the former from the latter. Consider, for example, a transient produced by Hewish and Brignell (Fig. 3) for injection through a grid, a case in which the actual charge injection waveform might only be guessed at. The model is provided with the current waveform and has a trial attempt at charge injection to match this. The resulting trial current is compared with the target value, and if the error is greater than 0.5% the model is reset and the quantity of charge injected reset. Fig. 4 shows the injected-charge curve and the resulting current transient for an assumed carrier mobility of 7.5×10^{-8} m^2/V.sec.

The deduced charge curve exhibits a form of oscillation and it was found that this was related to the mobility figure used. Fig. 5 shows the corresponding curves for a mobility of 6.75×10^{-8} m^2/V.sec which shows a more marked oscillation. The optimum value was found to be 9×10^{-8} m^2/V.sec; the curves are shown in Fig. 6. Although the physical meaning is not clear and is certainly dependent on the validity of the prior assumptions,

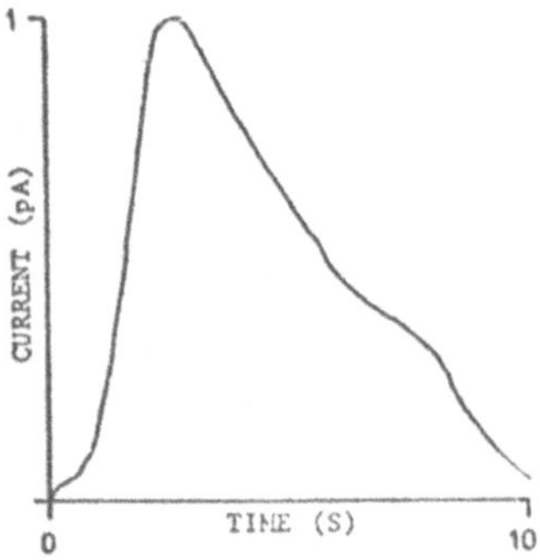

Fig. 3
Current transient
(Model input data)

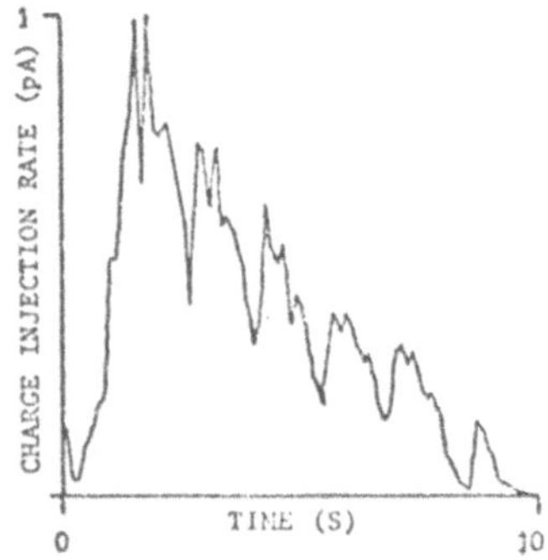

Fig. 4
Charge injection curve
obtained from model
(μ=7.5x10^{-8} m^{2}.(V.S.)$^{-1}$)

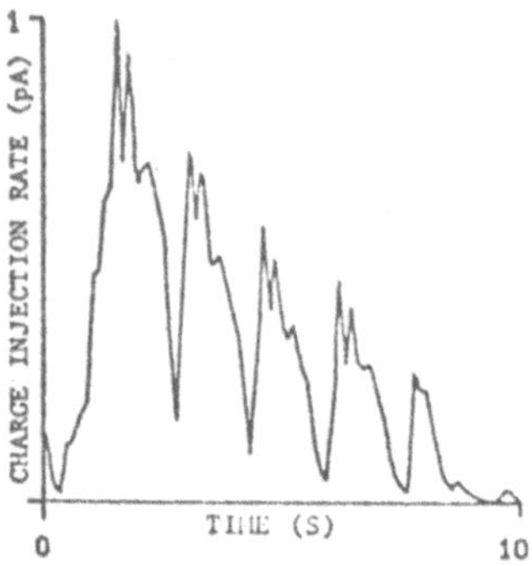

Fig. 5
As Fig.4 except μ =6.75x10^{-8} m^{2}.(V.S)$^{-1}$

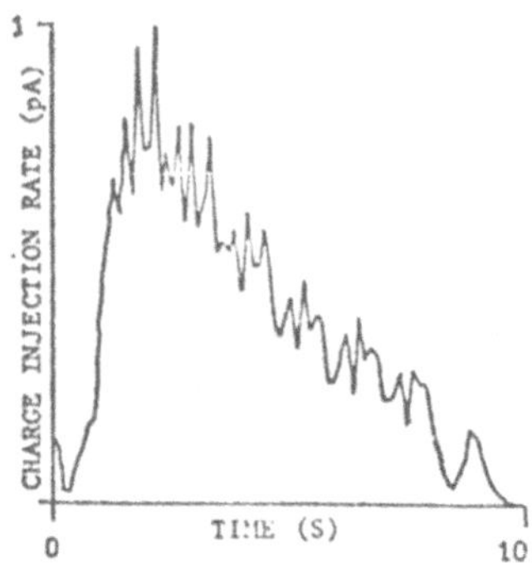

Fig. 6
As Fig.4 except μ =9.0x10^{-8}m^{2}.(V.S)$^{-1}$

this example does point the way to a possible powerful extension of such modelling techniques.

Discussion

The usefulness of a simulation technique is solely determined by the validity of the assumptions on which it is based. It is essential that it should be closely related to experimental work. In this case it was available while experiments were being conducted, and some of the results of the associated experiments[1] seem to contest some of the basic assumptions, and suggest possibilities such as the simultaneous exsitence of a whole range of carrier mobilities.

The technique of maximising the stability in the automated trial-and-error method is a very interesting one. The effects of an injected packet of charge persist for the whole of its transit time, and an incorrect value for a parameter such as mobility can mean that there is no possible charge curve to match the given current waveform, which accounts for the instability. However, it must be emphasised that such a result is only as good as the assumptions on which it is based.

Acknowledgements

The equipment and maintenance of one of us were provided by the U.K. Science Research Council. (JE).

References

1. Evison, J. Ph.D Thesis The City University,London. 1975.
2. Brignell, J.E. & Evison, J. 1973 J.Phys.D:Appl. Phys. 6 L55.
3. Many, A. & Rakavy, G. 1962 Phys. Rev. 126 1980.
4. Fleming, R.J. 1974 J.Appl.Phys. 45 4944.
5. Hartmann, G.C. & Lipari, N.O. 1973 J. Appl. Phys. 44 1676.
6. Batri, P. & Schechtman, B.H. & Keiji Kanazawa, K. 1970 Sol.Stat. Comm. 8 1433.

THE EFFECT OF TRAPPING AND DETRAPPING OF ELECTRONS ON THE NEGATIVE CHARGE CARRIER MOBILITY IN LIQUID TETRAMETHYLSILANE

J. M. Warman, M. P. de Haas and A. Hummel*

Of the various models proposed to describe the transport properties of excess electrons in molecular dielectric liquids that involving the temporary trapping of otherwise quasi-free electrons has received considerable attention in recent years.

$$e_f \underset{k_{-t}}{\overset{k_t}{\rightleftharpoons}} e_t \qquad (1)$$

Under conditions of thermal equilibrium between the free and trapped states of respective mobilities u_f and u_t (with $u_f >> u_t$) the overall electron mobility, u, will be given by the equivalent expressions (2)-(4)

$$u = u_f [e_f^-]/([e_f^-] + [e_t^-]) \qquad (2)$$

$$= u_f(1 + K_t)^{-1} \qquad (3)$$

$$= u_f(1 + \exp(-\Delta G_t^0/RT))^{-1} \qquad (4)$$

where K_t is the equilibrium constant for (1) and ΔG_t^0 is the standard free energy change involved in the trapping process.

It has been found that the temperature dependence of electron mobilities in molecular dielectric liquids can be described quite well by an expression of the form

$$u = A \exp(-B/RT) \qquad (5)$$

The parameters A and B, which may be obtained from the intercept and slope of ln u versus 1/T plots, have been frequently taken to be equal to the quasi-free mobility u_f and the mean trap depth $E_t = -\Delta H_t^0$ in the liquid respectively. This will be the case, as can be seen from equation (5), if the conditions $-\Delta G_t^0 >> RT$ and $|T\Delta S_t^0| << RT$ are fulfilled.

Recently it has been found by the present authors that the solute molecule biphenyl (Ph_2) behaves as a temporary trapping site for electrons in liquid tetramethylsilane (TMS) and from kinetic studies the thermodynamic parameters associated with the trapping process have been derived.[1] In pure liquid TMS the electrons are considered to be almost completely free, i.e. $u \simeq u_f$. This fact is supported by the rate constant found for electron capture by SF_6 in this liquid,[2] 2.1×10^{14} $mol.dm^{-3}s^{-1}$, which is greater than that found in the gas phase and close to the

*Interuniversitair Reactor Instituut, Mekelweg 15, Delft, The Netherlands

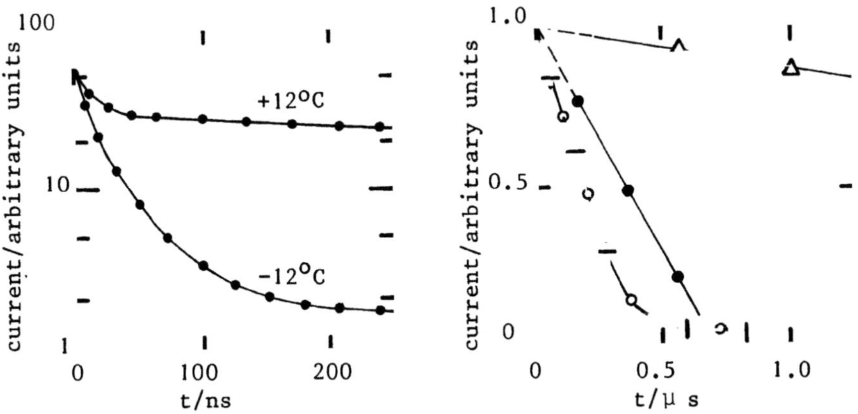

Figure 1. Electron decay in a solution of 3.3×10^{-6} mol/dm^3 Ph_2 in TMS

Figure 2. Time dependence of the current (normalized to 1 at t = 0) in pure TMS for a field strength of 7kV/cm at 23°C, +, and -35°C, o ; and for 3.3×10^{-6}mol/dm^3 solution of Ph_2 in TMS for a field strength of 8.5 kV/cm at 12°C, ●, and -12°C, △.

theoretical limit of 3.0×10^{14} mol.dm^{-3}s^{-1} for the attachment of quasi-free electrons at room temperature. By studying the effect of Ph_2 on the negative charge carrier mobility in TMS it appeared therefore that the applicability of equations (2)-(4) could be rigorously tested.

Excess electrons were formed in TMS by ionization of the liquid in a parallel plate conductance cell (50 Ohm stripline configuration) with 5ns pulses of X-rays produced by irradiation of a platinum target with 3 MeV electrons from a Van de Graaff accelerator. The time dependence of the electron concentration subsequent to pulse ionization was followed by monitoring the current flowing between the electrodes.

In pure TMS, at low field strengths (<1.5kV/cm) the electron half-life was 500 ns and was controlled principally by attachment to impurities in the liquid. As shown in Figure 1, on addition of 3.3×10^{-6}mol/dm^3 Ph_2 the electron concentration decreases initially much more rapidly than in pure TMS over the first 100 ns or so following the pulse to a slowly decaying "plateau" value due to establishment of the equilibrium

$$e^- + Ph_2 \underset{k_{-1}}{\overset{k_1}{\rightleftharpoons}} Ph_2^- \qquad (6)$$

As can be seen in figure 1, the equilibrium is displaced increasingly towards the right with decreasing temperature. From the kinetics of the relaxation to equilibrium the rate constants k_1 and k_{-1} have been determined as described previously[1] and the values of k_1 and the equilibrium constant $K = k_1/k_{-1}$ are listed in columns 2 and 3 of table I. It should be noted that the product $K[Ph_2]$, not K alone, is equivalent to the equilibrium

TABLE I

T (^oC)	$k_1 \times 10^{-12}$ ($mol^{-1}dm^3s^{-1}$)	$K \times 10^{-5}$ ($mol^{-1}dm^3$)	u ($cm^2V^{-1}s^{-1}$)	$(1 + K[Ph_2])u$ ($cm^2V^{-1}\ s^{-1}$)
24	8	1.0	74	98
12	11	3.1	49	99
1	14	10.6	25	112
-12	13	71	3.8	93

constant K_t since in the case of pure liquids the "trap" may be considered to be a property of the medium itself and does not appear as a reactive entity in (1).

Mobilities were determined from current versus time plots, as shown in Figure 2, at field strengths sufficiently high (>ca 7 kV/cm) that drift of the electrons to the electrodes was the dominant mechanism of electron removal and the current obeyed the linear relationship

$$i(t) = i_0(1 - uVt/d^2),\ t<d^2/uV \qquad (7)$$

(The "tailing" in the i versus t plots in figure 2 is due to the fringe fields surrounding the strip electrodes). From the data shown in Figure 2 for pure TMS at 22.5 and -35^oC, the mobility is seen to be almost independent of temperature as found by Allen and Holroyd.[3] Since the electrode separation distance in the present experiments (ca 0.3 cm) was not accurately known, the value of $u_{TMS} = 100$ obtained by the above authors was taken as a standard. As can be seen in Figure 2, in contrast to the pure TMS data the mobility in a $3.3 \times 10^{-6} mol/dm^3$ Ph_2 solution is markedly dependent on temperature. The values of u obtained over the temperature range 24 to -12^oC are listed in column 4 of Table I.

The ratio of trapped to free electrons in the biphenyl solution is given by $K[Ph_2]$ and if (2) holds, the overall negative charge carrier mobility should be given by

$$u = u_{TMS}(1 + K[Ph_2])^{-1} \qquad (8)$$

The product $u(1 + K[Ph_2])$ should therefore be independent of temperature and equal to the value of 100 for pure TMS. This is found to be the case, within the error of the present results, down to a mobility 25 times lower than for pure TMS, as shown in the final column of Table I.

In Figure 3 the mobilities determined in the Ph_2 solution are plotted as ln u versus 1/T as is customarily done with mobility data for pure liquids. The solid line in Figure 3 was calculated from the relationship

$$u = u_{TMS}\left[1 + [Ph_2]\exp(-\Delta G^0/RT)\right]^{-1} \qquad (9)$$

where ΔG^0 is the standard free energy change for attachment of electrons to Ph_2 in TMS and has been determined to be given by

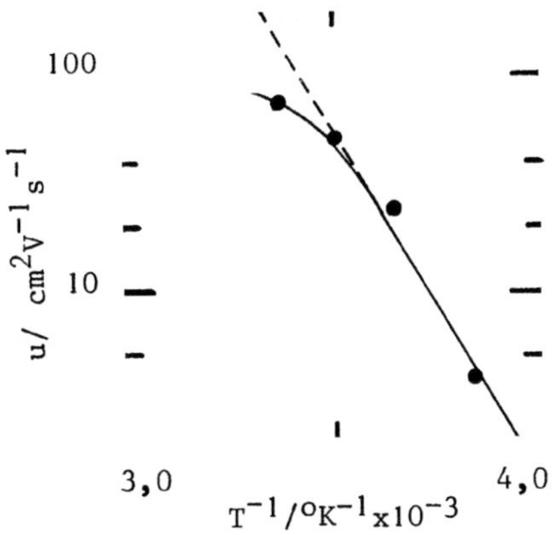

Figure 3. Dependence of the mobility in a 3.3 x 10^{-6}mol/dm^3 solution of Ph_2 in TMS on reciprocal temperature

$\Delta G^0 = \Delta H^0 - T\Delta S^0 = (-17.6 + 36 \times 10^{-3}T)$ kcal. With decreasing temperature equation (9) will tend to the simple exponential form

$$u = \frac{u_{TMS} \exp\Delta H^0/RT}{[Ph_2]\exp\Delta S^0/R} \qquad (10)$$

and a linear relationship between ln u and 1/T, given by the dashed line in figure 3, will be found. Equation (10) is seen to be a good approximation up to within 50% of u_{TMS} and it can be seen that a reasonably accurate estimate of ΔH^0 would be obtained from a straight line fit to the lowest three data points. Extrapolation of this straight line to 1/T = 0 gives as intercept u_{TMS} / $[Ph_2]\exp\Delta S^0/R$ (equivalent to A in equation (5) and it is apparent that u_{TMS} could not be determined from this extrapolation without knowledge of the term $[Ph_2]$ $\exp\Delta S^0/R$.

References

1) J. M. Warman, M. P. de Haas, E. Zador and A. Hummel, submitted for publication.

2) A. O. Allen, T. E. Gangwer and R. A. Holroyd, J. Phys. Chem., 79, 25 (1974).

3) A. O. Allen and R. A. Holroyd, J. Phys. Chem., 78, 796 (1974).

A CONTRIBUTION TO THE STUDY OF LOW-FIELD CONDUCTION BY IMPURITY PARTICLES IN DIELECTRIC LIQUIDS *

G.B. Denegri, G. Molinari, A. Viviani **

1 - Introduction

The strong influence of impurities on conduction processes in dielectric liquids is universally acknowledged, and shows itself in various forms, depending on experimental conditions. In the last years, attention has been given to the charge-convection mechanism by which impurity particles contribute to the current amount as macroscopic charge carriers, under the action of electro-dielectrophoretic forces [1,2,3,4]. Studies have been made of models of this phenomenon, especially in the case of simple conditions; for instance, to describe the behaviour of a single impurity particle between two electrodes [5,6].

Certainly, the observed phenomena present various and complicate aspects [7,8], so that it might be difficult for an analytical model to describe it completely. However, such a model can still provide some more information, even if it does not take into account non-linearities, space-charge, and so on. Therefore, starting from a previously developed theory [9], a more accurate description of the electro-dielectrophoretic field has been obtained [10]; applications to conduction processes are in progress.

2 - Electro-dielectrophoretic field theory

A theory of the electro-dielectrophoretic field has been developed for a spherical particle of radius R, constant permettivity ε_2 and conductivity γ_2, embedded in a homogeneous isotropic dielectric medium with constant permettivity ε_1 and conductivity γ_1, under general transient conditions. This has allowed us to deal also with charged particles having non-zero charge mean value in time.

In the dielectric, prior to the introduction of the particle, an unrestricted tridimensional time-varying electric field $\bar{E}$ is acting. If all magnetic field effects are neglected, the electric field $\bar{E}$ can be described in terms of an electric potential V. The Mac Laurin series of this potential in a Cartesian coordinate system x,y,z, centered in the position O which will be occupied by the center of the particle, gives:

$$V(x,y,z,t) = \sum_{1}^{\infty}{}_{n} \sum_{k+l+m=n} \frac{1}{k!\,l!\,m!} \left.\frac{\partial^n V(x,y,z,t)}{\partial x^k \partial y^l \partial z^m}\right|_{\substack{x=0\\y=0\\z=0}} \rho^n \cos^k\theta \sin^l\theta \sin^{k+l}\varphi \cos^m\varphi \qquad (1)$$

where ρ,θ,φ are spherical coordinates defined by $x=\rho\cos\theta\sin\varphi$, $y=\rho\sin\theta\sin\varphi$, $z=\rho\cos\varphi$.

Let us now introduce the impurity particle, on which an initial

* Work supported by C.N.R., Gruppo Alte Tensioni
** Department of Electrical Engineering, University of Genova
Viale F. Causa 13, 16145 Genova, Italy

superficial charge density distribution $\sigma(\theta,\varphi,0)$ is given by:

$$\sigma(\theta,\varphi,0) = \sum_{1}^{\infty} n \sum_{k+l+m=n} \frac{1}{k!\,l!\,m!}\, \sigma_{klm} \cos^k\theta \, \sin^l\theta \, \sin^{k+l}\varphi \, \cos^m\varphi \qquad (2)$$

By computing the distorsion produced by such a particle in the field, under the hypothesis that the particle does not affect the electric field appreciably for $\rho \gg R$, and by integrating the electromagnetic stress tensor on the particle, the instantaneous values of the force $\bar{F}$ and of the torque $\bar{C}$ are obtained in the form:

$$\bar{F}(t) = 4\pi\varepsilon_1 \sum_{0}^{\infty} n \frac{2n+1}{2^n\, n!} \frac{(2n)!!}{(2n+1)!!} R^{n+2} \sum_{k+l+m=n} \frac{1}{k!\,l!\,m!} \cdot$$

$$\cdot \left[n R^{n-1} V_{klm} * \left[f_n(t) - g_n(t) \right] - \sigma_{klm} h_n(t) \right] \operatorname{grad} V_{klm} \qquad (3)$$

$$\bar{C}(t) = 4\pi\varepsilon_1 \sum_{0}^{\infty} n \frac{1}{2^n\, n!} \frac{(2n)!!}{(2n+1)!!} R^{n+3} \left[(n+1)(2n+3) R^n \left[\sum_{k+l+m=n} \frac{1}{k!\,l!\,m!} \cdot \right.\right.$$

$$\left. \cdot \operatorname{grad} V_{klm} * f_{n+1}(t) \right] \times \left[\sum_{k+l+m=n} \frac{1}{k!\,l!\,m!} \operatorname{grad} V_{klm} * g_{n+1}(t) \right] +$$

$$\left. + \left(\sum_{k+l+m=n} \frac{1}{k!\,l!\,m!} \operatorname{grad} V_{klm} \right) \times \left(\sum_{k+l+m=n} \frac{1}{k!\,l!\,m!} \bar{S}_{klm} \right) \right] \qquad (4)$$

where:

$$V_{klm} = \frac{\partial^n V(x,y,z,t)}{\partial x^k \partial y^l \partial z^m} \quad ; \quad \bar{S}_{klm} = \sigma_{(k+1)lm}\bar{a}_x + \sigma_{k(l+1)m}\bar{a}_y + \sigma_{kl(m+1)}\bar{a}_z \qquad (5)$$

$$f_n(t) = \mathcal{L}^{-1}\left\{ \frac{\gamma_2 + s\varepsilon_2}{F(s)} \right\} \quad ; \quad g_n(t) = \mathcal{L}^{-1}\left\{ \frac{\gamma_1 + s\varepsilon_1}{F(s)} \right\} \quad ; \quad h_n(t) = \mathcal{L}^{-1}\left\{ \frac{1}{F(s)} \right\} \qquad (6)$$

$$F(s) = n\gamma_2 + (n+1)\gamma_1 + s\left[n\varepsilon_2 + (n+1)\varepsilon_1 \right] \qquad (7)$$

The symbol * represents the convolution product, $\mathcal{L}^{-1}$ is the inverse Laplace operator, and s is the complex frequency.

Eqs. (3) and (4) hold in Laplacian fields and for particles sufficiently far from the electrodes. If the particle is very close to a boundary where conditions are imposed on the potential or on the electric field, Eqs. (3) and (4) yield only approximate values.

3 – Model of the particle motion and of the conduction process

From the above theory of the electro-dielectrophoretic field, a model has been developed, enabling us both to compute the trajectory of a particle in a dielectric liquid under given field conditions, and to describe the charge convection mechanism due to macroscopic impurity particles.

In this model, the particle is subject to the electro-dielectrophoretic force and torque, as well as to an external field of forces (commonly the gravitational field). On the particle is present an initial charge distribution which is due to the contact with an electrode, and is generally approximated by taking into account, in (2), the terms for n=0 and n=1 only, i.e. σ_{000}, σ_{100}, σ_{010}, σ_{001}. Under the action of the forces, the particle translates and rotates, and a first short trajectory arc is computed on the basis of the liquid dynamic viscosity μ, of the particle gravitational mass m_g and inertial mass m_i, and of the initial values of the forces. Once the trajectory arc is determined, the time evolution of the applied field, due to its time dependence and to the motion, allows us to determine new values of σ_{000}, σ_{100}, σ_{010}, σ_{001}.

By iterating this procedure, it is possible to compute the trajectory with a precision depending on the elementary arcs length.

The initial values of σ are important parameters acting on the trajectory form. Such values are to be determined on the basis of a detailed model of the particle behaviour in the vicinity of the electrodes. Such a behaviour is not sufficiently known, and consequently we have introduced and compared some elementary and rough procedures to evaluate σ. They differ essentially in the dwell time that a particle spends in contacting an electrode. In a first simple model (A), the dwell time is zero: the particle rebounds elastically, and acquires a charge Q given as the minimum value between Maxwell's formula $Q = 2\pi^3 R^2 \varepsilon_1 E/3$ (proposed by many authors [5,6], and valid for a metallic sphere contacting a metallic plane), and the formula $Q = 4\pi\varepsilon_1 R V_0$, where V_0 is a limiting potential of the sphere, in conformity with the Felsenthal-Vonnegut hypothesis [11,7]. In a second simple model (B), the dwell time is not zero, and the assumption is made that there exists a limiting adherence force between the particle and the electrodes: the particle acquires just the charge Q sufficient to overcome this force. In this case, the charge increases as the field on the electrode surface decreases, and cannot be greater than a maximum value, fixed by either Maxwell formula or by Felsenthal-Vonnegut formula. Other models have also been used with intermediate characteristics between A and B. The values of σ_{000}, σ_{100}, σ_{010}, σ_{001} are computed from Q by means of suitable coefficients of theoretical derivation.

Some examples of particle trajectories computed by the above procedure in a sphere-plane electrode system are given in Fig. 1.

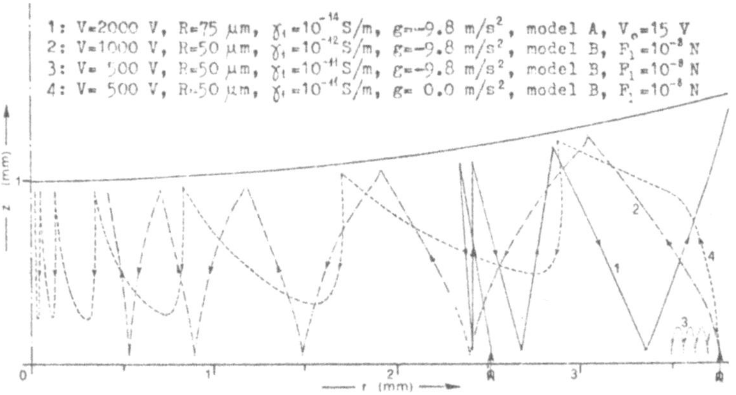

Fig.1 - Particle trajectories in a sphere-plane electrode system.
Parameter values defined in Sec. 4. - V, applied voltage - R, particle radius - g, gravitational field component parallel to the vertical axis - V_0, Felsenthal-Vonnegut potential - F_1, limiting adherence force - ⍋, starting points.

Under the only further hypothesis that the particle do not interact during their motion, it is possible to evaluate the current produced by a given initial particles distribution between two electrodes when a time-varying voltage is applied. The main difficulty in this procedure lies in the evaluation of the charge distribution induced on the electrodes by the particles, that might require frequent field computation. This onerous problem has been satisfactorily solved by using approximate equivalent RC circuits.

4 - Experimental tests

To verify the above model, experimental tests have been performed by using a sphere-plane electrode system, with a 1 mm gap and a 15 mm sphere radius. The liquid used for such tests has been n-exane, sufficiently purified by successive distillations, so that a conduction current of less than 10^{-10}A has been obtained under an applied d.c. voltage up to 5000 V. Then, the liquid has been polluted by means of metallic spherical particles, 130 + 150 μm in diameter. Precautions have been taken to avoid any other pollution form has far as possible. Then, the current has been measured after a step insertion of d.c. voltage, starting from an initial uniform distribution of particles on the flat electrode. These tests have been in good agreement with theoretical previsions. As an example, we report here some results concerning a test with an initial density of about 50 nickel particles per square millimeter and an applied d.c. voltage of 2000 V. In Fig.2 are reported, as time functions, the experimental values of the current which appears to be quite entirely due to the convection process. Under the action of the field, the particles have assumed a "to and fro" motion between the electrodes, and, under steady state conditions, such a motion has been concentrated on an annular region about 10 mm in radius. In the inner region, the number of particles moving or at rest has been negligible. In the outer region, no particle motion has occurred.

The computer approach by means of previous models has been carried out by introducing a particle diameter of 150 μm, a metal density of 8.9 g/cm^3, a liquid density of .663 g/cm^3, a viscosity of .312 centipoise, a d.c. voltage of 2000 V, and the constants $\varepsilon_1 = 1.89\ \varepsilon_0$, $\varepsilon_2 = \varepsilon_0$, $\gamma_1 = 10^{-14}$ S/m, $\gamma_2 = 10^7$ S/m; a gravitational field orthogonal to the flat electrode has been assumed. The electric field has been previously computed for the sphere-plane electrodes under linear conditions.

The best results have been obtained by using an intermediate model between A and B. Practically, the only difference with respect to A lies in an anelastic rebound of the particles, which retain only a quarter of their initial kinetic energy. The computed current is reported in Fig. 2 . In Fig. 3 are reported, as functions of time, the abscissas where particles of different start-points rebound on the flat electrode. Finally, a particle trajectory is plotted in Fig. 4. The only remarkable difference between theoretical and experimental results involves the tran-

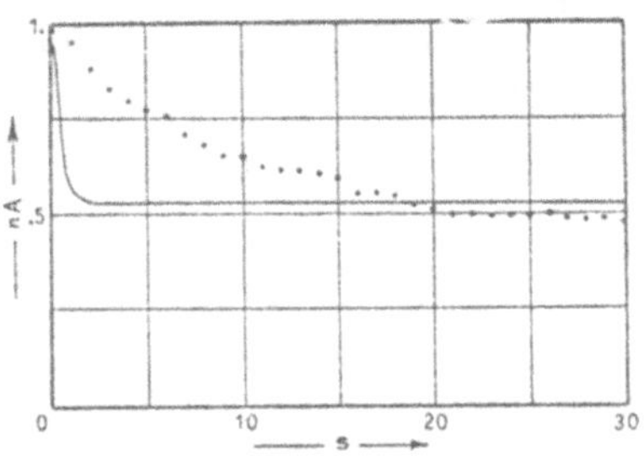

Fig. 2 - Plots of current versus time.
—— Theoretical ... Experimental

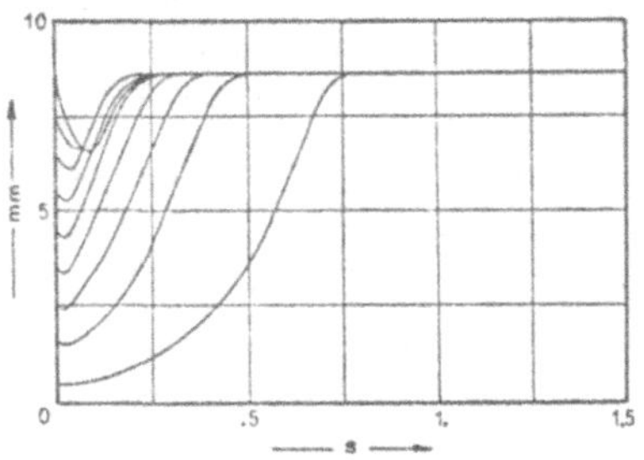

Fig. 3 - Abscissa where the particles rebounds on the flat electrode versus time.

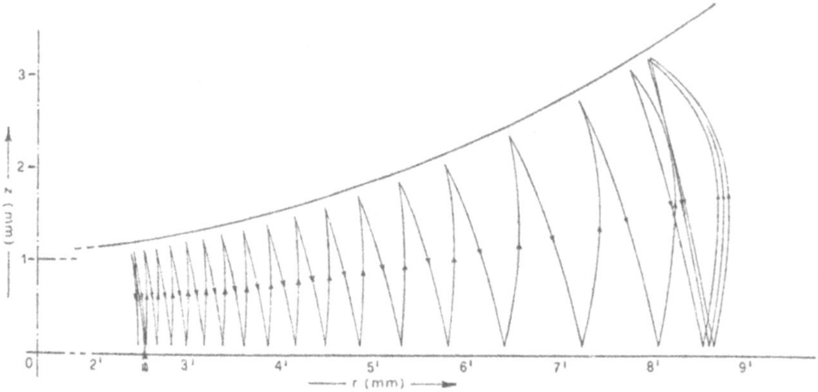

Fig. 4 - Particle trajectory computed under the experimental test conditions.

sient state in the current plot. This seems to be mainly due to the hypothesis of non-interaction among the particles.

5 - Conclusions

The above examples show that an accurate mathematical model of the charge convection mechanism due to macroscopic impurity particles in the conduction of dielectric liquids can give very good results, at least in a certain range of the relevant parameters. The description of the particles motion in the bulk seems to be already satisfactory, also if spatial charge distributions, interactions of particles, and other features of the phenomenon are neglected. On the contrary, further research on the particles behaviour in the vicinity of the electrodes is required. Such a behaviour seems to have most influence on the global evolution of the system, and the identification procedures of the involved electrical and mechanical parameters seem to be still inadequate. This is already true for the above particles size, and is especially true for lower-size particles, where the phenomenon may be more complicate. Studies on the model in this direction are in progress.

[1] H.House: Proc. Phys. Soc. B, 70, 913, 1957
[2] A.M.Sletten, T.J.Lewis: Brit. J. Appl. Phys., 114, 883, 1963
[3] G.M.Rhodes, J.E.Brignell: J. Phys. D, 4, L47, 1971
[4] Z.Krasucki: Proc. Grenoble Conf. on Dielectric Liquids, 1968
[5] Z.Krasucki: Proc. Dublin Conf. on Dielectric Liquids, 129-135, 1972
[6] S.Birlasekaran, M.Darveniza: Proc. Dublin Conf. on Dielectric Liquids, 124-128, 1972
[7] G.M.Rhodes, J.E.Brignell: Proc. Dublin Conf. on Dielectric Liquids, 116-119, 1972
[8] S.Birlasekaran, M.Darveniza: Proc. Dublin Conf. on Dielectric Liquids, 120-123, 1972
[9] A.Viviani: Proc. Dublin Conf. on Dielectric Liquids, 136-139, 1972
[10] G.Molinari, A.Viviani: Report SL-11-75, University of Genova
[11] P.Felsenthal, B.Vonnegut: Brit. J. Appl. Phis., 18, 1801, 1967

INFLUENCE OF SOLID IMPURITIES ON THE ELECTRIC STRENGTH OF TRANSFORMER OIL

F. ABGRALL AND J.M. CARDON *

Abstract

The reduction of the overall size of UHV transformers can be obtained by increasing the electrical field in the insulation, especially in the oil. Hence it is of paramount importance to know the breakdown voltage of the oil under various circumstances. In this paper, carefully controlled measurements of electric strength of degassed and filtered transformer oil are reported. The influence of the following parameters are studied : the volume of oil under stress, the size of solid particles impurities. With aluminum electrodes in uniform field configuration, the electric strength of the oil is low if it is not filtered between successive breakdowns. When the measurements are made with electrodes in a coaxial configuration and when the particles created between two breakdowns are removed by filtering down to 1 μm through a filter paper, very high electric strength is obtained. This value, 20 kV r.m.s/mm for an oil volume under stress of 300 cubic centimeter, is higher than the currently published values for the same volume under stress. The test results are explained on the basis of particles initiated breakdown.

Introduction

The insulating oil which filled the cooling ducts located between the conductor layers of a high voltage power transformer is one of the weak point of their insulation. On the one hand its electric strength is lower than that of the impregnated cellulosic paper. On the other hand its lower permittivity leads to higher electrical stress than in the paper insulation. In order to estimate the allowable or withstandable stress in the oil, it is of paramount importance to know its breakdown voltage distribution particularly in the low probability region. With this purpose we have undertook to measure the dielectric strength of a transformer oil versus the volume of oil under electrical stress in various conditions. In this paper we report the preliminary results we have obtained. We also measured the reduction of the oil dielectric strength due to the introduction in the oil of metallic and insulant solid particles. Indeed in a big transformer there are such particles : dust which penetrates during the construction and cellulosic fibres detached from the main insulation.

* Division Electrotechnique Générale
Laboratoires de Marcoussis
91460 - MARCOUSSIS (France)

Experimental apparatus

The breakdown tests are performed in three geometries of oil gap : plane, cylindrical and spherical electrode arrangement.

Plane electrode configuration

The aluminum plane electrodes are horizontal. The test surface is 65 cm^2 and the gap distance can vary from 0.5 cm to 2.5 cm. After each breakdown, the oil is circulated between the electrodes and through filters of 8 μm pore size.

Cylindrical coaxial electrode configuration

The aluminum coaxial cylindrical electrodes are vertically arranged. The diameter of the inner electrode is 4.4 cm with an area of 1250 cm^2. The oil gap is 0.85 cm. The volume of oil in which the electrical stress is greater than ninety percent of its maximum value is 315 cm^3. A circulation pump brings the oil between the electrodes and through a filter paper.We use three pore sizes for the paper filter : 5, 2 and 1 μm.

Spherical electrode configuration

This experimental apparatus is similar to the one used to measure the breakdown voltage of transformer oil following the C27-101 norm /1/. The diameter of the brass spherical electrodes is 12.5 mm and the gap distance is fixed to 2.5 mm. Like in the two preceding apparatus the oil is circulated between the electrodes. An agitating device is placed at the bottom of the vessel to prevent the settling of the particles.

Oil processing

Before the tests, the transformer oil is filtered through a filter paper of 5 μm pore size, degazed and dehydrated at 60° C under vacuum (0,1 torr). This industrial processing leads to an oil with a water content maintained in the range of 5 to 10 p.p.m. The breakdown voltage, measured according to the C27-101 norm /1/ is above 72 kV r.m.s.

Experimental results

The dielectric strength of the transformer oil is measured in these three experimental apparatus in various conditions. For each shot the alternative voltage is increased to a rate of 3 kV r.m.s/second. During all the breakdowns the liquid stays stationnary.

Results obtained with the first experimental apparatus

In these tests, the electrode gap was 0.6 cm and the oil was circulating after each breakdown. For a circulation time of two minutes, we measure an average dielectric strength of 8 kV/mm with a scattering of 17 % ($\frac{\sigma}{E}$). This value is multiplied by two when the circulation time is multiplied by ten. This result can be interpreted like following : each breakdown creates solid particles, metal ones from the electrodes and carbon ones from the oil. When the circulation time is short these particles are not completely eliminated from the space between the horizontal electrodes. One may conclude that these particles are responsible for the low dielectric strength observed and that their influence is of paramount importance since they can divide the electric strength by two.

Results obtained in the second experimental apparatus

In this experimental apparatus, the memory effect between two breakdowns is minimized by two facts : first of all, the electrode axis is vertical and secondly the liquid circulation is very effective because of the coaxial arrangement of the electrodes. Therefore the solid particles do not settle between the electrodes. After a rapid upward spark conditionning, the average value breakdown voltage remains constant and is independant of the circulation time between two breakdowns. This is true only when this time is longer than one minute. We measured a high electric strength, 20 kV r.m.s/mm with a scattering of about 15 %. This value, higher than the currently published values one for the same oil volume under stress /2/,/3/, is due to the effectiveness in eliminating the solid particles created by the breakdowns. We do not notice any improvement in the electric strength when decreasing the pore size of the filter paper from 5 m to 1 m.

We tried to fit the statistical series of breakdown voltage to the normal and to the extreme value laws. The curves on fig.1 show that the distribution of breakdown measurements is a close fit to the first asymptotic distribution of smallest values /4/ of which the cumulative probability is given by :

$$P\,(V) = 1 - \exp\left[- \exp \beta\,(V - u)\right]$$

This confirms the results obtained by many authors /2/,/3/,/5/.

Results obtained with the third experimental apparatus

We measure the reduction in electric strength of the transformer oil versus the average size of dehydrated alumina and aluminum particles. These particles are chosen to simulate dry and wet cellulosic fibres. When the breakdown voltage of treated oil is constant and near 72 kV r.m.s, the solid particles are vigourously scattered through the oil and the corresponding decrease in electric strength is then measured. The quantity of solid particles is always maintained equal to 6 mg/litre. Typical results obtained are presented on figure 2. Each point is the average of twenty breakdowns. The reduction of the breakdown voltage of the oil versus the average size of the metallic particles is higher than the one predicted by KRASUCKI /6/. The influence of insulating particles is less marked except when the alumina is not dehydrated. This reduction in breakdown voltage in the case of cellulosic fibres in association with moisture is already known /7/,/8/.

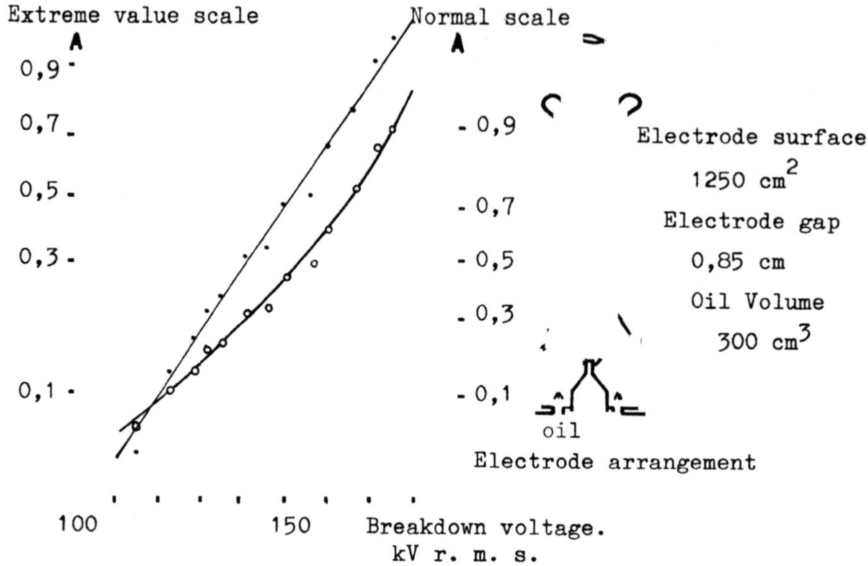

Fig.1. Cumulative probability for stationnary transformer oil.

• extreme value scale
o normal scale

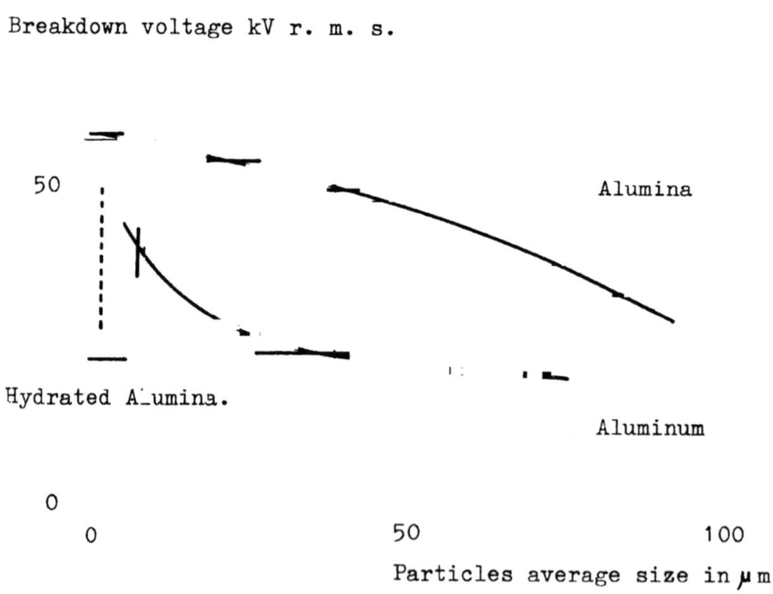

Fig.2. Dielectric strength of oil versus average size of alumina and aluminum particles.

In this apparatus, with an oil free of solid particles, the breakdown voltages obey to a normal distribution. If we compare it with the smallest value distribution obtained with the coaxial electrode arrangement we must conclude that there is an increase in the probability corresponding to the low breakdown voltage, when the volume of oil under stress increases. This confirms the results obtained by KAWAGUCHI and all /3/.

Conclusion

The electric strength of a transformer oil processed by industrial means is high. It is 20 kV r.m.s/mm with a scattering of 15 % for an oil volume under electrical stress of 315 cm^3. This high value can only be achieved if one eliminates the solid particles created by each breakdown.

The breakdown voltages of a processed transformer oil are found to be reduced by the introduction of solid particles. This reduction greatly depends on the average size of the particles. The reduction due to the conducting particles is much more pronounced than the one due to the non - conducting ones if these latter are well dried.

When the stress oil volume increases from 0.2 cm^3 to 315 cm^3, the form of the breakdown voltage distribution changes from a normal law to an extreme value distribution. This results in an increase of the probability corresponding to the low breakdown voltage. This point is very important from the point of view of the insulation design.

REFERENCES

/1/ NF C 27-101 - Huiles minérales isolantes sans produits d'addition -norme française.

/2/ J.K. NELSON, B.SALVAGE, W.A.SHARPLEY - Electric strength of transformer oil for large electrode areas.
Proc. IEE vol 118, n°2, Feb. 1971

/3/ Y.KAWAGUCHI, H.MURATA, M.IKEDA - Breakdown of transformer oil. IEEE Trans. on Power Apparatus and Systems, Jan/Feb.1972

/4/ E.J. GUMBEL - Statistics of extremes-columbia university press

/5/ K.H. WEBER, H.S. ENDICOTT - Extremal area effect for large area electrode for the electric breakdown of transformer oil.
T.AIEE, Vol.76, Déc. 1957 p. 1091-1098.

/6/ Z.KRASUCKI - Breakdown of liquid dielectrics
Proc. Roy. Soc. 1966, 294, 393.

/7/ P.K.WATSON, J.B. HIGHAM - Electric breakdown of transformer oil, ibid., 1953, 100, Pt II A, pp 168 - 174.

/8/ BUI AI., R.LACOSTE, J.F.MOREL, G.WIND
Etude de l'action des décharges partielles sur les isolants
R.G.E. Janvier 1969.

HIGH FIELD CONDUCTION IN HEAVILY CONTAMINATED TRANSFORMER OIL

B. Young *

Introduction

An earlier paper[1] has presented results for heavily contaminated transformer oil samples taken from the diverter switch compartment of a tap-changing transformer during the course of an extended life test. The present paper reports certain other test results derived from the same samples. The variation of conduction current with increasing voltage is discussed and a simple model proposed by recourse to statistical considerations. Various other results for the contaminated samples have been compared using multiple factor analysis in an attempt to obtain primary factors which may indicate some of the underlying processes involved in conduction and breakdown.

Experimental procedure

Readings of high direct field conduction current and breakdown voltage were obtained using a Skowronski cell. The electrodes were of stainless steel with radii of 20 mm and 40 mm. The cell wall was of glass with a diameter of 81 mm and height of 140 mm. Before each testing sequence the electrodes were polished by buffing and cleaned with n-hexane. The cell was then immediately filled with 200 mils of oil. The direct test voltage across the cell was increased automatically at a constant rate of 1.5 kV s^{-1}. The pre-breakdown conduction current was measured by a microammeter connected between the cell's lower electrode and earth and suitably protected by a spark gap. A metal ring encircling the cell just above the lower electrode and connected directly to earth provided a by-pass circuit for surface leakage current. Measurement of the test voltage was by a second microammeter and high series resistance connected between the cell's upper electrode and earth. Protection of the electrodes following breakdown was provided by a current limiting resistance of 40 MΩ and diverter unit. The time interval between successive breakdown measurements was 5 minutes. Readings of insulation resistance were measured at 25°C and at a direct field of approximately 3 kV cm^{-1} using a commercial megohm bridge. This incorporated a high input impedance detector consisting of a balanced electronic amplifier using electrometer valves. The test cell used with the bridge was fully guarded. Readings of power factor and permittivity were measured using a doubly screened Schering bridge operating at 2.5 kV r.m.s., 50 Hz.

Conduction current at high direct stress

The curves in Fig. 1 indicate the variation of conduction current during the application of linearly increasing direct fields up to the point of breakdown for an electrode separation of 4 mm.

*B. Young is with Thames Polytechnic, Woolwich, SE18.

Fig. 1 Variation of natural logarithm of current in microamperes with test voltage in kV.

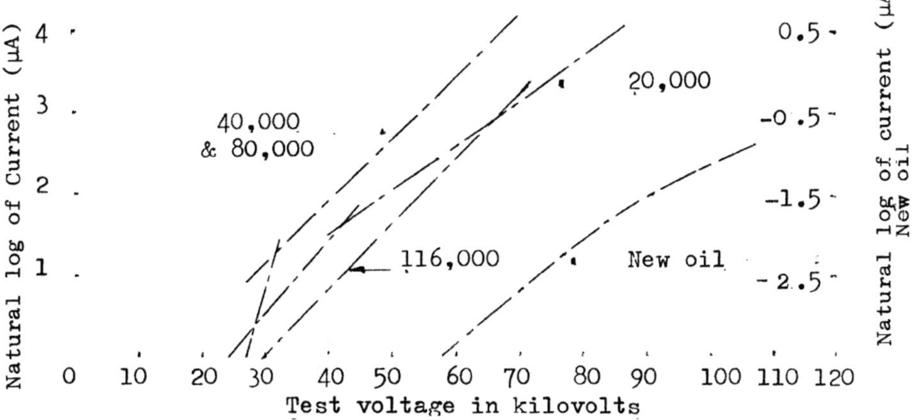

Similar curves were obtained when the gap length was 2 mm and 3 mm. At high direct voltages the readings for each sample, when plotted in the manner shown, fit closely to a straight line which implies a relationship of the form I = k exp (CV). The relationship at lower voltages is less well defined and in some cases there is a distinct break point between two linear regions.

For pure dielectric liquids high-field conduction is generally considered to develop at fields above 100 kV cm^{-1}. Since the samples involved here had field strengths at breakdown in the range 80 kV cm^{-1} to 160 kV cm^{-1} it is possible that the conduction processes operative before and at breakdown involved those normally associated with the high-field regime.

Particle count data show that the size-frequency distribution is of the Pareto form[2] and that in a volume of 0.1 mil the number of particles of size 1 to 5 μm is typically five million. By extrapolating the linear plot of the Pareto form into the sub-micron range as far as 0.1 μm the predicted number of particles in 0.1 mil of contaminated oil is 10^{10}. Assuming an initially uniform spatial distribution this implies an inter-particle separation of 2 μm. It has been suggested[3] that sub-micron particles of this kind can provide a quantitative explanation of high field conduction in terms of the charge transfer resulting from their oscillatory motion between electrodes. In contrast to this view other observations[4] have shown that charge transfer due to particle activity is not field dependent and that the dwell time of particles at the electrodes is large compared to the transit time. Particle motion[5] has also been shown to be inhibited when other contaminating particles are present. Although a model based directly on particle conduction has been questioned an indirect process involving particles is believed to affect the conductions level. It is suggested that positively charged particles at the cathode surface cause local field intensification which may be responsible for bursts of electron emission[6]. Examination of the Skowronski cell electrodes during testing revealed a heavy deposition of carbon particles, particularly on the cathode, and their presence may be a contributory factor in carrier multiplication in the present case.

The high level of contamination in the bulk of the liquid must also have a considerable effect upon the conduction. The moisture content of the samples has been reported and the effect of moisture upon the conductivity of transformer oil is well known[7]. Due to arcing of the diverter switch the samples were also expected to have a high gas content and the effect of absorbed air upon the conductivity of transformer oil has been considered by various workers[8].

The variation of the resistivity of transformer oil, including contaminated oil, has been observed at low and medium fields[9]. It has been shown that the resistivity is first constant, then increases steeply and at a certain critical field strength suddenly decreases. In the present case this abrupt change in the slope of the resistivity characteristic could be detected for some samples. Variation in transformer oil resistivity at low and medium fields has been explained by regarding the oil as a weak electrolyte and developing relevant differential equations. Due to the complex nature of the contaminated samples involved here a simple determin-istic formulation is unlikely to encompass all the operative factors. It may, however, be possible to assume that certain probabilities of charge carrier increase and decrease will apply. Probabilistic formulations of conduction processes in stressed oil prior to breakdown have previously been considered[10,11].

A concise description of the conduction model is now given. Because two carrier species are involved, one positive and one negative, the process is essentially two dimensional. As the test voltage is increased at constant rate there is the possibility of carrier withdrawal or generation in an interval Δt. Since the voltage is increasing it is assumed that the mean rates of carrier generation exceed the corresponding rates of withdrawal. Carrier withdrawal is considered to be associated with (i) the long residence time of charged particles at the electrodes (ii) large particle ejection from the maximum field region (iii) the deflecting action of interparticle forces and fluid motion. Carrier generation may arise from (i) further neutral particles entering the gap and charging at the electrodes (ii) positively charged particles at the cathode promoting bursts of electron emission (iii) negative ion space charge at the anode causing collision ionization[12] (iv) other multiplicative effects in the moist and gas saturated liquid bulk at higher stresses (v) possible ionization resulting from inter-mittent development and disruption of particle bridges.

In the simple model no distinction is made between carrier generation in the bulk liquid and that occurring at the electrode/ liquid interfaces. An overall rate of carrier generation, for one species, is assumed to apply to the system as a whole. As a further simplification the rates of generation of both positive or negative carriers are assumed to depend on the number of active negative carriers existing at time t in the entire region bounded by the electrodes. Thus, although the applied field is recognised as the primary cause of carrier multiplication, the mechanism of carrier generation is vested entirely in the carriers themselves for conceptual reasons. The validity of this assumption would appear reasonable for a collision ionization process in the liquid bulk, and in the region of the anode due to space charge formation. At the cathode negative carrier injection by burst emission may depend on the accumulation of positively charged carriers but these have been previously generated in the bulk of the liquid and fed back to the cathode surface.

The precise role of the macroscopic particles in the conduction process is less easy to define. A thick layer of particles on the cathode and a thinner film on the anode suggests that large numbers have acquired charge and may thus contribute to charge transport through the liquid. But the high concentration of particles may result in dielectric layers of coarse particles impeding neutralization of other particles in outer layers. Nevertheless, such a layer may be porous to the movement of sub-micron macroscopic carriers and those of molecular dimensions. The particle conduction at higher fields may therefore depend on particles of the finest size and carrier generation due to these impurities will largely come about from negatively charged particles by a process of electron breakaway[13] at high fields. Although the generation rates of both species have been assumed to depend on the number of negative carriers the withdrawal rates are assumed to depend on the respective numbers of positive and negative carriers.

Under the above assumptions the increase in conduction may be attributed to an underlying Markov process with discrete states in continuous time[14]. A forward differential equation may be formulated which when solved by a suitable generating function leads to a solution for the distribution of the numbers of positive and negative carriers in the system as a function of time. Since subgroups generated by different carriers are assumed mutually independent the conduction current will depend on the time functions specifying the means of these distributions. The solutions for the means in this case are of the form:

$$N_1(t) = n_1 \exp(at) \qquad \text{negative carriers}$$

$$N_2(t) = A \exp(at) + (n_2 - A)\exp(-bt) \qquad \text{positive carriers}$$

where n_1 and n_2 represent initial numbers of negative and positive carriers and a, b and A are constants depending on the rates of generation and withdrawal.

Since the preponderating terms are $n_1 \exp(at)$ and $A \exp(at)$ the total number of carriers involved in conduction increases according to the law

$$N_1(t) + N_2(t) \approx (n_1 + A) \exp(at)$$

Thus $i = a(n_1 + A) \exp(at)$ also increases exponentially, and since $V = Bt$ we may write:-

$$i = a(n_1 + A) \exp\left(\frac{aV}{B}\right) = k \exp(CV)$$

This is the solution for the current flowing between the electrodes. The current in the external circuit may be expected to follow with a slight delay due to the finite time assumed for particle neutralization at the electrodes. The proposed model is intended to be simple and tentative. A suggested modification would be to specify negative and positive carrier generation rates in a manner which makes the process dishonest. The population size then becomes infinite in a finite time and simulation of the breakdown process is possible.

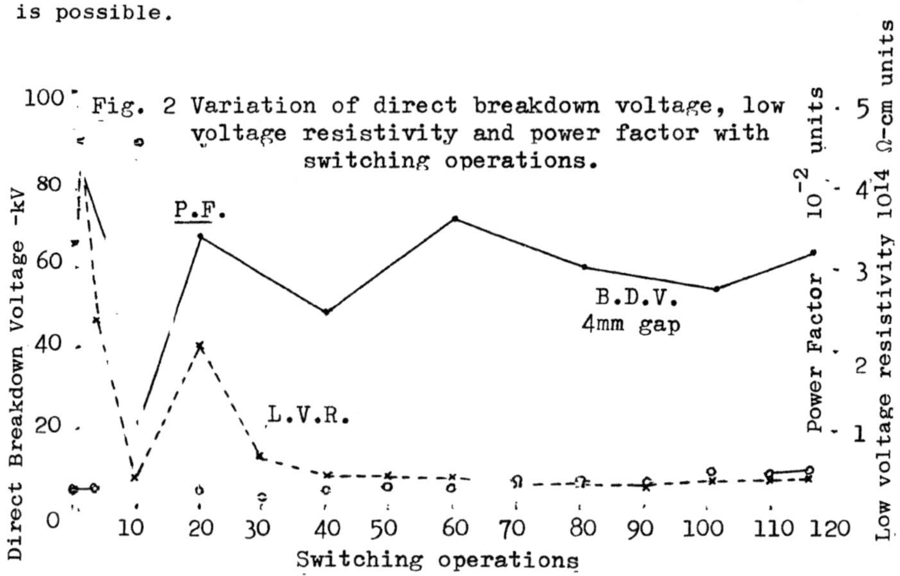

Fig. 2 Variation of direct breakdown voltage, low voltage resistivity and power factor with switching operations.

Additional test results

A relatively large number of independent electrical, physical and chemical tests were made on each oil sample. Fig. 2 shows the variation of direct breakdown voltage, power factor and low voltage resistivity for each of the samples tested. One method of condensing a large number of hybrid data is that of multiple factor analysis[15] which has previously been used in an investigation concerned with electrical insulating material. Some of the data in the present investigation were analysed in this way using the group centroid method. For the limited number of tests initially considered three primary factors were established. The statistical significance of this result would naturally be increased with a larger number of tests and work is in progress on a correlation matrix which includes all the test data. Initial results indicate that Factor I has almost constant loading in tests concerned with moisture, direct breakdown voltage and power factor, while Factor II indicates a link between low voltage resistivity and particle number.

References

1. Young, B., 1972, 4th International Conference on Conduction and Breakdown in Dielectric Liquids, Dublin.

2. Young, B., 1973, I.E.E. Conference on Diagnostic Testing of High Voltage Power Apparatus in Service, London.
3. Krasucki, Z., 1972, 4th International Conference on Conduction and Breakdown in Dielectric Liquids, Dublin.
4. Rhodes, G.M., and Brignell, J.E., ibid
5. Birlasekaran, S., and Darveniza, M., ibid
6. Sletten, A.M., 1960, Ph.D. Thesis, University of London.
7. Stannett, A.W., 1951, British J. Appl. Phys., 2, pp 110-114.
8. Sugita, K., Sato, T., and Toriyama, Y., 1960, ibid, 11, pp 539-542.
9. Harrison, N.L., 1968, Proc.I.E.E., 115, 5, pp 736-741.
10. Yamanaka, C., and Sinta, T., 1953, J. Phys. Soc. (Japan) 8, 277.
11. Megahed, I.Y., and Tropper, H., 1971, J. Phys. D, 4 pp 446-451.
12. Zein Eldine, M.E., et al., 1965, Proc.I.E.E., 112, 3, pp 580-585.
13. Lewis, T.J., 1959, Progress in Dielectrics I, Heywood, London.
14. Cox, D.R., and Miller, H.D., 1965, The Theory of Stochastic Processes, Methuen, London.
15. Thurstone, L.L., 1947, Multiple Factor Analysis, University of Chicago Press.

EFFECT OF ADDED ELECTRONEGATIVE SUBSTANCES (SF_6, I_2) ON ION MOBILITY IN MINERAL OIL

S. Sakamoto & S. Usuda *

1. Introduction

The relation between ion mobility and fluid motion in dielectric liquids (n-hexane or benzene) has been the subject for research and discussion by many investigators. This paper is concerned with ion mobility μ, fluid motion in mineral oil (transformer oil) under electric field, the effect of added electronegative substances (SF_6 and iodine I_2) on μ and relation between breakdown strength and the amount of the added substances.

2. Problems on ion mobility measurement

Many investigations into the experimental value of ion mobility of n-hexane as a representative dielectric liquid has been reported [(1)(2)], but there are some deviations from 10^{-4} to 10^{-3} $cm^2/V\cdot sec$ among the values. As the causes for these deviations, the following several factors are pointed out (1) difference between the methods of charge injection into drift field (2) difference between methods of measurement (3) effect of fluid flow induced by electric field (4) effect of impurities in dielectric liquid, etc. A direct discussion concerning the above considerations will not be done here. In this paper the measurement of μ is performed by the charge injection method [(2)] in which razorblade electrode is used as the emitter and also by the time-of-flight method. These are considered as the useful and convenient methods to get stronger current of dielectric liquids. For the research on the effect of fluid flow induced by the electric field, liquid velocity v_1 at various points between electrodes is measured by a laser-Doppler method [(3)(4)].

3. Experimental arrangement and results

An experimental arrangement is shown in Fig.1. As described above, 3 razor-blade electrodes are used as charge emitters and a pair of mesh electrodes is used as the grid and the collector electrode. Charge carriers drift between the two mesh electrodes with a certain drift velocity and the ion mobility μ is obtained from the measured drift time τ and drift field Ed.

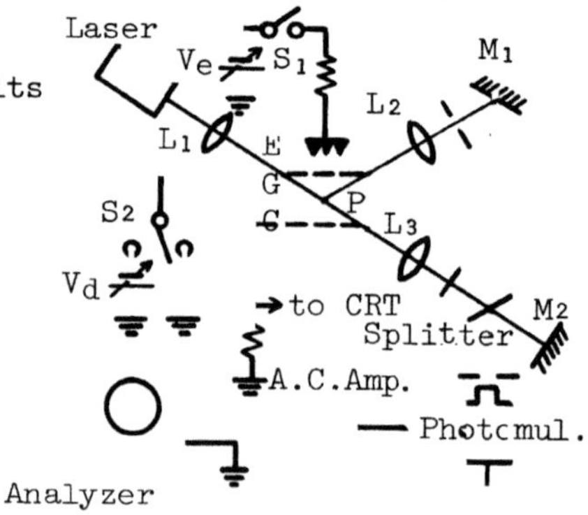

Fig.1.Experimental arrangement.

* Department of Electrical Engineering, Hokkaido University, Kitaku N13, W8, Sapporo, Japan 060

(3.1) relation between ion drift velocity and drift field

In Fig.2 relation between drift velocity and drift field Ed for several dielectric liquids are shown. V_i in the figure denotes applied emitter voltage. As V_i increases the ion drift velocity v_i increases. From Fig.2 the following relation among v_i, Ed and ion mobility μ will be obtained.

$v_i = v_o + \mu Ed$ (V_i=const.) ...(1)

where μ is obtained from the slope of v_i/Ed, and v_0 is v_i when Ed=0.

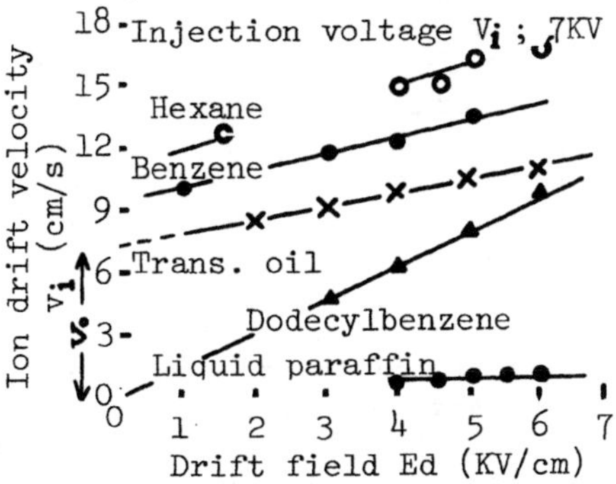

Fig.2. Ion drift velocity v_i vs. Ed.

This indicates that the electrons emitted from lazor-blade cathode drift with some speed toward the grid mesh electrode, but they are captured just after they leave the cathode and enter the drift field as negative ions with some initial speed v_0. In the drift field the negative ions are accelerated by Ed and drift toward the collector.

(3.2) effect of absorbed SF_6 in transformer oil

It is well known that SF_6 gas is electronegative and has a high dielectric strength. In transformer oil the effect of absorbed SF_6 gas on the ion drift velocity and the ion mobility is shown in Fig.3. SF_6 gas absorption treatment is as follows. First, the transformer oil is degassed under 0.05 Torr. Then it is kept more than 6 hours in SF_6 gas with 760 Torr. Negative ion drift velocity v_i-decreases markedly by this treatment (mark ▲), but remains unaltered after 24 hours under the atmospheric pressure (mark o in Fig.3). On the contrary, v_i- increases when compared with the non-treated oil by the degassing treatment. About conductivity κ it is obtained that κ of non-treated oil is 8.6×10^{-16}(℧/cm)degassed oil 2.6×10^{-16}, oil with absorbed SF_6 2.4×10^{-16}. It is assumed, however, from eq.(1) and Fig.3, that the ion mobility does not alter very much by absorption and the degassing treatment described previously. If the consideration as shown in eq.(1) is accepted, it would be concluded that the SF_6 absorption or degassing in the mineral oil affects mainly initial ion drift velocity v_0.

(3.3) effect of dissolved iodine I_2 in transformer oil

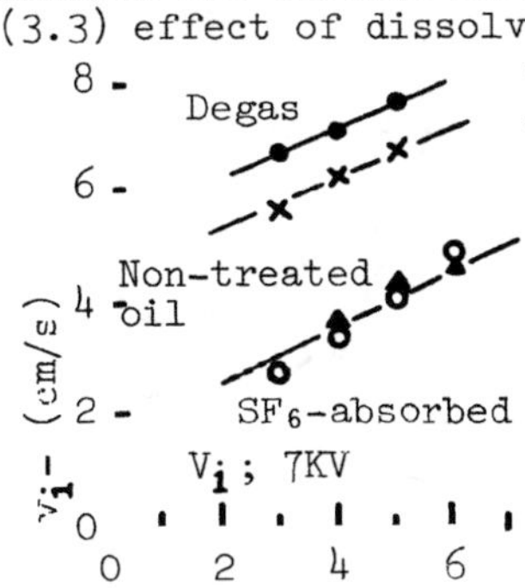

Fig.3. v_i- in trans. oil with dissolved SF_6 gas.

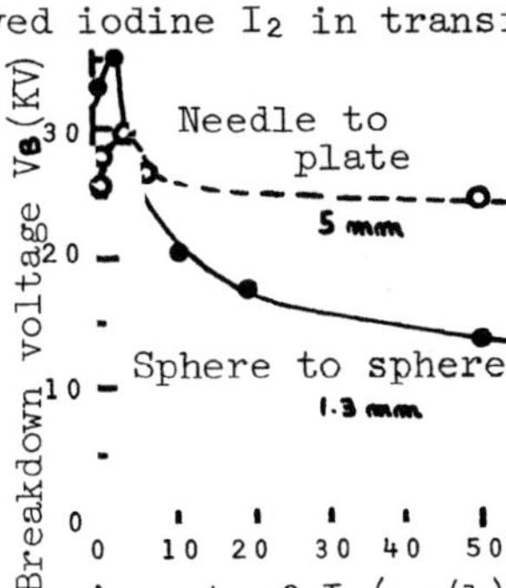

Fig.4 (a). Breakdown voltage in CCl_4 with additive I_2.

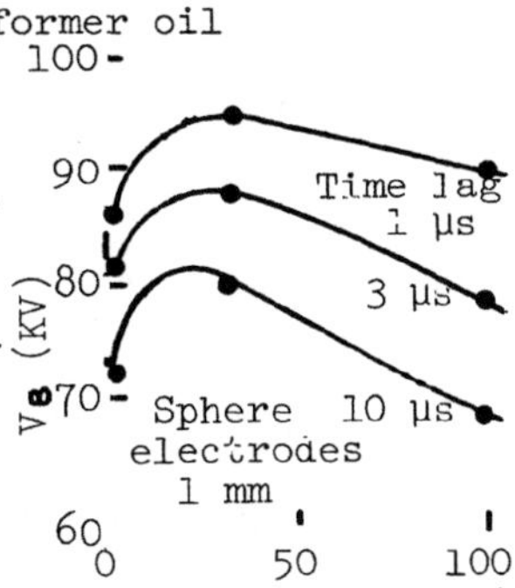

Fig.4 (b). V_B in trans. oil with additive I_2.[6]

I_2 is also electronegative substance, the same as SF_6 and an effect of dissolved I_2 in dielectric liquid (CCl_4, transformer oil) on its breakdown strength has been investigated [5][6]. The results are shown in Fig.4. It is seen that the breakdown voltage increases for proper concentration of dissolved I_2 (mg/l). In correlation with these investigations the ion drift velocities v_i- are measured at wide range of dissolved concentration of I_2 in transformer oil and the results are shown in Fig.5(a)(b). Several examples of transient current which are induced when the carge carriers drift between the mesh electrodes, are shown in Fig.5(a). From these results Fig.5(b) and Fig. 6 are obtained successively. In Fig.6 it is shown that the initial ion drift velocities v_0 take a minimum value at the amount of I_2 5~10 mg/l in transformer oil which agrees with the result shown in Fig.4(b). It appears that ion mobility μ does not alter as same as in (3.2). Conductivity of the transformer oil increases with the amount of dissolved I_2.

(3.4) fluid velocity in transformer oil under electric field

As described in introduction, it is important to check the effect of fluid motion under electric field for estimation of the value of ion mobility in dielectric liquid. For this purpose liquid velocities v_l at several points in drift field are measured in transformer oil using a laser-Doppler method.

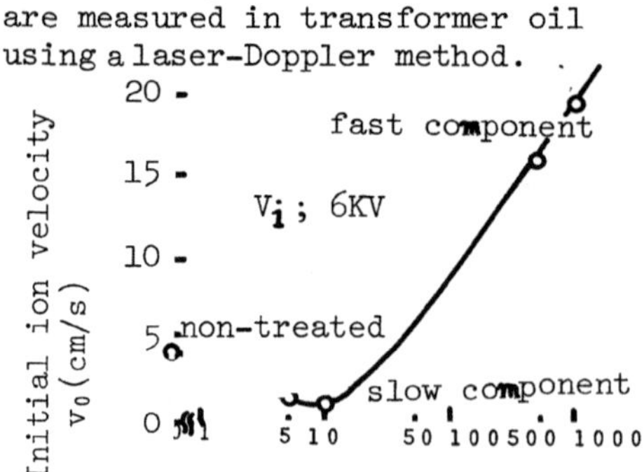

Fig.6. v_0 vs. amount of additive I_2 in transformer oil.

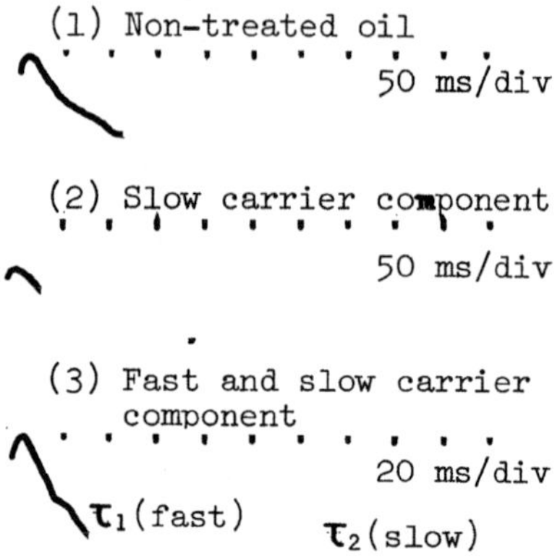

Fig.5 (a). Induced current in transformer oil with additive I_2.

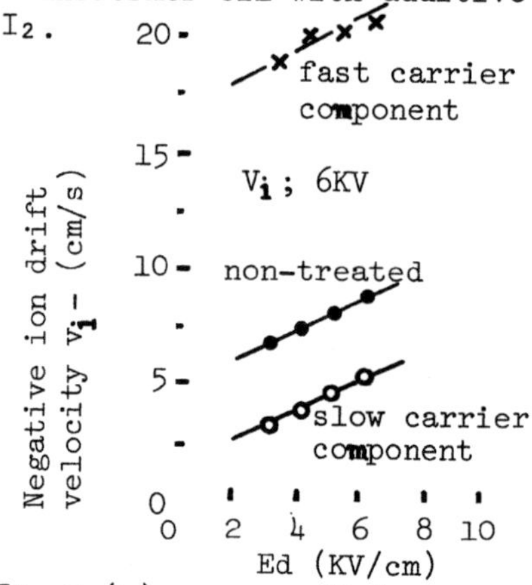

Fig.5 (b). v_i- vs. Ed in transformer oil with additive I_2.

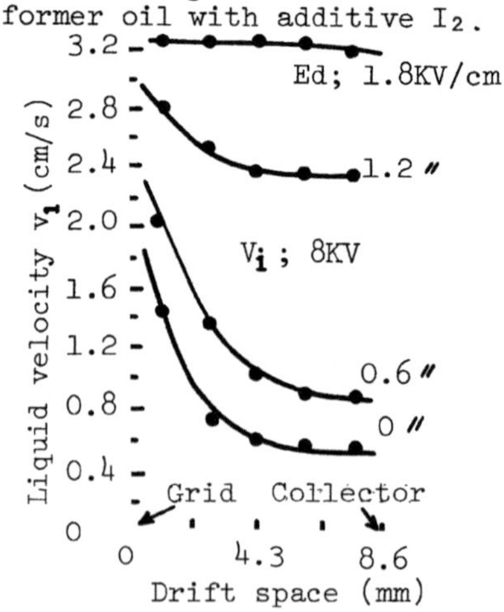

Fig.7. Distribution of v_l in drift field space in transformer oil.

Obtained results are shown in Fig. 7 and Fig.8. From these figures it is observed that v_l is built up in time of the order of second and reaches the maximum value in few seconds and then decreases. The ion drift time between the mesh electrodes is of the order of milli seconds. This means that v_l is much lower than v_i. It is also observed that the ion velocity decays rapidly in transformer oil.

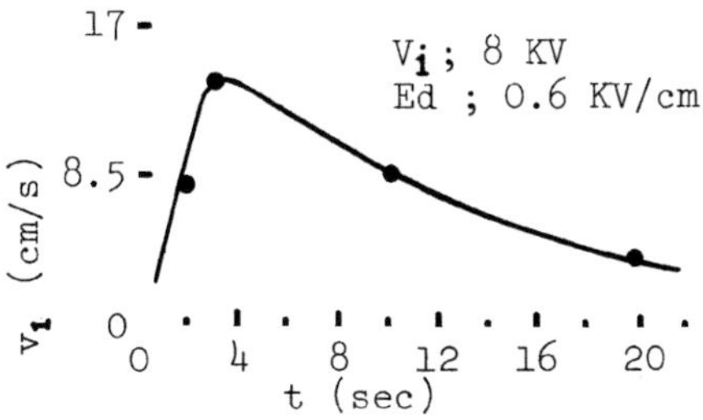

Fig.8. Variation of v_l with time after the voltage application in trans. oil.

4. Discussion and conclusions

(1) When a proper electronegative substances (SF_6, I_2) are added to transformer oil, the ion drift velocity decreases. It is assumed that electrons emitted from the cathode are captured just near the cathode by SF_6 or I_2 molecule in the liquid and enter drift field with an initial ion velocity v_0.
(2) It is concluded that the molecular weight of added substance has a great effect upon the value of v_0. From the next table it is considered that electron attachment to SF_6 or I_2 molecule produced heavy ions and decreases v_0. In this case it is assumed that the effect of molecular parachor is not much. The assumption is deduced from an experiment of $C_{12}H_{25}C_6H_5$ + I_2.

Molecule	H_2O	N_2	O_2	C_6H_6	C_6H_{12}	SF_6	$C_{12}H_{25}$-C_6H_5	I_2
Molecular wt.	18	28	32	78	84	146	246	254
Parachor	54.2	48.2	63.2	207.1	268.2	202.4	646.3	205.2

(3) The breakdown strength of the liquid increases by addition of a proper amount of SF_6, I_2 and the tendency agrees with the decrease of ion drift velocity.
(4) The liquid velocity v_l is negligiblely small in comparison with the ion velocity v_i in this measurement arrangement.
(5) The value of ion mobility μ is regarded as being constant in spite of the treatments as mentioned so far and it is considered that the value of μ depends on the individual property (mainly on the molecular weight) of the dielectric liquid.

References.(1) E.Gray & T.J.Lewis: Brit.J.A.P., 1965, 16, 1049
(2) V.Essex & P.E.Secker: J.Phys.D., 1968, 1, 63
(3)R.J.Taylor & H.House: Electronics Letters, 1971, 7, 404
(4) T.Sato & S.Sakamoto: J.Phys.D., 1971, 4, L53
(5) Y.Toriyama & S.Sakamoto: E.Tech.J. of Japan., 1941, 5
S.Sakamoto: Tech.Rept.Hokkaido Univ., 1953, 8, 131
(6) H.Yamashita & T.Mori: IEEJ, 1973, 93-A, 369
(7) S.Usuda & S.Sakamoto: IEEJ, 1974, 94-A, 507

MONITORING OF INJECTION PROCESSES IN DIELECTRIC LIQUIDS

S. Barret, F. Gaspard, F. Mondon★

Introduction

Many experiments and theoretical works have been devoted during the last years to the conduction in polar dielectric liquids. In spite of some discrepancies between the results great progress has been made in clarification of the problem. One of the main of them, i.e. the explanation of the current vs voltage characteristics, is now well understood. A typical current-voltage curve shows 3 parts (1)

- an ohmic behaviour at very low voltage
- a saturation plateau at medium voltage
- a fast increase of the current at higher voltage.

Physical chemistry provides explanation for this behaviour. The saturation of the current is due to the removal of ions from the dielectric by electrolysis, the rate of which is fast enough to displace the dissociation-recombination equilibrium. The saturation is conveniently studied by means of electrodialysis between ion exchange membranes (2). The rapid increase of the current when the voltage is raised takes its origin in a new carriers generation at one or both electrodes : new ions are injected into the dielectric liquid owing to electrochemical reactions. The measured current results from super imposition of bulk conduction and injection conduction. The injection induces space charges, field distortions and electrohydrodynamic instabilities (3). The study of the injection regime is a hard problem because the phenomena are not controlled. The nature of injected ionic species is generally unknown and the intensity of the current is unstable. Monitoring the injection processes is thus necessary and could lead to the same success as obtained with electrodialysis in the study of bulk ionic conduction.

1 - Injection as an electrochemical process

The existence of electrochemical reactions at both electrodes is a necessary condition for the circulation of the d.c. current in an ionic conductor. Oxidation occurs at the anode ($A \rightarrow A^{ox} + e^-$) and reduction at the cathode ($B + e^- \rightarrow B^{red}$).
Two cases have to be distinguished :

- Species A and B are ions (A^- and B^+), then the oxidized and reduced species (A^{ox} and B^{red}) are neutral, ions have been withdrawn from the liquid by electrochemical reactions. This leads to the bulk ionic purification of the medium by electrolysis.
- Species A and B are neutral species, then A^{ox} and B^{red} are ions (A^{ox} = cation, B^{red} = anion) then the result of electrochemical reaction is an injection of new ionic carriers in the liquid.

Monitoring of injection thus requires the knowledge of electrochemical reactions occuring in a dielectric liquid, this is the field of organic electrochemistry.

★ Université de Grenoble CERMO BP 53 38041 Grenoble Cédex (France).

2 - Monitoring of injection by use of suitable additives

The ability of a specie to be oxidized (or reduced) is called electro activity. If several species are present near an electrode, the most electroactive one will undergo an electrochemical reaction prior to the others. This first statement is the key of our problem but others specifications must be fullfilled in order to obtain a valuable result. The ideal electroactive additive (which is not an ionic specie and does not dissociate into ions) is defined as follow. Let us take the example of an oxidizable specie.

- high electroactivity (low oxidation potential)
- the oxidized form is a charge carrier ($X \rightarrow X^+ + e^-$)
- chemical stability of the oxidized form X^+
- reversibility of electrochemical behaviour : the oxidized form X^+ must yield back the initial specie X when reduced. This behaviour is sketched on figures 1 and 2

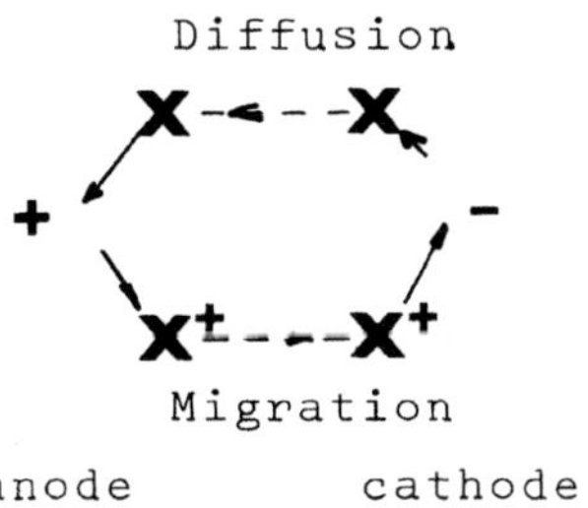

Fig 1

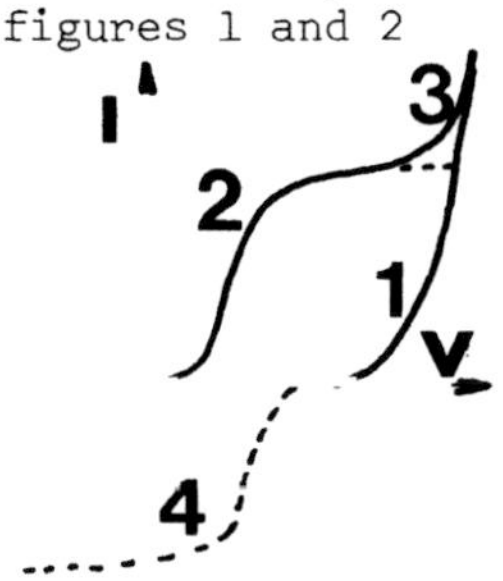

Fig 2

1 oxidation of the solvent.
2 oxidation of active species alone.
3 oxidation of active species + solvent.
4 reduction of oxidized form of active species.

This additive leads to a strong anodic injection of the oxidized form X^+ only. This is because the potential drop at anode is kept very low (the voltage drop beeing stabilized by the reaction). No injection occurs at the cathode for the same reason, the only reaction is the discharge of X^+. The use of an additive easily reduced and oxidized (or the use of two additives) would result in an ambipolar injection.

3 - Example of controlled injection of $HMDPA^+$ in chlorobenzene

Hexamethoxy - diphenylamine (HMDPA) is readily oxidized and yields the free radical $HMDPA^+$. The half-wave potential is 0,04 V vs Ag/Ag^+ in acetonitrile for oxidation of HMDPA and for reduction of $HMDPA^+$ too. This potential is very weak as compared to half-wave potential of most organic compound which is generally higher than 1 V. Reversibility and chemical stability are checked by means of polarography and cyclic voltametry. HMDPA is used in concentration of 10^{-3} M/l and 5.10^{-4} M/l to study injection in chlorobenzene.

Resistivity measurements

The resistivity of several samples of pure chlorobenzene is in the range 1.5 - 3.0 10^9 $\Omega \times$ cm (25°C and 10 k Hz). The resistivity of doped samples beeing $3 \times 10^7 - 1.5 \times 10^8$ $\Omega \times$ cm. Contamination of the liquid by impurities during preparation of doped samples is the cause of this decrease in resistivity as HMDPA does not dissociates into ions.

After some stay in the cell under moderate voltage the low field d.c. resistivity of doped samples is raised to 4×10^8 $\Omega \times$ cm.

Current intensity measurements

Current intensity vs voltage characteristics for pure chlorobenzene shows a saturation plateau at field strendth oxtending from 10^4 V/cm to 10^5 V/cm (electrode spacing 300 μm). The saturation current density is 5.10^{-8} A/cm^2. At higher field, the current increases rapidly. In doped chlrobenzene, the most evident fact is the considerable increase of the current, at 10^4 V/cm the current density is now $2.16\ 10^{-5}$ A/cm^2 (electrode spacing 280 μm). The, difference of three orders of magnitude is due to a strong injection (the increase of a.c. conductivity is only of one order of magnitude for the same sample). A perfect relation $i = CE^2$ is obtained (fig 3). Assuming a real space-charge limited current regime and the validity of the formula $j = 9/8\ \varepsilon\ \mu_e V^2/l^3$ the deduced electrohydrodynamic mobility is $\mu_e = 6.2.10^{-4}$ cm^2.V^{-1}S^{-1}.

Transit time measurements

The transient behaviour of injection is now studied, single rectangular pulses are applied to the cell. The current intensity vs time characteristic shows a peak wich indicates the transit time τ (4). The electrohydrodynamic mobility computed from τ is voltage dependent (fig 4). Extrapolation to zero field provides the value of the true (electrochemical) mobility of the radical cation HMDPA$^+$ in chlorobenzene ($\mu = 3.2\ 10^{-4}$cm^{2}V^{-1}S^{-1}). This value has to be compared to those obtained from conductivity measurements in chlorobenzene : $\mu(C_6H_9)_4H^+ = 2.54\ 10^{-4}$cm V^{-1}S^{-1}, $\mu I^- = 5.47\ 10^{-4}$cm V^{-1}S^{-1} (5). The transient mobilities of HMDPA are twice the steady value given above, this difference is usual (6).

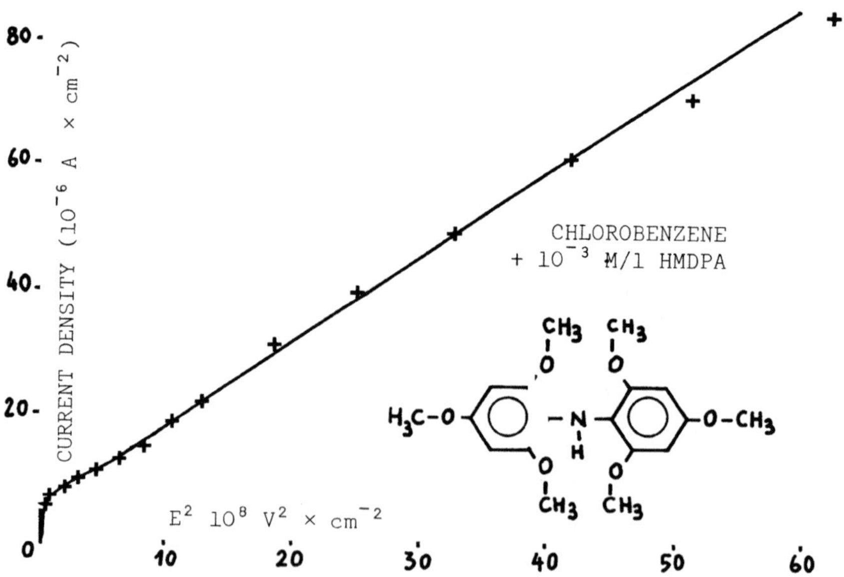

Fig 3 : steady injection current variation

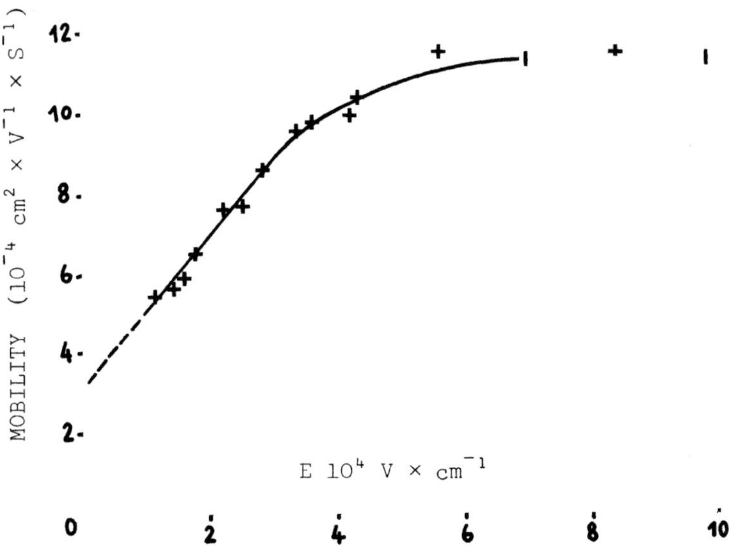

Fig 4 : transient mobility

Conclusion

The monitoring of injection process is possible in a medium of weak permittivity as chlorobenzene. This will be a powerful tool for the study of electrodydrodynamic phenomena. Moreover the method is of general interest as it allows the control of electrodes reactions in many situations. Successfull results were obtained in a similar way in liquid crystals displays. A rapid degradation, caused by non reversible electrolysis, occurs when a pure liquid crystal is submitted to a d.c. voltage (7) (as in dynamic scattering displays). A liquid crystal doped with HMDPA is protected against electrolysis and the life-time of the display is increased ten fold. Application of the method to more polar dielectric liquid does not present any difficulty. Electrochemical reactions need only the build-up of an ionic double layer on the electrode and enough ions are always present in a polar liquid. Extension to non polar liquids such as hexane and benzene is investigated. However these liquids are not dissociating solvents and the presence of enough ions is less evident. Checking the validity of the method will previde informations about electrochemical reaction in these media where no electrochemistry experiments have ever been done.

(1) G. Brière, F. Gaspard Chem. Phys. Lett. (1968) 1 706
(2) G. Brière, F. Gaspard, R. Hérino J. Chim. Phys. 1971 68 845
(3) N.J. Félici Direct Current 2 147
(4) P. Atten, J.P. Gosse J. Chem. Phys. 51 2804 (1969)
(5) F. Gaspard, H. Seinera to be published
(6) S. Barret, F. Gaspard J. Chim. Phys. (1974) 71 255
(7) G. Brière, R. Hérino, F. Mondon, Mol. Cryst and Liq. Cryst. 19 757 (1972)

CHARGE INJECTION IN DIELECTRIC LIQUIDS BY PHOTOELECTRIC EMISSION

W. Tauchert and W. F. Schmidt*

Introduction

The investigation of the transport properties of excess electrons in dielectric liquids requires also measurements of the energy of the conducting state relative to the vacuum level. This quantity V_o can be obtained by measuring the work functions of a metal in vacuum Φ_{vac} and in the liquid Φ_{liq}. V_o is then given by

$$V_o = \Phi_{liq} - \Phi_{vac} \tag{1}$$

Positive V_o represents the energy necessary to bring an electron from the vacuum into the liquid, negative V_o corresponds to the energy which is released during this process.

Measurements of the work function can be performed by means of the photoelectric effect. The photocurrent from a metal electrode is studied as a function of the frequency of the incident light. After Fowler [1] the photocurrent density $j(\nu)$ in vacuum is given by a universal relation

$$j(\nu, T) = \alpha A T^2 F(x) \tag{2}$$

or

$$\ln \frac{j(\nu, T)}{T^2} = \ln \alpha A + \ln F(x) \tag{2a}$$

where T is the absolute temperature, α the fraction of emitted electrons per absorbed photon. A is a universal constant given by

$$A = \frac{4 m k_B^2}{h^3} \tag{3}$$

with m denoting the electron mass, k_B the Boltzmann constant and h Planck's constant. The frequency dependence is given by the Fowler function

$$F(x) = \sum_{n=1}^{\infty} (-1)^{n+1} \frac{e^{nx}}{n^2} \tag{4}$$

with

$$x = \frac{h\nu - \Phi}{k_B T} \tag{5}$$

* Hahn-Meitner-Institut für Kernforschung Berlin GmbH, Bereich Strahlenchemie, 1 Berlin 39, Germany

The experimental data are plotted as $\ln j(\nu)/T^2$ versus $x' = \frac{h\nu}{k_B T}$. The work function Φ is then obtained by shifting the experimental curve in x-direction until it coincides with the Fowler curve according to eq. 2a. Holroyd was the first to show that eq. 2 also describes the photoeffect of a metal electrode in a dielectric liquid [2]. In this case an additional shift of the exponential curve in y-direction is necessary in order to account for the reduced efficiency due to electron back scattering by the atoms or molecules of the liquid.

Measurements of V_o for liquid hydrocarbons were reported by Holroyd et al. [2-4] and by Schiller et al. [5]. Data for liquid argon and helium were reported by Lekner et al. [6] and by Sommer [7].

A correlation between V_o and the electron mobility in liquefied rare gases has been derived by Springett et al. [8]. The interaction between the excess electrons and the atoms of the liquid consists of the long range polarization and a short range repulsion. Domination of the polarization results in $V_o < 0$ and large electron mobilities $> 10^2 cm^2V^{-1}s^{-1}$. In helium which exhibits only a small polarizability repulsion is prevalent and $V_o > 0$; the electron mobility is small $\sim 10^{-2}$ $cm^2V^{-1}s^{-1}$. In liquid hydrocarbons V_o values of -0.45 eV for neopentane to + 0.1 eV for n-hexane and mobilities of 70 $cm^2V^{-1}s^{-1}$ to 0.1 $cm^2V^{-1}s^{-1}$, respectively have been reported. A phenomenological relation between μ and V_o

$$\mu = A e^{-BV_o}$$

was suggested by Holroyd [3].

Experimental

Our experimental set-up was similar to that of Holroyd [2] and is shown schematically in figure 1. The light of a xenon arc (1) was focussed on the entrance slit of a grating monochromator (2). A

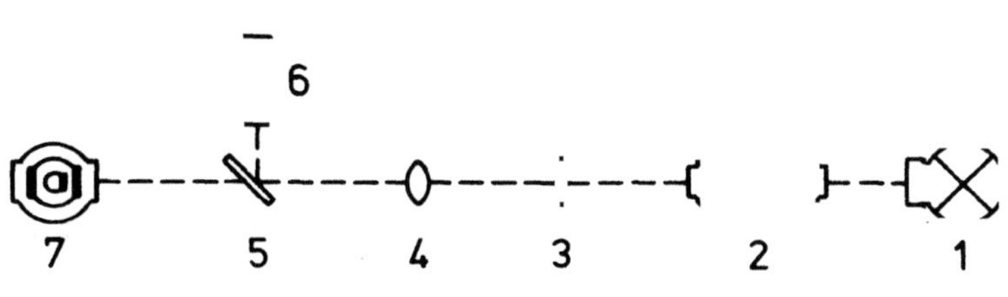

Fig. 1 - Schematic of experimental set-up (see text)

filter (3) served to remove scattered light and harmonics of the selected wave length. A beam splitter (5) reflected about 10 % of the light to a photo multiplier (6) which monitored the light intensity. The light beam was focussed (lens 4) on the cathode of the measurement cell which was enclosed in a cryostat (7). The temperature was measured by a thermistor and by the vapor pressure of the liquid in the measurement cell. All optical windows were made of suprasil. The cathode was brass coated with zinc. The anode consisted of a tungsten wire mesh with an optical transparency of 80 % separated by d = 2.5 mm from the cathode. The photo currents i as a function of the wave length were measured with a Keithley electrometer. The relative light intensity I at each wave length was determined with the photo multiplier, the response of which had been determined with a calibrated photodiode (United Detector Technology PIN 10).

All samples were prepared from high purity substances which were purified further by passage through columns of activated 3 Å molecular sieve and charcoal maintained at dry ice temperature. The measurement consisted of the following cycle: First the cell was filled with the liquid to be investigated. Then, the liquid was removed again and the cell evacuated to a pressure below 10^{-5} torr and the work function was measured in this vacuum. Measurement in the liquid followed and finally the cell was evacuated and the vacuum work function was determined again. The photocurrents in the vacuum were measured at 400 V cm^{-1} while field strengths from 2 to 8 kV cm^{-1} were applied in the liquids. Typical values of the photocurrents ranged from 0.1 pA to 100 pA depending on the wave length. The dark current was of the order of 0.01 pA and was subtracted from the photocurrent. The illuminated electrode area was approximately 1 cm^2.

The data $i(\nu)/I(\nu)$ as a function of $x' = \frac{h\nu}{k_B T}$ were fitted to eq.2 by a non-linear least square program on a Siemens 4004 computer.

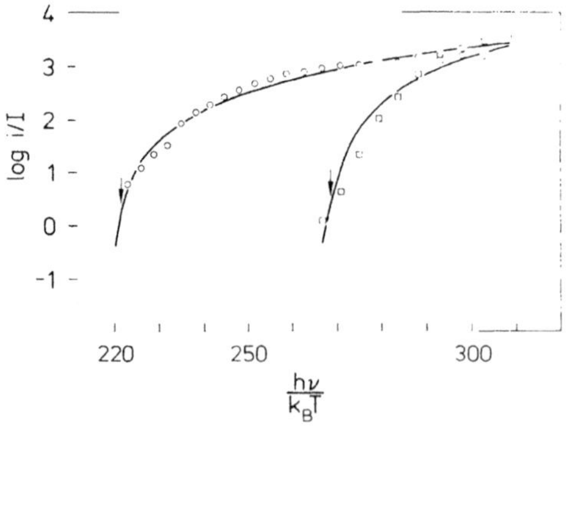

Fig.2: Fowler plot of photocurrents in liquid xenon at T = 161 [K] ○ and vacuum □.

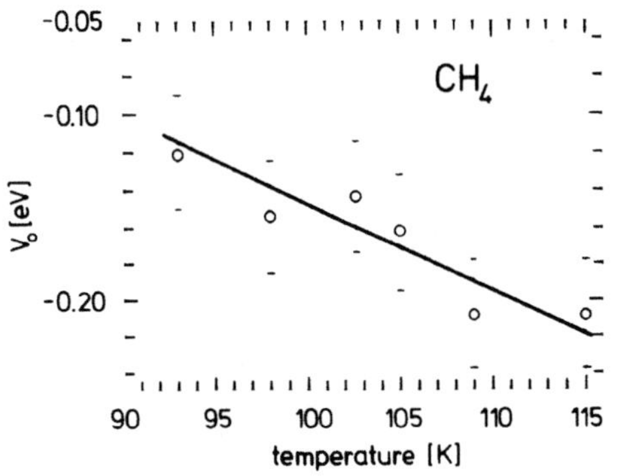

Fig. 3: V_o in liquid methane

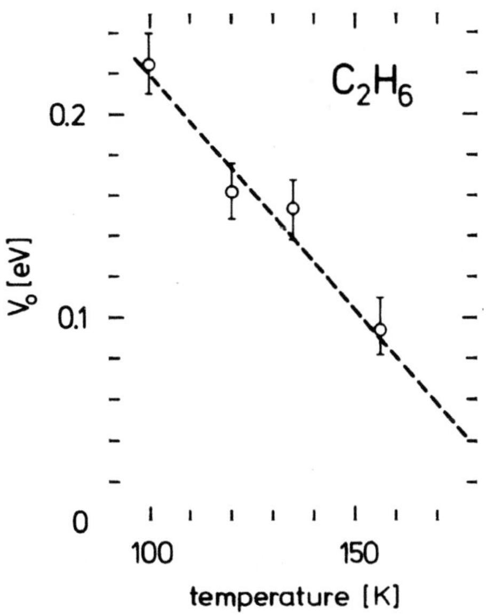

Fig. 4: V_o in liquid ethane

Results and Discussion

Typical data of the relative photocurrent i/I as a function of the frequency measured in liquid xenon and vacuum are shown in figure 1. The arrows define the work functions. A value of Φ_{vac} = 3.72 for the zinc electrode was obtained. Literature data scatter from 3.1 eV to 4.3 eV [9] which indicates the influence of surface properties. Our value remained fairly constant even for freshly deposited zinc layers. In liquid argon and xenon the measurements could be carried out only over a small temperature range. The results are summarized in table 1. The values V_o obtained in liquid methane and ethane as a function of the temperature are shown in figure 3 and 4. In both liquids V_o decreases with increasing temperature.

	T[K]	V_0[eV]
argon	84	-0.20 ± 0.02
	87.5	-0.21 ± 0.02
xenon	161	-0.67 ± 0.05
	165	-0.61 ± 0.05

In accuracy of the V_0 measurement is decisively influenced by the fact that V_0 is obtained as the difference of two large numbers (eq. 1). The work functions measured in different runs did not vary on the average by more than ± 1 %. This represents in absolute terms ± 0.035 eV and the influence on the accuracy of the V_0 value depends on the magnitude of V_0.

The temperature dependence of V_0 in methane, ethane, and argon can be explained by the semicontinuum model of Springett et al. [8]. The governing factor seems to be the change in density. The temperature variation of V_0 in xenon is not understood.

Acknowledgement: We are grateful to "Deutsche Forschungsgegemeinschaft" for financial support.

References: [1] R.H.Fowler, Phys.Rev. 38, 45 (1931); [2] R.A. Holroyd and M.Allen, J.Chem.Phys. 54, 5014 (1971); [3] R.A. Holroyd and R.L.Russel, J.Phys.Chem. 78, 2128 (1974); R.A. Holroyd and W.Tauchert, J.Chem.Phys. 60, 3715 (1974); [5] R. Schiller, Sz.Vass, and J.Mandics, Int.J.Radiat.Phys.Chem. 5, 491 (1973); [6] J.Lekner, B.Halpern, S.A.Rice, and R.Gomer, Phys.Rev. 156, 351 (1967); [7] W.T.Sommer, Phys.Rev.Lett. 12, 271 (1964); [8] B.E.Springett, J.Jortner, and M.H.Cohen, J.Chem.Phys. 48, 2720 (1968); [9] G.L.Weissler, in "Handbuch der Physik", Vol. 21, 347 (1956)

FACTORS CONTROLLING ION INJECTION BY METALLIC ELECTRODES IN DIELECTRIC LIQUIDS

N. Félici, B. Gosse, J.P. Gosse *

Introduction

The conductive behaviour of resistive liquids when subjected to strong D.C. fields puzzled physicists for many decades. Having in mind Thomson's results on X-ray-ionized gases, which should apply to liquids whatever the ionizing agency, they could not figure any reason for the apparently erratic conductivity of liquids like nitrobenzene or even hydrocarbons. Nikuradse's attempt to resort to collision ionization as an explanation was unwarranted, and it was not before the '60s that the fundamental similarity of electrode effects in resistive liquids with the electrode processes of electro-chemistry came to light.

The understanding of the proper role of electrochemical concepts in such matters was long obfuscated by the undue reliance of many chemists on the "electroneutrality rule" which seems to preclude any applicability to the cases when strong space charges pervade the medium and Ohm's law fails completely. In fact, what happens at the electrode/liquid interface, across a layer a few molecular diameters thick, controls the whole system. In the bulk of the liquid, one has to deal with the familiar effects of electrolytic dissociation and ionic drift, usually attended by space charge formation and turbulence when electric fields are strong enough, these secondary phenomena being negligible when higher concentrations of electrolytes provide for "electroneutrality".

A chief feature of electrochemical processes is the fact that electron exchanges at either electrode occur both ways, i.e. ions can be formed as readily as they can be neutralized, in contradistinction to gases where the second process is largely dominant, and this explains why "electrical purification" by D.C. voltages between metallic electrodes is not feasible. Another essential feature is the existence of a "double-layer" at the interface, due to the gathering of undischarged ions having sign unlike that of the electrode. The potential drop across this layer entirely controls the electron exchanges between ionic or neutral species from the liquid and the electrode itself. The part of the macroscopic field nearby remains absolutely negligible as long as it does not overstep 10^6 V/cm, for the field across the layer is 10^7 to 10^8 V/cm.

Our recent work has shown that the behaviour of the double-layer and its catalytic effect on electron exchange

* Laboratory for Electrostatics, Centre National de la Recherche Scientifique, B.P. 166, Centre de Tri 38042 GRENOBLE CEDEX, (France).

remains unaltered for bulk resistivities up to 10^8 Ωcm at least and other evidence, though indirect, leaves no doubt that this applies whatever the resistivity (1). Thus, provided the field does not exceed 10^6 V/cm, electrochemical concepts will describe conductive processes entirely (2).

Various electrochemical mechanisms

We investigated the influence of the following factors :
- Amount of residual ions available to build up the double layer (closely related to the low-voltage conductivity of the liquid and to the geometry of electrodes)
- Oxidation or reduction potentials of neutral molecules, which give off ions after electron exchange
- Nature of the compounds created by electrode reactions.These compounds, if liable to electrolytic dissociation, are an additional ion source.
- Other important factors : the ionizing power of the liquid and the passivation of metals.

I - Influence of residual ions

This influence was studied in propylene carbonate (PC) either between two plane parallel electrodes or in a coaxial wire-cylinder arrangement (3).

In the first case, the amount of residual ions available to build up the double layer was made to vary by modifying the liquid resistivity, or the gap : the wider the gap, the larger the volume from which ions can be drawn, and consequently, the higher the ion amount.

In the second case, we used wires of various diameters :
For a given volume of liquid, the charge density that can build up on a wire electrode is the higher as the wire surface is smaller. In both cases we observed that the injection current density increases as the ion flux from the liquid is becoming higher.

II - Oxidation or reduction of neutral compounds

This influence was clearly observed in the same wire-cylinder arrangement. With a wire made of silver, an easily oxidised metal, the positive injection current at a given voltage is about three times as high as with a wire of the same diameter made of platinum.

In this latter case, it may be surmised that a neutral impurity from the liquid is responsible for injection. Actually, if the PC is saturated with hydrogen the positive injection current on a platinum wire is much larger (2). It can reach values corresponding to the SCLC limit, 15 times higher than without H_2.
In this case, the current is less with the thinner wires. This, because the reacting molecules (H_2) cannot reach the electrode surface save by diffusion, an inefficient process when the wire surface is very small.

III - Nature of the reaction products at the electrodes

We leave aside ion neutralization which, if alone, would reduce conductivity, and we only consider cases when neutral molecules are involved at the start.

1°) Reversible reactions

A few examples :

- silver oxidation

$$Ag \rightleftarrows Ag^{+} + e$$

- nitrobenzene reduction

$$NB + e \rightleftarrows NB^{\bar{\circ}}$$

- hydrogen oxidation

$$H_2 \leftrightarrow 2H^{+} + 2e$$

In all these cases the ion formed at the injecting electrode neutralizes as it reaches the collecting one, and the primary product is regenerated. The final result of such a reaction is the mere transfer of an electron or hole through the liquid. It does not entail any chemical alteration of the liquid itself, not even concentration gradients, which are prevented by the ubiquitous turbulence. With easily oxidized or reduced products in sufficient amount the current can reach comparatively high values, only limited by space charge effects or diffusion.

2°) Irreversible reactions

This type of reaction is the most commonly observed and results in the irreversible destruction of the starting product. Many examples can be given :

- residual water oxidation

$$H_2O \rightarrow 2H^{+} + {}^{1}/{}_{2}O_2 + 2e$$

- residual water reduction

$$H_2O + e \rightarrow OH^{-} + {}^{1}/{}_{2}H_2$$

- PC reduction

$$PC + 2e \rightarrow CH_3CH = CH_2 + CO_3^{--}$$

- water reduction in PC

Water reduction is followed by basic hydrolysis of PC (1), the total reaction is :

$$PC + 2H_2O + e \rightarrow CH_3CHOHCH_2OH + CO_3H^{-} + H_2$$

In all those cases ions are formed along with neutral products that can, however, be more or less dissociated themselves. This liability to dissociation is very important for the actual behaviour of the liquid. For instance, the products resulting from PC reduction are very little dissociated, if any, and the resistivity of PC remains very high (10^{11} Ωcm) even under high electric fields (10^5 Vcm^{-1}).

At the opposite, water oxidation in CH_3CN gives protons as well as ammonium acetate. Formation of the latter leads to a strong increase in conductivity :

$$CH_3CN + 3H_2O \rightarrow CH_3CO_2NH_4 + 2H^{+} + 2e + {}^{1}/{}_{2}O_2$$

$$CH_3CO_2NH_4 \leftrightarrow CH_3CO_2^{-} + NH_4^{+}$$

Another interesting case was encountered in the study of the liquid crystal p-Methoxybenzilidene p-butylaniline (MBBA). Hydrolysis occurs spontaneously in MBBA and gives p-butylaniline and p-anisaldehyde. The p-butylaniline is oxidised at the anode, giving off protons which neutralize at the cathode, but at the same time coloured polymers are formed, which are stronger electrolytes, with considerable conductivity. Thus D.C. passing through MBBA brings forth an increasing degradation of the liquid, which makes it unsuitable for applications (4).

IV - Other important factors

1) For the same concentration of electrolytes, the residual ion concentration largely depends on the solvating and ionizing properties of the liquid, i.e. the strength of the bonds its molecules can form with ions of a given kind and this is not always reflected by the magnitude of dielectric constant alone. For instance, nitrobenzene (ε = 36) is a weakly ionizing solvent as compared to methanol (ε = 32).

2) The importance of the metallic electrode must not be overlooked. We have already mentioned the case of an easily oxidized material. There are also materials on which a protective film is formed by the current (passivation). The actual current under high field may be cut tenfold be merely using aluminium.

Conclusion

All the factors discussed above which control injection phenomena and, as a consequence, high field conductivity have been studied in highly polar liquids like propylene carbonate and some mildly polar ones such as MBBA. There is no doubt that the same factors are operative in all polar liquids and even in some non-polar ones like benzene where solvation of ions is still favoured by the displacement of pi-electrons. In alcanes, where the solubility and dissociation of electrolytes are almost nil in the absence of polar impurities like water, the formation of a double layer becomes problematical. Nevertheless, it has been observed that their conduction is higher when they contain easily reacting species like oxygen. Thus it seems plausible that even in this extreme case the conductive behaviour of the liquid is still controlled by some sorts of electrochemical reactions, at least for fields which do not considerably exceed 10^6 V/cm.

(1) B. Gosse J. Electroanal. Chem. 1975

(2) B. Gosse Thesis Grenoble 1974

(3) N. Félici, B. Gosse, J.P. Gosse Conference on Dielectric materials, measurements and applications I.E.E. Cambridge 1975

(4) N. Félici, B. Gosse, J.P. Gosse 5th International Conference on conduction and breakdown in dielectric liquids Delft 1975

CONDUCTION IN INSULATING LIQUIDS DUE TO CARRIER INJECTION: NATURE OF CARRIERS INJECTED AND REGENERATION THEREOF BY IONIC POLYMERS.

R. Tobazéon* and M. Sauviat*

I - Introduction . For both fundamental research and applications in the field of semi-conductors and dielectrics, it is necessary to have electrodes working in well-defined conditions. Solid-state physicists usually speak of "contacts" : ohmic (injecting), blocking, neutral (1), and they know how such contacts are difficult to contrive. In liquids, we know to-day that two classes of phenomena may happen at a metal/liquid interface : i) oncoming ions can neutralize, but not always with a fast kinetics, or remain blocked. ii) injection of new charge carriers can occur by electrochemical processes (i.e. oxidoreduction of the liquid itself or its impurities). Both phenomena may be interrelated, and metallic electrodes often behave in an impredictable and erratic way. Covering the metal with an electrodialytic - or ionic - polymer (ion exchange membranes or varnishes) allowed a noteworthy stride in the use and the study of strongly polar liquids. Thus, electrodialysis gives proof of the active and efficient role of the ionic polymers which facilitate the discharge of oncoming ions and are the best electrodes to carry out electrical purification to obtain polar liquids of very low conductivity (2)(3)(4). When the voltage is reversed with respect to the polarity used for electrodialysis, the active role of the polymer appears even more clearly : the metal/polymer assembly becomes a "strong" injector into nitrobenzene (5)(6) (it works like an ohmic contact). In strongly polar liquids, injected currents are remarkably stable and reproducible with several ionic polymers, and, on the contrary, injection is poor and erratic with other polymers or in weakly polar liquids (7). Membrane electrodes are so effective for deionization that the resistivity limits of several self-ionized liquids have been actually reached (8). Moreover, under higher fields, the saturation current of electrolytic dissociation has been measured in ethanol for various electrode separations ; nevertheless, saturation plateaus - due to either self-dissociation or dissociation of impurities - are always followed by an increase of the current if the field is raised further (9). This must be ascribed to an unwanted but unavoidable behaviour of membrane electrodes under very high fields. The purpose of this paper is to give a first idea of the various mechanisms that provide for generation and neutralization of charge carriers in ionic polymers.

II - Experimental . Injection is performed in a recirculated liquid deionized in an auxiliary electrodialytic cell. In the test cell, the electrode separation can be varied from 0.1 mm to 8 mm. The injection strength is appreciated from current-voltage curves and field measurements (studied by KERR technique). To ascertain the nature of the mobile ion, commercial membranes (MB) were regenerated

* Laboratory for Electrostatics, Centre National de la Recherche Scientifique, B.P. 166, Centre de Tri
38042 - GRENOBLE CEDEX, France

by soaking during 24 hours in appropriate solutions, and then cleaned with distilled water and ethanol. Varnishes were prepared with techniques described elsewhere (4). To determine the H^+ and Na^+ concentrations into the membranes, these were soaked in KCl solutions, which were dosed by acid-base titration or flame emission spectrometry.

III - Introducing remarks . i) If we turn our attention to the composite electrodes used, there is a priori a major difference between a membrane covering a metal and a varnish stuck to the metal. Another fact, revealed by experiments, is that only MBs are good collectors keeping their qualities even under a high flux of ions ; varnishes are not : they peal or are destroyed very quickly because of electroosmotic effect dragging the solvent to the metal. In most cases, MBs under Na^+ form (collecting positive carriers) or Cl^- (for the collection of negative ions) were used ii) Possible mechanisms of carriers generation have been examined :

1) Metal/polymer interface. a) anodic oxidation (metal dissolution) can take place when the metal is anode. (No similar mechanism possible at the cathode). b) oxidation (at the anodic side) or reduction (at the cathode) of the liquid itself (swelling the polymer) or its impurities.

2) Inside the polymer . dissociation of an ionizable species, possibly enhanced by the electric field of the ion-exchange groups.

3) Polymer/liquid interface . Supposing an electronic conductivity of the polymer, the same processes than in 1b) can take place.

4) The above-mentioned possible mechanisms can happen simultaneously or be interrelated.

IV - Some results and discussion about positive injection into NB

1) Soluble anode . A silver electrode covered with an AMF C100 membrane is a strong injector ; in the injecting and collecting membranes, Na^+ ions are replaced by Ag^+ : The polymer here plays the role of a simple "filter". The mechanism involved is 1a) : the perenniality of the injection is secured, the metal being an inexhaustible source of ions. The easy oxidation of silver has been described in (10). A polyimide varnish peals quickly, probably because of the disappearance of its metallic foundation.

2) Insoluble anode . We used platinum sheets covering massive brass electrodes ; varnishes were deposited on stainless steel.

a) AMF C100 as injector . Fig. 1 shows the progressive replacement of Na^+ ions by H^+ in the injecting and collecting membranes, and the amount of deposited carbonated sodium hydroxide at the interface MB/metal cathode : which is in rough agreement with Faraday's law (11). Thus H^+ ions have been identified ; it has been also shown that their production seems to be inexhaustible because the current is stable even though the amount of electricity passed though the system is several times the one corresponding to the exchange capacity of the membranes. About the origin of H^+, it comes immediately to mind that water may be involved but :

- mechanism 3) is to be rejected because in the injecting membrane, Na^+ would not be replaced by H^+
- 1b) is in agreement with the kinetics of replacement of Na^+ by H^+. Nevertheless, if we cover the anode with 4 membranes, and if the amount of electricity passed corresponds to the capacity of only one MB, the acidity of the 4 sheets is roughly the same (This experiment however lasts hours and H^+ can diffuse).

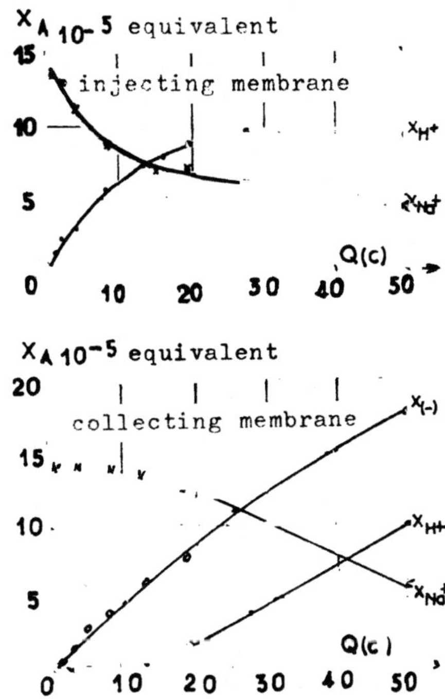

Fig. Positive injection into NB by AMF C100 MBs-X : quantities of dosed ions. X(-) : quantity of carbonated sodium hydroxide.

- this latter experiment supports mechanisms 2), but against it, we can cite the following one : an injecting assembly is made of 1 AMF C100 (Na^+) in contact with the liquid and 3 AMF A100 (under Cl^- form) between the metal (anode) and the C100 membrane ; in this last one, Na^+ are replaced by H^+ ; but the A membranes become brown coloured, and EPR spectra typical of the presence of the nitrobenzene anion NB^- are obtained : the density of this radical decreases from the C membrane to the platinum. We may conclude from this, that water has been oxidized at the interface A/C.

b) PII varnish as injector . This ionic polymer is under H^+ form ; in the collecting MB there is a progressive replacement of Na^+ by H^+ (in rough agreement with Faraday's law). 1)2)3) are possible.

Remarks 1) We have tried to dehydrate MBs and PII (under vacuum at 70°C on P_2O5), but our results are inconclusive (we observe generally that the injection ability is unaffected -in disagreement with (12)- mainly because the NB even carefully treated contains moisture (several tens of ppm) and may swell very quickly the MB.

2) We have tried to collect under the injecting MB oxygen which may arise from water oxydation ($H_2O - 2e^- \rightarrow 2H^+ + 1/2\ O_2$). A gas is collected, but it is air introduced probably by some EHD phenomena.

3) Against the mechanism 2) comes also the fact that polyvinyl alcohol (PVA) varnishes are poor injectors in NB either under Na^+ or H^+ form. However PVA is known as a strong hydrophylic polymer.

4) The use of an auxilliary lateral compartment filled with salted water does not increase the injection ability of an AMF MB C100.

V - Factors affecting the injecting ability of polymers

1. Nature of the polymer. In NB we have tried several polymers containing strong exchange groups (sulfonic acid or alkali metal sulfonate) : polyethylene (AMF C60-100, RAI 5010), fluorinated polymers (RAI 1010-4010), aromatic polyimides and polyamide -imides, polyvinyl alcohol. Only PVA varnishes were poor injectors whereas PVA MBs were strong ones.

2. Influence of the liquid. The AMF MBs are strong injectors into highly polar liquids such as propylene carbonate (PC), ethanol, acetonitrile. This seems by no means surprising because the liquid swells the polymer and ionizes the sites. Nevertheless, this is not observed with PVA varnishes. What is much surprising is that AMF 60 MBs are often good injectors into Aroclors (13) or chlorobenzene, and that AMF 100 are very feeble ones whereas their characteristics are very similar.

3. Nature of the exchange groups. RAI P 1030 (teflon weak acrylic acid in K^+ form), and P 3193 (polyethylene with the same groups) are poor injectors into NB. Polyacrylic acid varnishes are also poor injectors in NB, but Sodium polyacrylate ones are better injectors, as expected from the fact that these groups are much more ionizable. 4. Influence of the mobile ion. With the monovalent ions Na^+, H^+, K^+ in AMF strong injection is observed ; with Ca^{2+}, Ba^{2+}, and with Al^{3+}, Fe^{3+} the injection strength is considerably reduced.

VI - Negative injection . With MBs owning strong groups, strong and durable injection is observed into NB and PC. With NB, the injecting membranes become coloured and EPR signal testifies the presence of $\dot{NB}^-$: this reaction insures the durability of the injection. In PC and in Aroclors, the perenniality of the injection has been insured; but if the reduction or dissociation of water are possible mechanisms, it has not been verified that the injecting or collecting MBs become in OH^- form : in PC the aspect of the mb is not modified, but the injector becomes purple in Aroclor (and it gives an EPR signal indicating the presence of a radical). PI injectors are strong injectors at the start, but are progressively altered, and destroyed (6). PVA injectors are weak-injectors, even in highly polar liquids. In weakly polar liquids, it is also observed that AMF 60 MBs are efficient injectors (whereas AMF 100 are poor ones).

VII - Conclusion . It has been proved that ionic polymers may behave as simple filters (injection of Ag^+ in NB, collection at low fields), and more generally, when they work as injectors, as active media allowing efficient regenerative mechanisms. A great deal of facts are in favour of an enhanced dissociation of certain impurities inside the polymer, for instance water (what also would explain an unwanted small injection at high fields during electrodialysis) ; however, it is by no means excluded that the polymer be involved in some oxydoreductive mechanisms, implying an electronic conduction of the material.

Acknowledgements. Thanks are given to Dr. M. RINAUDO (CERMAV,Grenoble) and to Drs. M. LAMBLA and R. VAROQUI (CRM, Strasbourg) for helpful discussions and technical assistance.

1. Lampert M.A. and Mark P. "Current Injection in Solids", Academic Press, New York and London 1970
2. Brière G. and Félici N. Comptes Rendus, Paris, 259, 1964, 3237
3. Brière G., Félici N. and Filippini J.C., ibid. 261, 1965, 5097
4. Filippini J.C., Sauviat M. and Tobazéon R., ibid. 271, 1970, 600
5. Brière G. and Gosse J.P., J. Chim. Phys., 65, 1968, 1341
6. Félici N. and Sauviat M., Proc. of the 4th Intern. Conf. on Cond. and Breakd. in Diel. Liquids, Dublin, Inl. 1972
7. Tobazéon R. and Sauviat M., Contract Report, DGRST, Sept. 1974
8. Barret S., Brière G. and Pierre G., J. Chim. Phys., 63,1966, 725
9. Tobazéon R., Doctoral Thesis, Grenoble 1973
10. Gosse B. and Gosse J.P., Proc. of the Conf. on Diel. Mat. Meas. and Appl., Lancaster, Jul. 1970
11. Sauviat M., Thesis, Grenoble 1974
12. Barret S. and Gaspard F., J. Chim. Phys., 71, 1974, 255
13. Lacroix J.C. and Tobazéon R., Proc. of the 4th Intern. Conf. on Cond. and Breakd. in Diel. Liquids, Dublin, Jul. 1972

PHOTO INJECTION INTO N-HEXANE

T.R. Hewish & J.E. Brignell*

Introduction

Originally the method of injecting electrons into a non-polar liquid through a vapour gap was developed to extend the range of injection currents[1,2]. This was necessary to obtain comparison with the interesting results which were emerging from studies with polar liquids, which are much more amenable to high injection rates. Furthermore, electronic shuttering of the primary current by means of a grid in the vapour gap was faster and more versatile than mechanical shuttering of the U.V. light. A disadvantage of the original technique was that the field strength in the liquid could only be estimated in a rather doubtful way; so a second grid was placed in the liquid surface to establish this field.

Experimental results with this system[3] revealed a defect which could have been anticipated, that in contrast to the familar case of ballistic motion, under drift conditions the low interstitial field in the grid greatly increased the diffusion error. Nevertheless, work with this system has been continued as it seemed to offer the best means of studying high injection phenomena such as EHD stability etc.

The measurements were made by on-line computer, and it was comparatively easy to include a program to translate the practical variables into a form due to Atten and Moreau[4] which gave a measure of the degree of space charge pertubation and the EHD stability.[5] An obvious next step was to provide some practical means of monitoring this stability.

Apparatus

In extending the experimental technique to include the direct observation of mechanical instability we have sought a method which combines high-sensitivity with objectivity, i.e. a schlieren method independent of the limitations of the human eye. The tetrode test-cell previously described was modified by incorporating two windows level with the drift space. A simple schlieren system was built, but instead of a screen, the image was focussed onto a pair of light-sensitive diodes connected to a high-gain differential amplifier. The common mode rejection of this system obviated problems with hum etc. and provided a substantial signal wherever motion occurred in the liquid. It was certainly more reliable than the human eye in identifying the occurrence of schlieren disturbances and hence motion. The signals from the electrometer amplifier and the light detector

* Department of Electrical and Electronic Engineering, The City University, London, England.

were fed to the computer via an ordinary fast multichannel A-D converter or the transient recorder.[6,7] Once in the computer they were available for observation on the VDU or as hard-copy graphs from the incremental plotter.

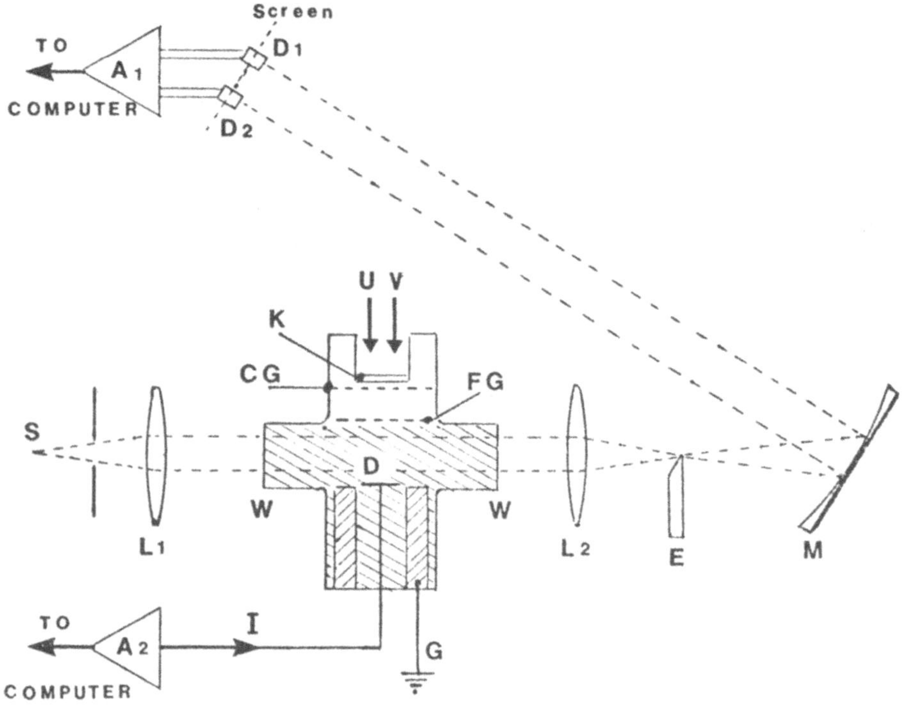

Fig. 1 Tetrode Test Cell modified for schlieren observations

Fig. 1 shows the complete experimental system. The light from source S, either a helium neon laser or ordinary incandescent bulb, is focussed on to the knife edge E and the schlieren image is focussed on a plane containing the diodes D_1, D_2.

Results

The first results with the schlieren system were somewhat surprising in that they revealed schlieren disturbances occurring for injected currents much smaller than might have been expected. Fig. 2 shows the sort of result obtained. The upper curve shows the current following the application of one-second pulses to the control grid (the edges of the pulses are seen as break-through spikes). The waveform is similar to that shown previously for this cell construction[3]. The lower trace shows the amplified differential signal from the light sensitive diodes. The unbalanced apparent in the waveform is due to the diodes straddling the edge of the schlieren field, an arrangement which was found to give high sensitivity.

It will be seen that the initiation of the schlieren disturbance

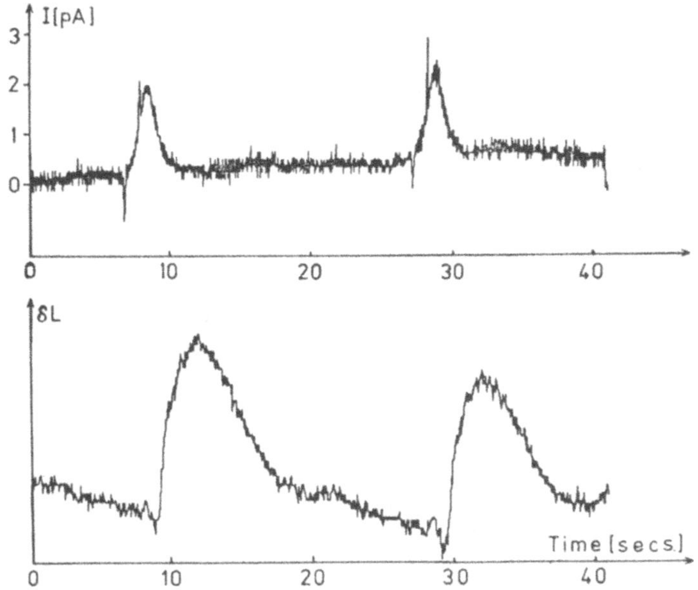

Fig. 2 Upper Trace: Current response to 1-second pulses applied to the control grid.

Lower Trace: The corresponding schlieren disturbance signal from the diodes.

occurs after the current pulse with a consistent time-lag. A preliminary investigation of this time-lag has revealed behaviour somewhat similar to that described by Lacroix and Tobazeon[8] following their work on Anoclor, though the orders of magnitude of current and time are rather different. In particular, the time-lag decreases approximately as the square of the applied field.

The disturbance front has been found to move with a velocity of the order of 10^{-3}m.S^{-1} (at an applied field of 1 kV.cm^{-1}). This means that the required correction to the apparent mobility for this reason is relatively small - about 15%.

It must be emphasised that a disturbance of the schlieren image was detected at very low values of injected current and applied field (e.g. 1 pA at 150.cm^{-1}), in transient conditions of step and pulse injection.

Discussion

These results cast doubt on some of the previous attempts to investigate EHD instability by electrical observations alone[1]. Whether they are in fundamental disagreement with the EHD stability theory[4] is not quite so clear, as the stability criterion is very dependant upon the actual value of mobility (the critical voltage varies as the mobility, but the critical current as the cube of the mobility[5]). The response is not ideal, and the assumptions made in the experimental determination of a mobility value may ultimately prove to be erroneous. A further theoretical problem is the nature of a grid as an EHD boundary. It is normally considered to be equivalent to a solid wall[9], in which case the theory should apply to this method of injection.

Acknowledgment

This work was made possible by an equipment grant from the U.K. Science Research Council.

References

1. Brignell, J.E., and Buttle, A.J., 1971, J. Phys. D: Appl. Phys., 4, 1560.

2. Buttle, A.J., and Brignell, J.E., 1969, Electron. Lett., 5, 687-8.

3. Hewish, T.R., and Brignell, J.E., 1972, Proc. 4th Int. Conf. on Conduction and Breakdown in Dielectric Liquids, Dublin, 5.

4. Atten, P., and Moreau, R., 1970, C.R. Acad. Sc. 270A, 415.

5. Hewish, T.R., and Brignell, J.E., 1972, J. Phys. D: Appl. Phys., 6, 747.

6. Buffam, C.J. and Brignell, J.E., Communication to the present conference.

7. Brignell, J.E., and Rhodes, G.M., 'Laboratory On-Line Computing', Intertext, London, 1975.

8. Lacroix, J.C. and Tobazeon, R., 1972, Proc. 4th Int. Conf. on Conduction and Breakdown in Dielectric Liquids, Dublin, 93.

9. Atten, P., private communication.

FREE ION YIELDS OF SOME DIELECTRIC LIQUIDS AND STUDY OF BINARY MIXTURES OF THOSE DIELECTRICS IRRADIATED BY ^{60}Co AND ^{137}Cs γ RAYS

J. CASANOVAS, R. GROB et D. BLANC*

I. Determination of free ion yield G°_{fi} for pure liquids

The resistivity of the alkanes we have used is higher than 10^{16} Ω.m under and electric field strength of 2,5 $MV.m^{-1}$ (distance between electrodes : $1,65.10^{-3}$ m), the carbon tetrachloride (CCl_4) and the oxyperfluoro-octane cyclic ($c\text{-}C_8F_{16}O$) possessed respectively a resistivity of 10^{14}Ω.m and of 5.10^{15} Ω.m. In order to measure the energy absorbed in the sensitive volume of the cell, we have used a special chamber with plane and parallel beryllium electrodes filled with hexane or with trimethyl-2,2,4 pentane (1). The G°_{fi} values were determined by the extrapolation technic to zero field strength we have shown in effect (1) that in taking some experimental precautions the obtained G°_{fi} values are in agreement with the clearing field ion yields (2). We give in table 1, the G°_{fi} values at 23°C for the different liquids irradiated by ^{60}Co and ^{137}Cs γ rays, the activation energy values for G°_{fi} and the theoretical and experimental of the slope/intercept values P.

II. Interpretation of expérimental $G_{fi} = f(E)$ (^{60}Co-γ) curves by making use of ONSAGER theory

In order to bring into reality the $G_{fi}=f(E)$ curves for hexane(t=23°C $0 \leq E \leq 16$ $MV.m^{-1}$), for methyl-2 pentane (t= 25°C, $0 \leq E < 10,5 MV.m^{-1}$), for trimethyl-2,2,4 pentane (t=25°C, $0 \leq E \leq 14 MV.m^{-1}$) and for CCl_4 (t=23°C, $0 \leq E \leq 22 MV.m^{-1}$) by employing the ONSAGER theory (3) on the geminated recombination of ions, we have calculated the expression of the probability $\phi(r,E)$ for an ion pair with an initial separation distance r, subjected to an electrical field strength E, to escape geminate recombination, in other words $\phi(r,E) = \int_0^{\infty} \varphi(r,E)F(r)dr$ where $\varphi(r,E)$ is the expression published by TERLECKI and FIUTAK (4) and F(r) is the distribution function for initial separation distances, r, of ions. This function is not known in a liquid, in order to compare the experiment with the theory, we must arbitrarily establish a form of distribution to see if the experimental results are realized or not. We have used the four following distribution functions : the Gauss function(G.) (2), the Gauss modified function (P.G.)(5), the exponential function (E.)(6) and the exponential modified function (P.E.)(7). Table 2 gives the value of b for which optimal adjustement between theoretical curve $\phi(r,E) = f(E)$ and experimental curve $G_{fi}=f(E)$ is obtained and gives the corresponding value of $G_{tot} = G^E_{fi} / \phi(r,E)$. The distribution functions which correctly renders an account of experimental results (electrical field effect on G_{fi} and good G_{tot} value) are the exponential modified function for hexane and trimethyl-2,2,4 pentane, the exponential function for

*Centre de Physique Atomique, Université Paul Sabatier, TOULOUSE, (France)

Table 1

Compounds	G^0_{fi} * 23°C ^{60}Co-γ	G^0_{fi} * 23°C ^{137}Cs-γ	ΔE** 10^{-2}eV ^{60}Co-γ	ΔE** 10^{-2}eV ^{137}Cs-γ	$P_{obs.}$*** $10^{-6}m.V$ ^{60}Co-γ	$P_{obs.}$*** $10^{-6}m.V^{-1}$ ^{137}Cs-γ	$P_{calc.}$*** $10^{-6}m.V^{-1}$
Hexane	0,148	0,141	8,9	8,9	0,62	0,65	0,59
Octane	0,136		8,5		0,60		0,57
Dodécane	0,129		7		0,60		0,55
Méthyl-2 pentane	0,173		9,7		0,62		0,59
Méthyl-2 heptane	0,144		8,8		0,60		0,57
Diméthyl-2,2 butane	0,445	0,408	5,7	5,7	0,69	0,76	0,59
Diméthyl-2,2 hexane	0,24		7,3		0,61		0,57
Triméthyl-2,2,4pentane	0,39	0,355	4,3	5,6	0,62	0,67	0,57
CCl_4	0,116		3,4		0,51		0,50
c-$C_8F_{16}O$	0,032		8,8		0,60		0,59

* Précision of G^0_{fi} values ± 6%

** Précision of ΔE values ± 5% , values calculated between 10°C and 60°C.

*** Values given at 23°C

Table 2

Compounds	t°C	r_c(nm)	distribution	b(nm)	G_{tot}	$\overline{G_{tot}}$ literature liquid	gas
Hexane	23	29,9	G.	8,8	2,5	4	4,2
			P.G.	7,4	3		
			E.	4,8	7,6		
			P.E.	4,4	5,4		
Méthyl-2 pentane	25	29,9	G.	9,1	2,7		4,4
			P.G.	7,2	3,7		
			E.	4,8	9		
			P.E.	4,3	6,6		
Triméthyl-2 2,4pentane	25	29	G.	13,2	2,8	4,5	
			P.G.	11,9	3,1		
			E.	8,7	6,2		
			P.E.	8,5	4,9		
CCl_4	23	25,3	G.	8,5	1,5		4
			P.G.	7,3	1,6		
			E.	5,1	3,7		
			P.E.	4,8	2,6		

Table 3

Scavenger	Alkane	t°C	E $MV.m^{-1}$	G_{fi}	b(nm) P.E.	Lowering induced in% of G^0_{fi}	réf.
$c\text{-}C_8F_{16}O$	Hexane	23	0	0,148	3,1	11,8	
G^0_{fi}=0,032		23	3	0,411	3,1	10	
23°C	Triméthyl-2,2,4pentane	23	0	0,39	7,3	54,1	
		23	3	0,975	7,3	49,4	
CCl_4	Hexane	23	0	0,148	3,1	11,8	
G^0_{fi}=0,116	Diméthyl-2,2 butane	23	0	0,445	8,5	62,6	
23°C		23	3	1,147	8,5	59,2	
	Triméthyl-2,2 4 pentane	23	0	0,39	7,3	61,3	
	Diméthyl-2,2 propane	23	0	0,86		72,8	(2)

CCl_4 and the Gauss modified function or the exponential modified function for methyl-2 pentane.

III. Lowering of G_{fi} values in the alkane-scavenger(CCl_4, c-$C_8F_{16}O$) mixtures

The table 3 gives the lowering of G_{fi} values(in % of G_{fi} values of pure alkanes) induced by a concentration of 3% in volume of scavenger and gives the value of b for each alkane. The efficiency of the two scavengers is that much better than the b value of the alkane is higher, this indicates a scavenger efficiency depending on initial separation distances of ions in alkanes. We think that those scavengers interact with "subvibrational electrons" (prethermalization regime), this results in an higher geminate recombination probability for ions, that is in a lowering of the G_{fi} values of the mixture.

References

(1) CASANOVAS (J.) Thèse de Doctorat ès sciences, n°653,TOULOUSE (1975)

(2) SCHMIDT (W.F.) and ALLEN (A.O.) J.Chem.Phys. 52, n°5,2345(1970)

(3) ONSAGER (L.) Phys.Rev.,54,554,(1938)

(4) TERLECKI (J.) and FIUTAK (J.) Int.J.Radiat.Phys.Chem., 4 ,469, (1972)

(5) DODELET (J.P.), SHINSAKA (K.), KORTSCH (U.) and FREEMAN (G.R.) J. Chem.Phys. , 59, n°5 , 2376 , (1973)

(6) ABELL (G.C.) and FUNABASHI (K.) J. Chem.Phys. , 58 , n°3 , 1079, (1973)

(7) DODELET (J.P.) and FREEMANN (G.R.) J. Chem.Phys. , 60 , n°12 , 4657, (1974)

ELECTROHYDRODYNAMIC STABILITY OF LIQUIDS SUBJECTED TO UNIPOLAR INJECTION

P. Atten*

Introduction

The hydrodynamic stability of a liquid layer subjected to a unipolar injection is a fundamental problem in electrohydrodynamics (EHD) to which important work has been devoted for some years. The first theoretical studies (1,2) gave values of the critical number T_c : $\varepsilon V/K\eta$ (ε : permittivity ; V : applied voltage ; K : ionic mobility ; η : viscosity) ; the lower bound, in the space-charge-limited current (SCLC) case, is T_c = 161 (two rigid boundaries). A detailed discussion of the first experiments showed a discrepancy between theoretical and experimental values (3) which has been confirmed by careful measurements giving $T_c \simeq$ 80-90 (4). In order to explain this disagreement, a criticism has been undertaken of the main approximations of the theoretical treatment : i) diffusion is negligible ii) injection is autonomous iii) the liquid is perfectly insulating and also of the implicit assumption in linear stability analysis that the disturbances which will be amplified have an infinitely small amplitude.

Influence of liquid conduction on stability

Considering only a stationary state of conduction leads to a model where the bulk conductivity proceeds from the weak dissociation of a chemical species. Without injection, such a conducting liquid is stable for all not too high values of the parameter $C_o = en_o d^2/\varepsilon V$ (e : carrier charge ; n_o : density of pairs of ions at thermodynamic equilibrium ; d: gap width). The analogy with the fluid layer heated from above shows the stabilizing character of a dissociation-recombination process (5,6). With a superimposed unipolar injection, the effect of conductivity is more intricate and depends on the level of injection $C = q_o d^2/\varepsilon V$ (q_o : charge density on the injector) and on $\beta = C_o/C$ roughly measuring the ratio of conduction and injection currents (6). The main results are (fig. 1) :

a) Instability sets in as a convective motion of growing amplitude for $\beta < \beta_1$ (β_1 from 0.2 to 0.5) and as an oscillating motion (overstability) for $\beta_1 < \beta < \beta_2$

b) When injection is not strong ($C \leqslant C_t \simeq 2.5$) the criterion T_c increases with β, which is consistent with the stabilizing influence of conductivity.

c) for strong injections ($C > C_t$) the oscillating mode of instability is characterized by an increase of T_c versus β and the existence of a lower bound $T_c \simeq$ 400 obtained for $\beta \simeq 0.28$, C $\simeq$

* Laboratory for Electrostatics, Centre National de la Recherche Scientifique, B.P. 166, Centre de Tri, 38042 GRENOBLE CEDEX France.

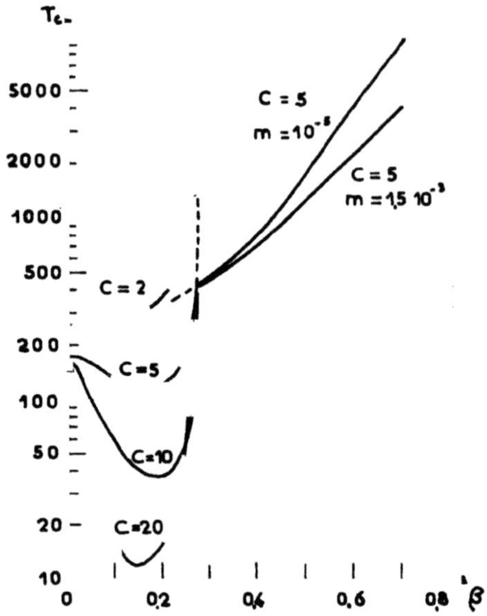

Fig. 1 - The critical number T_c versus $\beta = C_o/C$.

5.5 and all usual values of $m = 1/M^2$ ($M = \sqrt{\varepsilon/\rho}/K$, ρ : density).

d) For $C > C_t$ conductivity has an indirect destabilizing influence on the convective mode of instability and $T_c \leqslant 161$ for $0 \leqslant \beta \leqslant 0.25$, with a quadratic variation in β for $\beta \ll 1$ (fig. 1). The latter property shows that the disagreement between the theoretical criterion and experimental results obtained in conditions such that $C \gg 1$ and $C_o \ll 1$, cannot be attributed to the bulk effect of residual conductivity.

The indirect destabilizing effect of conductivity can be understood as follows ; in the case of pure unipolar injection, the space charge perturbation becomes negative in the vicinity of the injector (2). When ions of unlike polarity are present, they provide the negative part of the space charge perturbation and the perturbation of injected ions which plays the driving role in putting the liquid into motion is in the mean much more important, leading to a decrease of the criterion. (6).

This theoretical study can be used to predict the behaviour of liquid layers bounded by two metallic plane and parallel electrodes. At low voltages, the ions coming from the liquid contribute in part to the polarisation of both electrodes. At a given voltage depending on the nature of the liquid as well as of the metallic electrodes, an electrochemical reaction occurs, giving a unipolar injection. Increasing the voltage further results in a decrease of C_o, an increase of C and consequently a decrease of $\beta = C_o/C$; when β takes values about 0.25, the critical voltage becomes lower than the voltage exerted upon the liquid layer and instability sets in. Thus for high enough C_o values, instability would follow triggering of injection. This mechanism might also play the major role in putting the nematic liquid crystals in motion (6).

Non-linear analysis for $T \sim T_c$

In certain problems, such as the stability of Poiseuille flow, finite amplitude motion can be sustained under subcritical conditions, i.e. under conditions for which all infinitely small perturbations are damped. It is indeed the case for unipolar injection, as was pointed out by Félici (7) in the case of weak injection.

For strong injections an analysis has been carried out with the same assumptions as in the infinitely small amplitude

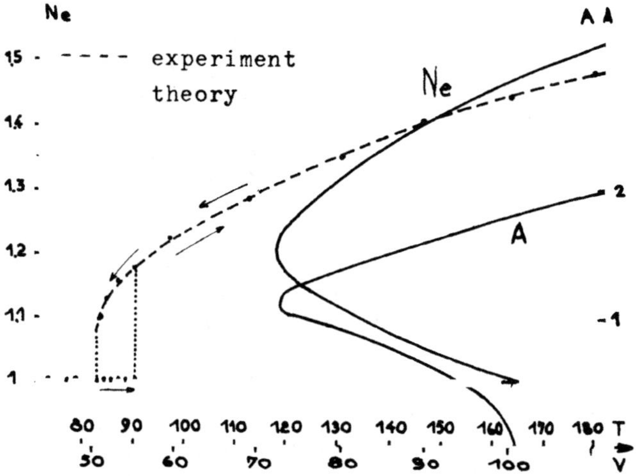

Fig. 2 - Variations of A and Ne versus T for C = 10 and comparison with experiment in pyralene 1460 (d = 0.5mm, resistivity ≃ 6.10^{13} Ωcm)

case (8). Introducing the number A as the ratio of the maximum fluid velocity to the mean ionic velocity KV/d, a solution can be obtained for 0 < A < 1 only if T < T_c (fig. 2) ; further considerations (7) indicate that this solution is unstable. For A > 1, i.e., for a maximum fluid velocity exceeding the ionic velocity, there are locations in the bulk which are charge free. An approximate calculation gives solutions corresponding to stable steady convective motions and leads to the criterion for finite amplitude perturbations T_f (T_f ≃ 118 for C = 10). Thus the system can exhibit a finite amplitude movement under subcritical conditions unlike the analogous Bénard problem. Physically, the non-linear behaviour originates from the mobility character of the electric bulk conduction in contrast with the diffusive nature of heat conduction.

For T_f < T < T_c there exist two stable solutions characterized by A = 0 (no motion) and A > 1. This implies a hysteretical behaviour which has been observed experimentally when strongly injecting with permselective membranes into weakly polar liquids (Aroclors) of sufficient ionic purity (8). The voltage is increased or decreased by steps and the measured stationary current I is compared to the SCLC I_0 which would exist without motion (Ne = I/I_0). For a certain voltage range two values of Ne are obtained depending on the way voltage is varied (fig. 2).

Influence of diffusion on stability

The above considerations show that i) the measured criterion T_c = 80-90 is the linear one ii) bulk effect of conduction cannot be invoked to explain the disagreement between experiment and theory. The two other approximations upon which the EDH stability analysis has been based refer to the working of the injector, which is in most cases a permselective membrane. In such a membrane faced with a liquid of very low conductivity, only ions of a definite sign can move, the charge carriers of unlike polarity being fixed. The concentration of mobile ions being generally very high (fraction of mole/litre) the injection arrangement composed of two identical membranes bounding a well purified liquid layer bears a strong analogy to a p-i-p structure obtained with semi-conductors. Therefore, diffusion plays an important role in layers immediately adjacent to the membranes, even though its influence on the SCLC is very weak. In particular, in the

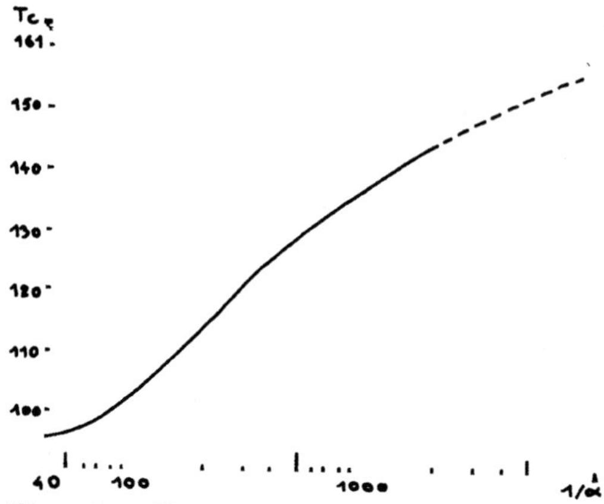

Fig. 3 - The critical number T_c versus $1/\alpha$ for very strong unipolar injections from electrodialytic membranes.

diffusion layer of the injecting membrane, the electric field has direction opposite to that in the bulk of the liquid. The Coulomb force qE (E : electric field) which is nearly uniform in the core, decreases to zero and then takes very high opposite values when approaching the injector. This situation may be more instable than that characterized by a uniform force density distribution obtained when neglecting diffusion.

This expectation is confirmed by an analysis taking diffusion into account and assuming that the liquid is perfectly insulating. Another non-dimensional number is then introduced :

$$\alpha = D/KV = kT/eV \simeq 1/40 \text{ V}$$

(D = diffusion coefficient ; k : Boltzmann constant ; T : absolute temperature). For very strong injections ($C > 10^4$) the results (fig. 3) show that the lower the critical voltage, the greater will be the departure from the asymptotic criterion value $T_C = 161$. ($\alpha \to \infty$). For nitrobenzene $T_c = 124 \pm 2$ and for aroclors $T_c = 145 \pm 3$. Although these values are significantly lower than 161, they do not satisfactorily agree with the experimental results ($T_c \simeq 80\text{-}90$).

Conclusion

The study on the influence of conductivity strongly suggests that in a stationary state of conduction, instability can set in only when injection is present. Thus, it is of primary interest to know the different factors which interfere to make the criterion lower than 161 in the case of pure unipolar injection. The results presented show how the situation is intricate but nevertheless support the hope that combining the effects of both conductivity and diffusion will lead to a clear apprehension of this stability problem.

(1) J.M. Schneider, P.K. Watson, Phys. Fluids, 19, p.1948 (1970)
(2) P. Atten, R. Moreau, J. Mécan. 11, p. 471 (1972)
(3) P. Atten, Phys. Fluids, 17, p. 1822 (1974)
(4) J.C. Lacroix, P. Atten, E.J. Hopfinger, J. Fluid Mech. (1975) To be published.
(5) P. Atten, Proceed. 4th. International Conference on Conduction and breakdown in dielectric liquids, Dublin, 1972, p. 85
(6) P. Atten, J. Mécan., 14, N° 3 (June 1975). To be published.
(7) N. Félici, Rev. Gén. Electr., 78, p. 717, (1969)
(8) P. Atten, J.C. Lacroix, C.R. Acad. Sc., 278 B, p. 385 (1974)

DOUBLE INJECTION WITH RECOMBINATION : EHD STABILITY AND CHARGE TRANSPORT

J.C. Lacroix*, P. Atten*

Introduction

The phenomena related to the fluid motion induced by the electric field acting upon space charges have been mainly investigated in the unipolar injection case (1,2). By using permselective membranes, we can obtain reproducible stationary states of strong double injection into weakly polar liquids (pyralènes 1460 and 1500). We present results of both theory and experiment which clearly show the role of the recombination of the two injected ionic species and also confirms the features of the electrohydrodynamic (EHD) role of unipolar injection.

Theoretical study of EHD stability

The linear stability analysis has been carried out with the following assumptions : i) the fluid is perfectly insulating ii) diffusion is negligible iii) the injector is autonomous iv) the recombination products remain undissociated. The computations have been limited to symmetric double injections $C_+ = C_-$ of ions of equal mobility $K_+ = K_- = K$ (K : mobility, $C_\pm = q_\pm d^2/\varepsilon V$ with $q_\pm$: charge density on the injector, d : gap width, ε : permittivity, V : applied voltage). An important parameter is $A = k_R/k_{RL}$ the ratio of the recombination constant to its limiting value obtained by Langevin $k_{RL} = e(K_+ + K_-)/\varepsilon$ (e : carrier charge).

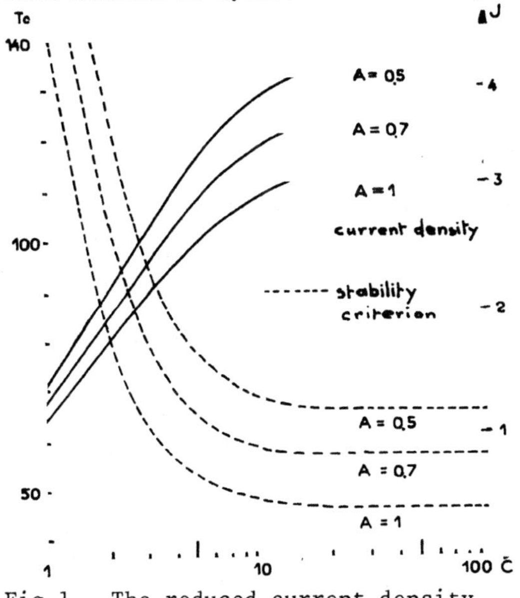

Fig.1 The reduced current density and the stability criterion versus C, the reduced density of symmetric double injection

With the above assumptions, an analytic expression of the solution without motion can be obtained in two cases only (3) i) for $C \to \infty$ and ii) for A = 1. Fig.1 shows the variations with C and A of the reduced current density $j = Jd^3/\varepsilon KV^2$ (J : current density) obtained by computation. The stability analysis leads to a criterion of same form as for unipolar injection (4) :

* Laboratory for electrostatics, Centre National de la Recherche Scientifique, B.P. 166, Centre de Tri
38042 - GRENOBLE CEDEX, France

$T_c = \varepsilon V/K\eta$ (η : viscosity), the values of which are plotted fig.1 (5).

Experimental procedures

The experiments were carried out with pyralène 1460 denoted P 1460 ($\varepsilon = 5.9\ \varepsilon_o$, η = 10 cp, density ρ = 1,41 g/cm^3) and pyralène 1500 denoted P 1500 ($\varepsilon = 6,3\varepsilon_o$, η = 50 cp, ρ = 1,35 g/cm^3) permanently deionized by electrodialysis and recycled in a temperature-controlled teflon circuit (20°C). The working resistivities were about 8.10^{13} Ω.cm (P1460) and 6.10^{13} Ω.cm (P1500). We verified that the two membranes AMF A60 and C60 inject Cl^- and Na^+ ions by collecting on them cristallized sodium chloride which has been proved to be quite unsoluble in these liquids (6). This fact combined with the observation that at fixed voltage the current does not vary with time for $t > 2\ T_t$ ($T_t = d^2/KV$: transit time) warrants the assumption of non-redissociation of the recombination product. Preliminary experiments gave the mobility values (6) :

P 1460 : $K_{Cl^-} \simeq 3.4\ 10^{-5}$ cm^2/V.s, $K_{Na^+} \simeq 3.10^{-5}$ cm^2/V.s

P 1500 : $K_{Cl^-} \simeq 7.5\ 10^{-6}$ cm^2/V.s, $K_{Na^+} \simeq 7.10^{-6}$ cm^2/V.s

As for the analogous unipolar injection experiments (7), varying the electrode separation has been made necessary by the important extent of investigated voltages ranging from volts to tens of kilovolts. At low voltages, small gap widths are required to make negligible the conduction current and at high voltages large d values are used to limit the field strength below a value which could cause a breakdown.

Experimental results

The current-voltage characteristics show four different régimes of charge transport :

a) Under the critical voltage V_c ($24V < V_c < 25V$ for P 1460 and $23V < V_c < 24V$ for P 1500) the current I varies like V^2/d^3 ; no motion can be detected from Schlieren observation and the liquid remains at rest. This space-charge-limited-current I_o which is of course also limited by recombination, is taken as reference for defining the electric Nusselt number $Ne = I/I_o$.

b) For $V_c < V < V_1$ ($V_1 \simeq 50V$ for both liquids) I approximately varies as V^3/d^3 (fig.2). Schlieren observations with transparent electrodes

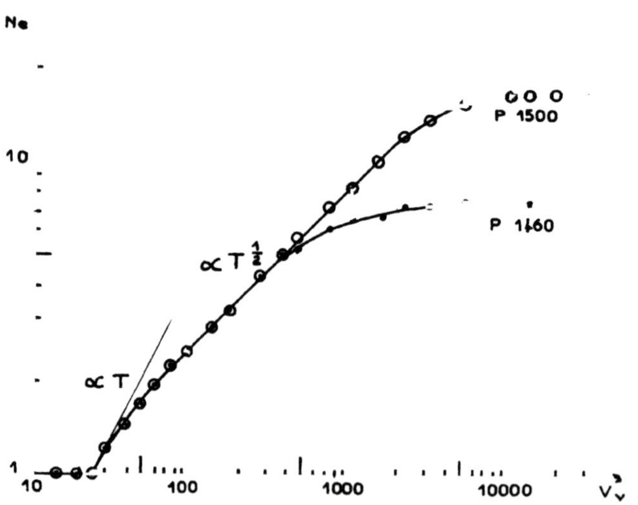

Fig.2 Variations with voltage of the electric Nusselt number Ne for two pyralènes.

a

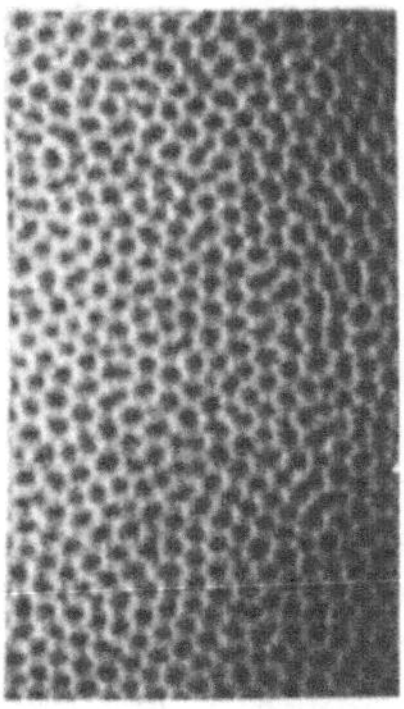

b

Fig.3 Schlieren photographs in pyralene 1460. a) V=26V b) V=50V

(see (2)) show just above V_c a roll pattern (fig. 3a) analogous to the Williams striations in nematic liquid crystals (7). As voltage is increased these elongated rolls progressively break up into smaller cells and at $V \simeq 50V$ a stable very regular hexagonal pattern is obtained (fig. 3-b) which is undistinguishable from that observed at the same voltage with unipolar injection.

c) For $V_1 < V < V_T$, I roughly varies like $V^{5/2}/d^3$ (fig.2). The pattern progressively looses its stability and its regularity as V is increased. From observation of light diffusing charge free particles, the maximum liquid velocity w is seen to follow the law $wd/KV \simeq 1.3V/V_c$ as long as the motion remains laminar ($V \lesssim 400$ V).

d) At voltages above V_T, I again varies like V^2/d^3 whereas the liquid is very agitated. Ne takes the saturation values $Ne_s \simeq 7.5$ (P 1460 - see fig.2) and $Ne_s \simeq 16$ (P 1500).

Discussion

For $V < V_c$ the current obeys the theoretical law of variations with respect to V and d. Assuming very strong injections and taking the mean mobility value make it possible to estimate the recombination constant. By comparison with the calculated current densities (fig.1), we obtain $A \simeq 0.9$ for P 1460 ($j \simeq 3.33$) and $A \simeq 0.8$ for P 1500 ($j \simeq 3.5$). For both liquids the recombination constant k_R is close to the Langevin value k_{RL}, a result consistent with measurements in non-polar liquids (8).

The experimental value of T_c is about 40 for P 1460 and 36 for P 1500 whereas theory predicts for strong double injection : $46 \leqslant T_c \leqslant 53$. The disagreement may be ascribed to the same reasons as in the unipolar case (1) : the model does not take into account the physical peculiarities of injection by electrodialytic membranes.

The variations of Ne versus V (fig.2) bear an analogy with the similar curve in the unipolar case (2). For $2\ V_c < V < V_T$ the phenomena can be accounted for with the picture of columns (laminar cellular flow) in which the velocity exceeds the ionic velocity and which contain alternatively positive and negative charges so that recombination is confined into boundary layers only. The system is then expected to roughly behave as if there were an addition of the effects of two independent unipolar injections, which is consistent with the observations upon the rough doubling of the fluid velocity (with respect to the unipolar case (2)), the $T^{1/2}$ dependence of Ne and the identity in both cases (unipolar and double injections) of the hexagonal pattern.

The picture fails, however, for the range T_c to $2\ T_c$ where Ne varies like T, because the fluid velocity is not high enough to produce charge separation in the whole core. As voltage is increased, the fluid velocity increases, which causes an extent of the regions of no recombination effect. Therefore the current increases faster than $V^{5/2}$ because in addition to the effect of velocity, more charge carriers are transported on paths or increasing length.

For high voltages $V > V_T$, the motion is turbulent and from theoretical considerations we can expect the saturation of the r.m.s. reduced velocity and of Ne. The two experimental values suggest that $Ne_s \simeq M^{1/2}$; this law, analogous to the one obtained in the unipolar injection case, is consistent with the form of the transition number $(MR)_T \simeq 20$. Indeed the latter is related to a Reynolds number $Re_T \simeq 10$ which appears as a universal transition number between viscous and non-linear regimes of such structured flows.

Conclusion

Although simplified, the theory gives stability criteria not far from the observed ones. This result confirms the measurement of k_R from the current density values in the stable régime and we have therefore an original way for determining the recombination constant with a reasonable accuracy. This method nevertheless cannot be used with strongly polar liquids as nitrobenzene for example, in which no stationary double injection can be observed (the current grows quasi exponentially with time). Such a behaviour difference between weakly and strongly polar liquids arises from the different orders of magnitude of the dissociation constants which are extremely sensitive on the dielectric constant.

Finally, in this particular case of double injection into an insulating liquid, the convective charge transport is seen to be still more important than in the unipolar case ($Ne_s \simeq M^{1/2}$ whereas in the unipolar injection case $Ne_s \simeq (M/3)^{1/2}$. This once more shows the importance of the EHD phenomena induced by the injection processes.

1. P. Atten, Communication to the present Conference "Electrohydrodynamic stability of liquids subjected to unipolar injection"

2. J.C. Lacroix, P. Atten, Communication to the present Conference "The influence of induced motion on unipolar charge transport"

3. R.H. Parmenter, W. Ruppel, J. Appl. Phys., 30, p. 1548 (1959)

4. P. Atten, R. Moreau, J. Mécan., 11, p. 471 (1972)

5. J.C. Lacroix, P. Atten, C.R. Acad. Sc., 278B, p. 689 (1974)

6. J.C. Lacroix, Doct. Thesis, Grenoble University 1 (1975)

7. R. Williams, J. Chem. Phys., 39, p. 384 (1963)

8. A. Hummel, A.O. Allen, J. Chem. Phys., 44, p. 3426 (1966)

THE INFLUENCE OF INDUCED MOTION ON UNIPOLAR CHARGE TRANSPORT

J.C. Lacroix and P. Atten*

Introduction

It is now well known that new charge carriers are created at metallic electrodes when a polar dielectric liquid is subjected to a DC field (1). This injection phenomenon, electrochemical in nature, generally induces a space charge q which under the action of the electric field E can put the liquid into motion. Two questions then arise concerning i) the conditions under which the liquid cannot remain at rest and ii) the influence of motion on charge transport. A considerable amount of work has been devoted to the stability problem : instability sets in when the number $T = \varepsilon V/K\eta$ (ε : permittivity ; V : applied voltage ; K : mobility ; η : viscosity) exceeds a critical value depending on the characteristics of conduction (2). The electroconvection problem has been investigated more recently (3) and we present here results in the case of the stationary state of space-charge-limited unipolar injection into an insulating liquid (plane parallel electrodes). Without motion the Coulomb force qE is uniform, a situation analogous to the problem of the fluid heated from below, but the self-induced charge transport does not obey the same laws as heat transfer. In particular, the electric Nusselt number Ne defined as the ratio of the effective current I to the space-charge-limited current (SCLC) I_o which would exist if there were no motion, exhibits a saturation behaviour for high voltages (4).

Theoretical analysis

An approximate treatment has been carried out neglecting diffusion and assuming that the liquid is perfectly insulating for two asymptotic states, one for which viscosity dominates throughout the space, the other corresponding to a fully developed turbulence.
a) It is assumed in the first case that there is a quasi-cellular laminar motion. Two spatial regions may be distinguished : A layer adjacent to the injector where the drift transport is dominant and the core where the convective transport dominates : Introducing the number $A = \max|w| d/KV$ (w : velocity component perpendicular to the electrodes ; d : gap width), it can be shown (5) that Ne varies as $A^{1/2}$. Balancing the forces leads to the relation $A \sim T$ so that $Ne \sim T^{1/2}$ .As Ne = 1 for $T < T_c$ (no motion), we obtain $Ne \approx (T/T_c)^{1/2}$ for $T > T_c$.
b) For high voltages such that the motion is fully turbulent, we distinguish a boundary layer of thickness z_g where a strong gradient of mean charge exists from a core of weak mean charge gradient. An energy argument shows that the r.m.s. value of reduced fluid velocity $w' = (\overline{w^2})^{1/2}$ varies roughly as $M\overline{E}(z_g)$, with $M = (\varepsilon/\rho)^{1/2}/K$ (ρ : density), $\overline{E}$ denoting the mean reduced value of E.

*Laboratory for Electrostatics, Centre National de la Recherche Scientifique, B.P. 166, Centre de Tri, 38042 GRENOBLE CEDEX France.

It has been established (5) that Ne ~ $\bar{q}(1 - \bar{q}/2)(1 + \beta M^{1/2})$, β being a constant of order of 1. A maximum dissipation principle leads to a predicted reduced mean charge $\bar{q} = 1$. For sufficiently high M values Ne varies as $M^{1/2}$ which no longer depends on the voltage : we thus obtain the saturation of both the electric Nusselt number and the r.m.s. reduced fluid velocity.

c) Transition between these two regimes is due to the change from a viscous to a non-linear flow. For the analogous Bénard problem (heat transfer) this occurs at a Reynolds number of about 10. Taking this value leads to the expression $(MR)_T \simeq \sqrt{10 T_c} \simeq 30$ with $T_c \simeq 90$ (5) $(R = T/M^2 = \rho K V/\eta)$.

Experimental procedures and results

The experimental apparatus and the purification technique are those generally utilized (5,6). Eight different liquids have been studied whose dielectric constant ranges from 5.6 to 69 and viscosity from 0.6 to 90 cp. Most liquids have a sufficient Kerr constant to permit an accurate determination of the electric field distribution using a rotating analyser.

Fig. 1 shows the variations of the current versus voltage for various gap widths in pyralene 1460 ($\varepsilon = 5.9\varepsilon_0$, $\eta = 10$ cp at 20° C). Three states of charge transport are visible :

a) For applied voltages smaller than the critical value $V_c \simeq 50$V, the current I varies as V^2/d^3 ; no motion can be detected by Schlieren observations and within the experimental errors $I = I_0$, the SCLC calculated using the true ionic mobility ;

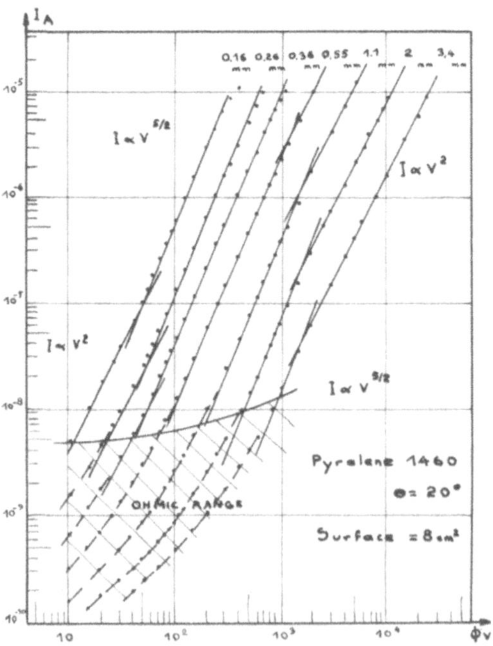

Fig. 1 - Current-voltage characteristics for various distances

b) for voltage values somewhat above V_c, I varies as $V^{5/2}/d^3$; Schlieren observations made in pyralene 1460 contained between two transparent electrodes consisting of glass coated with a conducting layer and ion-exchanger varnish, show just above V_c a stable very regular hexagonal pattern which is only slighly perturbed at $V = 2V_c$ (5) ; furthermore, direct observation of particles moving with the liquid show that the parameter A characterizing the maximum fluid velocity in the cells follows the law $A \simeq T/T_c = V/V_c$ (7).

c) For applied voltages above V_T ($V_T \simeq 1000$ V) I varies again as V^2/d^3 ; an apparent mobility can then be defined which is electrohydrodynamic in nature because the liquid is in a turbulent state of motion

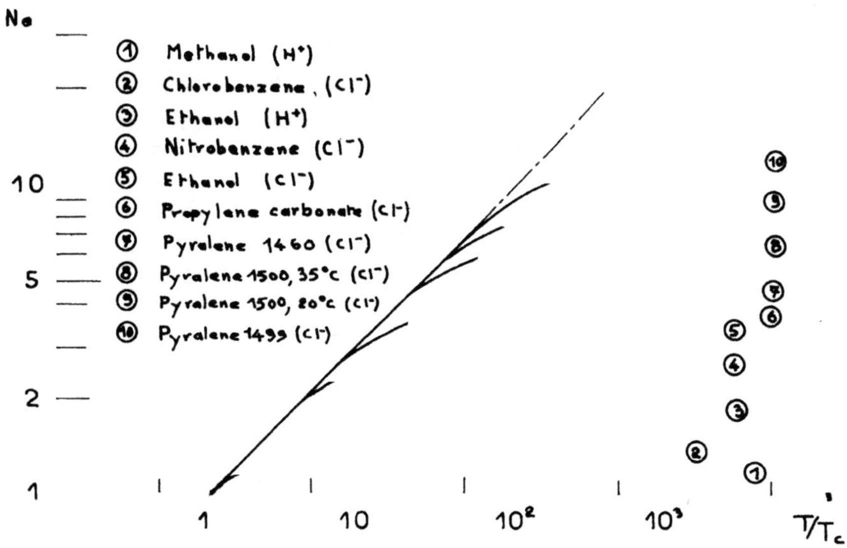

Fig. 2 - Electric Nusselt number versus the stability parameter T/T_c.

as is shown by Schlieren observations.

The results of the study performed on eight liquids are summarized in fig. 2. The experimental values of the criterion T_c lie between 80 and 90 (5). Above T_c all curves exhibit first a $\sqrt{T/T_c}$ behaviour as predicted by the analysis in the laminar case and then tend asymptotically to a saturation value Ne_s which obeys the law $Ne_s = \sqrt{M/3}$. Equating the two expressions of Ne at the transition leads to $(MR)_T \simeq 30$, in noticeable agreement with the prediction. Electric field determinations also give interesting indications. At voltages well above V_T (case of low viscosity liquids) the electric field is apparently linear throughout the space. The slopes give values about 1.2 - 1.3 for $\overline{q}$, not far from the rough prediction $\overline{q} = 1$. With high viscosity liquids, the working voltages are only a few times V_T and two boundary layers are visible (fig. 3), whose thickness depends on viscosity and voltage. As voltage increases the fully turbulent core where $\overline{q} \simeq 1.2$ progressively invades the whole space.

Discussion and conclusion

The experiments give the first precise determination of the stability criterion of unipolar injection under stationary conditions obtained from the jump in the slope of I versus V. The departure from the theoretical criterion is discussed elsewhere (2).

A very important consequence from a practical viewpoint arises from the expression of the saturation law $Ne_s \simeq \sqrt{M/3}$. Because $M = \sqrt{\varepsilon/\rho}/K$ and $K\eta \simeq C^{te}$ from Walden's rule, the more viscous the liquid, the greater will be the magnification by convective transport of the SCLC I_q. Consequently, when injection is present the benefit of using a liquid of high viscosity is only proportional to $\eta^{1/2}$ and not to η .

The value M = 3 plays a very important role. As $Ne \gtrsim 1$,

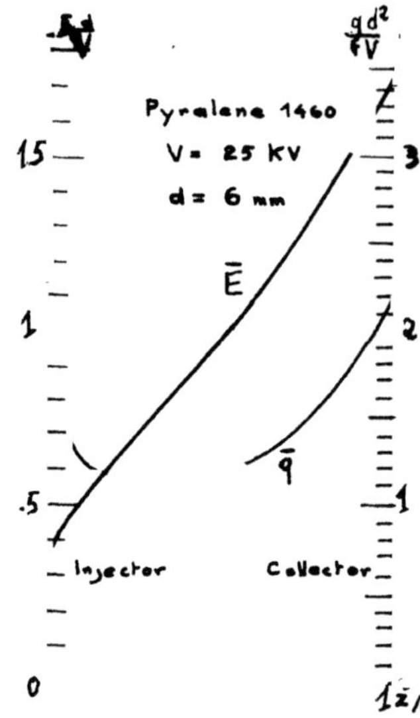

Fig. 3 - Electric field and space charge density distributions.

the saturation law does not hold for $M < 3$. This is supported by two remarks :
i) the r.m.s. fluid velocity can be estimated in the turbulent regime (5) : $w' \simeq (M/3)KV/d$ and for $M = 3$ approximately equals the mean ionic velocity
ii) at the critical voltage the value of $(MR)_c$ is $T_c/M \simeq 90/M$ and for $M = 3$ takes the transition value $(MR)_T \simeq 30$. Two classes must be distinguished in the set of insulating fluids :
1°) $M < 3$: this class includes all gases and a few liquids of both low permittivity and viscosity. The induced fluid velocity will be smaller than the ionic velocity and the effect of convection is negligible on charge transfer and also on the charge and electric field distributions. At $V = V_c$ we can expect a direct passage from rest to a turbulent fluid motion.
2°) $M > 3$: this is the case commonly encountered with liquids. There the induced motion plays the major role in charge transport and the fluid velocity takes values higher than the mean ionic velocity KV/d. Motion also strongly affects the charge and field distributions which exist at rest.

Comparison with the analogous problem in heat transfer shows that despite very different laws concerning the dependence of the absolute fluxes and Nusselt numbers there is a common relation reflecting the role of convection in transport : Nu and Ne roughly varies as $A^{1/2}$ in both regimes (5). It can also be seen that the number M takes the role of the Prandtl number.

It is not easy to generalize the above results. In the case of unipolar injection superimposed on bulk conduction, it appears plausible to predict that convection will significantly contribute to the charge transport only if the conduction current does not exceed αI_0, α depending, of course, on the number M.

(1) N. Félici, Direct Current, 2, p.90 (1971)
(2) P. Atten, Communication to the present conference "Electrohydrodynamic stability of liquids subjected to unipolar injection"
(3) E.J. Hopfinger, J.P. Gosse, Phys. Fluids, 14, p.1671 (1971)
(4) E.J. Hopfinger, J.C. Lacroix, Proceed. 4th International Conference on Conduction and breakdown in dielectric liquids Dublin 1972, p.97
(5) J.C. Lacroix, P. Atten, E.J. Hopfinger, J. Fluid Mech., 1975 (to be published)
(6) J.C. Lacroix, P. Atten, Communication to the present conference "Double injection with recombination : EHD stability and charge transport".
(7) J.C. Lacroix, Doct. Thesis, Grenoble University 2 (1975).

THE ELECTROVISCOUS EFFECT AND ITS EXPLANATION

T. Honda and P. Atten*

Introduction

It has long been known that external application of an electric field transverse to the flow of a liquid causes an increase in apparent viscosity (1). This phenomenon, called electroviscous effect has been extensively studied in dielectric liquids (2) and colloidal suspensions (3). We restrict ourselves here to the case of isotropic liquids for which several different mechanisms have been proposed such as transport of momentum by electrophoresis of ions, orientation of the polar molecules (4), cluster formation in the vicinity of the electrodes (2), space charge and hydrodynamic effects (5). However, because of the lack of reproducibility and even contradictions in results of different workers, the basic phenomenon was not clear.

Electrohydrodynamic nature of electroviscous effect

All previous authors working in isotropic liquids point out a certain correlation between apparent viscosity increase and the passage of an electric current through the liquid. It is therefore necessary to take into account the different phenomena involved in conduction. We now have precise ideas upon electrical conduction of polar liquids : charge carriers are created through i) bulk dissociation of a chemical species (an impurity or the liquid itself) and ii) injection by the metallic electrodes, which has been proved to be a very general phenomenon controlled by electrochemical reactions (6). Under the action of an electric field E the charge carriers migrate and therefore build up space charges q which under appropriate conditions can extend to the whole interelectrode space. The subsequent Coulomb force qE can put the liquid into motion. For a strong injection into a liquid at rest, this occurs at low D.C. voltages and under high voltages the motion becomes turbulent (7). If an electric field is applied transverse to a Poiseuille flow, the Coulomb force may induce a secondary flow which disturbs the velocity profile ; for an electrically induced turbulence the profile becomes quite flat. Momentum is then transferred not only by molecular process (viscosity) but also by convection. Therefore higher pressures will be required in order to maintain a constant flow rate. This is formally equivalent to an increase of mean viscosity.

Experiments performed in carefully controlled conditions completely confirm this explanation.

Experimental procedures

The lack of reproducibility of previous electroviscosity measurements is due to the erratic nature of injection phenomena at

* Laboratory for electrostatics, Centre National de la Recherche Scientifique, B.P. 166, Centre de Tri, 38042 GRENOBLE CEDEX France

metallic electrodes. Using electrodialytic membranes we can obtain very well defined conditions (unipolar injection, electrodialysis). The experiments were carried out with pyralene 1460 (permittivity $\varepsilon = 5.9\ \varepsilon_0$, viscosity $\eta = 10$ cp, density $\rho = 1.41$ g/cm^3) and nitrobenzene ($\varepsilon = 35\ \varepsilon_0$, $\eta = 2.0$ cp, $\rho = 1.2$ g/cm^3) permanently deionized and recycled in a temperature-controlled teflon circuit. The flow rate is controlled by a flow-metre and we measure the electric current and the pressure drop. The electrodes (1 cm x 5 cm) are separated by teflon plates of thickness d ranging from 300 µm to 1000 µm.

Effect of a pure D.C. unipolar injection

The characteristics of the electroconvection induced by a unipolar injection are now well known (7,8) and make it possible to give precise interpretation of the phenomena observed with a transverse flow. Three different hydrodynamic regimes are clearly visible (pyralene 1460, fig. 1) :

a) For applied voltages V less than V'_c, the current I varies as V^2/d^3 and the pressure drop P_E equals the pressure drop P_o of the Poiseuille flow without field. As $V'_c \approx 55$ V is close to the critical voltage V_c of the liquid at rest, it is probable that no secondary flow exists in this regime.

b) For $V'_c < V < V_1$, P_E increases with V (fig. 1). As I roughly varies as $V^{5/2}/d^3$, we expect that the resulting liquid flow is viscosity controlled.

c) For $V > V_1$, P_E increases more strongly with V and then exhibits a saturation behaviour whereas I approximately varies as V^2/d^3. The P_E/P_o versus V curves (fig. 1) show a change in slope for $V_1 \simeq 300 - 400$ V close to the value at which the electric Nusselt number Ne departs from the law $\sqrt{T/T_C}$ (7,8) (the same

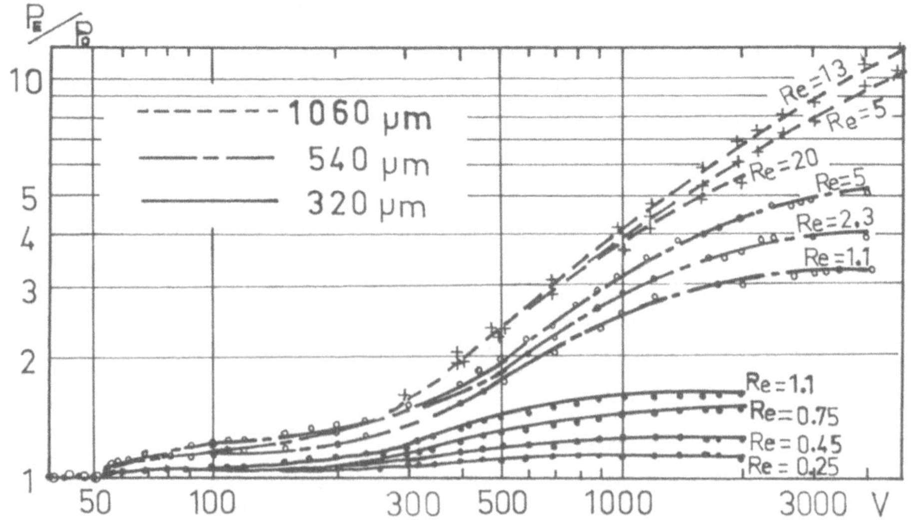

Fig. 1 - Variations of P_E/P_o versus voltage in pyralene 1460 ($\rho \simeq 5.10^{13}$ Ω.cm) subjected to unipolar injection for different gap widths d and Reynolds number Re.

agreement is also found with nitrobenzene).

The variations of P_E/P_o, with distance and with the Reynolds number of the longitudinal flow are not so easily understood. A decisive factor seems to be the ratio B of the mean electrically induced velocity to the longitudinal mean velocity. As B is greater than 1 (low Reynolds numbers Re) the ratio P_E/P_o increases with Re for fixed V and d values. Conversely for B significantly lower than 1, we should observe a decrease of P_E/P_o with Re reflecting the decreasing relative importance of the secondary flow. With regard to the variations with distance (V and Re fixed) no simple argument can account for the strong influence of d.

Effect induced by electrodialysis under D.C. voltage

As a weakly conducting liquid flows between two electrodes coated with purifying membranes (no injection), there is a progressive separation of charge carriers such that space charges appear in two boundary layers. The magnitudes of both the force and the mean layer thickness are decisive for the occurrence of the "electroviscous effect". Thus with very pure liquids such as nitrobenzene of resistivity $\rho > 10^{12}$ Ω.cm and also with sufficiently conducting liquids (for nitrobenzene $\rho < 10^{8}$ Ω.cm), the space charges are very small and no pressure drop increase can be detected whatever the applied voltage. In both cases experiment clearly confirms the hydrodynamic origin of the electroviscous effect in isotropic liquids.

For intermediate resistivity values, a secondary flow is induced which gives an effect of lower magnitude as in the pure injection case. The reduced charge density is of order C_o in the layers ($C_o = en_o d^2/\varepsilon V$, e : carrier charge, n_o : number of pairs of ions at thermodynamic equilibrium). As C_o also measures the ratio of transit time to relaxation time, a maximum driving force and consequently electroviscous effect are expected to occur for $C_o \sim 1$. For nitrobenzene this corresponds to a resistivity $\rho \sim 10^{10}$Ω.cm in the case V = 1000 V, d = 1 mm ; this value is in reasonable agreement with the observations (fig. 2).

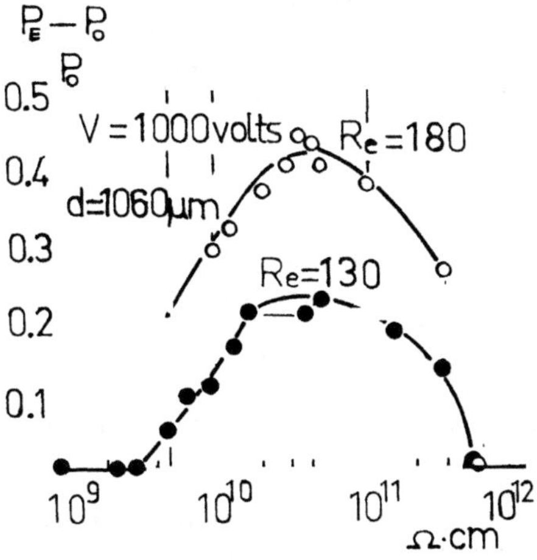

Fig. 2 - Relative pressure drop increase versus resistivity for electrodialysis in nitrobenzene.

Discussion and conclusion

We have proved that under D.C. voltages, the apparent viscosity increase results from the electrically induced secondary flow which disturbs the initial Poiseuille flow (moderate voltages). For high voltages a turbulent motion takes place which entirely destroys the laminar flow.

The two fundamental components of the electrical conduction of weakly conducting dielectric liquids cannot be distinguished when using A.C.

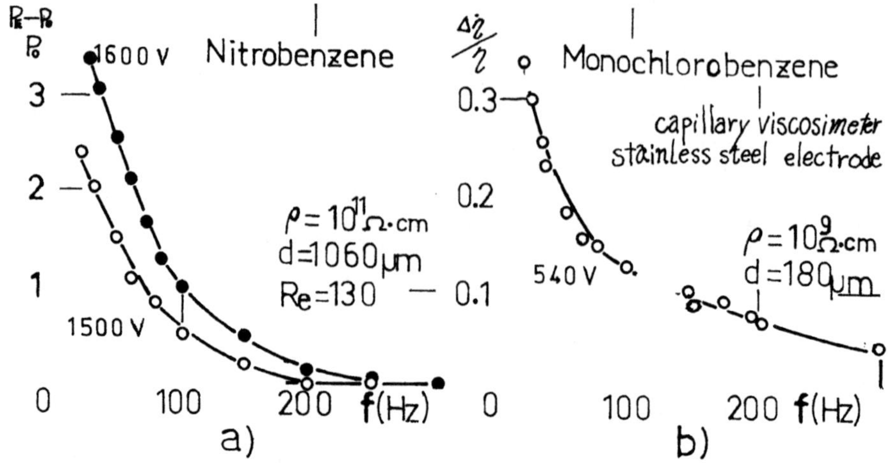

Fig. 3 - Relative pressure drop increase versus frequency
a) two identical permselective membranes b) metallic electrodes

voltages. It is nevertheless possible to give a picture of what happens with metallic electrodes. At low frequency both mechanisms i.e. bulk conduction and injections are present. Above a certain cut-off frequency f_c depending on resistivity and electric field, the injections disappear because during each half alternance the electrode polarisation can no longer reach the values characterizing electrochemical reactions (9). The electroviscous effect relative to the injection processes will be cancelled for a lower cut-off frequency because the onset of a secondary flow requires a minimum unipolar layer thickness. This is illustrated by an experiment carried out using two identical electrodialytic membranes which inject immediately after the reversal of the electric field (fig. 3-a). Although it is not possible to study only the bulk conductive component under A.C. conditions by entirely eliminating injections, it seems plausible to attribute the sharp (f < 50 Hz) and the slow decreases in fig. 3-b (two metallic electrodes) to, respectively, the injection process and the bulk conduction.

(1) A.W. Duff, Phys. Rev., 4, p. 23 (1896)
(2) E.N. Andrade, C. Dodd, Proc. Roy. Soc., 187 A, p. 296 (1946)
(3) D.L. Klass, T.W. Martinek, J. Appl. Phys., 38, p. 67 (1967)
(4) E.D. Alcock, Physics, 7, p. 126 (1936)
(5) P. Sokolov, S. Sosinski, Acta Physicochim. URSS, 5, p. 691 (1936)
(6) N. Félici, B. Gosse, J.P. Gosse, Communication to the present conference "Factors controlling ion injection by metallic electrodes in dielectric liquids".
(7) J.C. Lacroix, P. Atten, Communication to the present conference "The influence of induced motion on unipolar charge transport"
(8) J.C. Lacroix, P. Atten, E.J. Hopfinger, J. Fluid Mech.,(1975)
(9) N. Félici, Direct Current, 2, p. 90 (1971)

ELECTROLYTIC CONDUCTION AND E.H.D. TURBULENCE

N. FELICI*

Introduction

Electric currents passing through resistive liquids are usually attended by hydrodynamical turbulence, which often gives an important contribution to charge transport. This effect can be compared with the convection of heat. In a vessel filled with water and heated from below, instability develops for a certain temperature drop across the liquid layer, and the ensuing convection both facilitates the heat transport and is maintained by it, as a consequence of a positive feedback between them.

It is tempting to postulate that instability is always linked with some sort of facilitation, as in the preceding example, and positive cases abound. Nevertheless, exceptions do exist, and it may not be said that a system under constant voltage (or pressure, etc) always tends to the state of maximum entropy production, although it is very often so (1).

In the case of D.C. conduction through a resistive liquid, it is well known to-day that whenever charge injection is the dominant process, instability occurs very easily (at voltages as low as a few volts) and hydrodynamical charge transport (in contradistinction to transport by ionic drift) becomes the dominant process.

The situation when electrolysis is the only conduction mechanism (i.e. separation of charges in the bulk of the liquid and collection thereof by the electrodes) is not yet fully understood.

If separation of ions is achieved beforehand, as in a strong electrolyte, there is little doubt that instability and facilitation are present. The neutral core of the system may be likened to a charge injector of both signs, while the electrodes act as mere collectors. Turbulence enhances transport and reduces space charge effects by mixing. Experiment entirely confirms this qualitative description (2).

Steady-state electrolysis

When steady-state electrolysis is contemplated, however, things are vastly different. Ions are now created by dissociation at the same rate as they are neutralized

* Laboratory for Electrostatics, Centre National de la Recherche Scientifique, B.P. 166, Centre de Tri, 38042 - GRENOBLE CEDEX, France

at the electrodes. Elementary reasoning, together with a large body of elaborate theoretical results, suggests that stability can be absolute (3). Nevertheless, turbulence appears to be prevalent in practice, and this raises the question whether or not superimposed (and hardly detectable) injection is responsible for the observed instability.

In a previous paper (4), I tried to find a way out of this difficulty by resorting to the picture of a stratified liquid conductor. Whenever conductivity is discontinuous between adjacent layers mechanical instability ensues, and the overall resistance decreases. It may be inferred that this very situation is created by electrolysis, and that the resulting motion prevents the liquid from reaching its stable steady-state condition.

This problem has been reexamined and conclusions are considerably different.

To begin with, there was a flaw in the derivation of the second differential $\delta^2 W$ of the work of the forces, as compared with that of the dissipated power $\delta^2 P$. Although δW and δP always have like signs, this is not necessarily true of $\delta^2 W$ and $\delta^2 P$. Thus, while P always increases (under constant voltage) when the strata are disturbed, it is not true of W. This need not surprise us, for a general law of maximum entropy production under constant voltage does not exist.

Calculations for a uni-dimensionally stratified liquid conductor having conductivity $\gamma(x)$ at rest show that $\delta^2 W > 0$ in any volume element where a certain criterion involving the first and second derivatives of $\gamma(x)$ is satisfied. This makes for a fundamental difference with $\delta^2 P$, which only depends on $(\gamma'(x))^2$.

As a rule, disturbances of short wavelength λ and (or) developing slowly (in a time $t \gg \tau = \varepsilon/\gamma$) favour $\delta^2 W > 0$.

A few examples of the criterion for $\delta^2 W > 0$: short λ and $t \gg \tau$: $\gamma'^2 + \gamma\gamma'' > 0$; short λ and $t \ll \tau$: $-2\gamma'^2 + \gamma\gamma'' > 0$, or $E''(x) < 0$, $E(x)$ being the undisturbed field ; long λ (whatever t) : $-3\gamma'^2 + \gamma\gamma'' > 0$.

In all cases save the first $\gamma'' > 0$ is necessary.

Now how do matters stand when steady-state, symmetrical electrolysis occurs in a liquid whose conductivity is large enough to be described by a single function $\gamma(x)$? Diffusion must be taken into account, and the reduced conductivity $g(x) = \gamma(x)/\gamma(0)$ ($\gamma(0)$ being the value at the middle plane) is given by the second-order differential equation : $g''(x) = -C + Rg^2$; C and R being positive constants ($C > R$) related to the creation and recombination of ions, respectively, and to K (ionic mobility), D (ionic diffusion coefficient), $q(0)$ (ionic charge density at the middle plane), and $E(0)$. For symmetry reasons, $g'(0) = 0$. For a very low voltage applied, $C \neq R$ and $g(x) \neq 1$ everywhere. If voltage is increased, $g(x)$ remains nearly constant in the middle region, decreases slowly towards the electrodes and falls abruptly in their immediate vicinity over a distance $\sim U/E$, where $U = D/K = 0.025$ volt. As to $g''(x)$, it is negative everywhere, since $C > R$ (creation must exceed recombination). Thus, instability is not possible, save in the first case (short λ, that is $\lambda << U/E$, and $t >> \tau$) which will be now elaborated, the criterion for $\delta^2 W > 0$ being $g(x) < 0.52$.

Let us denote the x-component of the local displacement by δx. The work of the electrostatic forces per unit volume $\delta^2 W = A\varepsilon E^2(\delta x)^2/a^2$, where A is a dimensionless number of the order of unity related to λ and to the $\gamma(x)$ distribution, and $a \sim \gamma(x)/\gamma'(x)$. Here, obviously, $a \sim U/E$. (In other parts of the field, where γ is near unity, $A < 0$).

The growth time t of the disturbance may be controlled by either inertia or viscosity. If $\lambda << a$, A is largest, and viscosity is dominant. By equating the forces, one obtains : $A\varepsilon E^2 \delta x/a^2 \sim \eta u/\lambda^2$ (u = velocity, η = viscosity) $t \sim \delta x/u \sim (\eta/A\varepsilon E^2)(a^2/\lambda^2)$.
The growth, however, will be checked if diffusion can blurr the distorsion of the conductivity distribution in the time t, that is, $Dt > \lambda^2$ or $(D\eta/A\varepsilon E^2)(a^2/\lambda^2) > \lambda^2$

Since $a \sim D/KE$ this can be rearranged by introducing the specific mass ρ and letting (η/ρ)(kinematic viscosity) and $\sqrt{(\varepsilon/\rho)}$ (hydrodynamical mobility) appear.
One finds $(\eta/\rho)/D > (A(\varepsilon/\rho)/K^2)(\lambda^4/a^4)$. The electric Prandtl number $(\eta/\rho)/D \sim 10^4$, and $(\varepsilon/\rho)/K^2 \sim 10^2$ to 10^3. Thus, a disturbance with $\lambda << a$ cannot grow.

One may wonder whether the disturbance could be inertia controlled, thus giving a different growth time t. A simple derivation, however, shows that for inertia to be dominant, λ^2 must be larger than $\eta a/E\sqrt{(A\varepsilon\rho)}$. Since $\lambda < a$, one must have $(\eta/\rho)/D < \sqrt{A}\sqrt{(\varepsilon/\rho)}/K$, a most impossible condition.

It may be concluded that the large value of the Prandtl number together with the fact that the depth of the active region ($A > 0$) is diffusion controlled render instability unlikely.

If the liquid was moderately stirred by means of an external agency, the region where $g(x)$ departs from unity would be extended, but its distribution being smoother, the criterion $gg'' + g'^2 > 0$ could no longer be satisfied, as it is seen from model distributions like $g(x) = 1 - x^2/d^2$ subjected to the condition of giving the same integrated conductivity $\int g(x)\,dx$ as the original $g(x)$. Thus, turbulence could not maintain itself.

Conversely, if the electrodes inject charge carriers, injection being superimposed to the above considered electrolysis, g"(x) can easily be positive, and the various criteria are then satisfied in most of the field, not merely in a part of the diffusion controlled region. Moderate mixing does not change this, and turbulence can be self-sustaining. Moreover, the curvature of g"(x) being related to the relative intensity of injection and dissociation currents, voltage is not important, and instability can set in at low voltages provided E is large enough to give significant forces.

Closing remarks

The validity of the preceding discussion is of course restricted to the case when the conductive behaviour of the liquid may be described everywhere by a space and time-dependent single function γ. The criterion for that can be derived as follows : Let c be the creation rate of ions by dissociation, i the current density. Because of the shape of the g(x) functions, dissociation and recombination balance each other almost exactly, save in the diffusion region whose depth a $\sim$ U/E. Thus, i must arise there, and ca > i. Remplacing i by γE or 2KqE, one gets $E^2 < cU/\gamma$ or $cU/KqE^2 > 2$. If $c = 10^{-5}$ A/cm^3, $\gamma = 10^{-10}\Omega^{-1}cm^{-1}$, E < 50 V/cm. Thus, it may be said that the turbulence arising in resistive liquids under low voltages ($\sim$ 10 volts or even less) cannot be explained by the conductivity gradients due to pure, steady-state electrolysis. Conversely, when injection is superimposed, it may easily provide for instability, for there is no threshold voltage as far as the $\gamma(x)$ description holds.

(1) N.J. FELICI, C.R.Acad. Sciences Paris, t. 278, Série B - 151 to 153, 28 janvier 1974, and Série B - 807 to 809, 29 avril 1974.

(2) J.P. GOSSE, Journal de Chimie Physique, 1970, 67, n° 3, p. 630.

(3) P. ATTEN, Proceedings of the 4th Intl. Conf. on Conduction and Breakdown in Dielectric Liquids, Dublin, 1972, p. 86.
also P. ATTEN, paper to be published in Journal de Mécanique, 14, n°3, June 1975.

(4) N.J. FELICI, Proceedings of the 4th Intl. Conf. on Conduction and Breakdown in Dielectric Liquids, Dublin, 1972, p. 69.

ON THE ROLE OF IONS AND DIPOLES ON ELECTROCONVECTIVE PROCESSES IN INSULATING LIQUIDS

H. Vanderschueren•, R. Coelho••

Introduction

We have previously shown ([1]) ([2]), by use of Schlieren optics and by bolometric measurements, that thermo-electroconvection in a divergent electric field is quite sensitive to the presence of ionic or dipolar impurities in dielectric liquids. We principally used a horizontal platinum wire (20 µm) in the axis of an annular electrode (diameter = 3 cm). The wire is grounded at one end, and a high voltage up to a few thousand volts, is applied to the other electrode. The system is supported in a glass cell with two optical windows. Remarkable Schlieren sensitivity is achieved when the wire, perpendicular to the optical beam, is heated by a current step, and the boundary between cold and warm liquids appears as a highly contrasted line rising above the wire. The light from a normal iodine lamp is sufficient to permit direct observation or photography, or to use a position sensitive detector (P.I.N. "optical potentiometer" from U.D.T. for instance) for recording the boundary motion. In addition, the wire being heated by a regulated current, we have performed a bolometric measurement of its resistance during the application of voltage between the electrodes. Finally the current between the electrodes is recorded and the current voltage characteristics compared with the other informations derived from the two other methods.

Experimental findings

With an applied voltage of a few hundred volts, the thermal convection out of the wire is impeded by the impurities moving towards the wire. The boundary between warm and cold liquids, straight during natural convection (fig. 1a), is now distorted (fig. 1b). With a higher applied voltage, quadratic electroconvection ([3]) outbalances the first effect (fig. 1c). The general behaviour is summarized on figures 1d and 1e. Figure 1d shows, for a given voltage, the boundary rise after application of a current step.

• Université de Liège, Institut Electrotechnique Montefiore, Rue Saint-Gilles, 33, 4000 Liège, Belgium.

•• Laboratoire de Génie Electrique de Paris et L.C.I.E. 33, Aven. du Gén. Leclerc, F.92 Fontenay-aux-Roses.

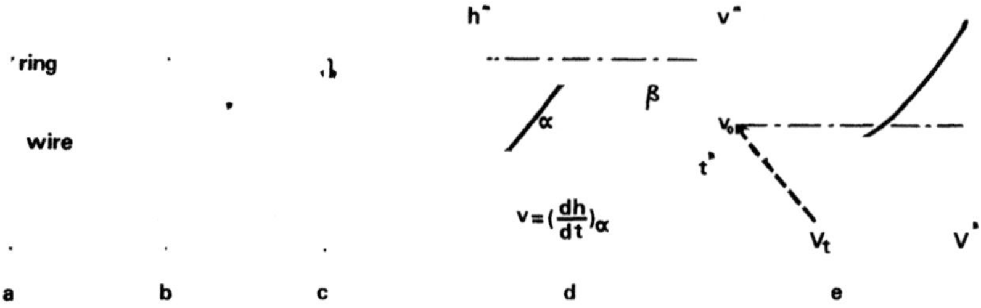

Fig.1

Figure 1d shows another particularity. After rising linearly (α), the boundary is decelerated until it becomes blocked (β) at a distance from the ring which depends on voltage, polarity and impurity content. On figure 1e, we have reported the rising velocity v, defined on 1d, as a function of the applied voltage. For moderate impurity content (one ppm of sulfonate in C_6H_6 for instance), the curve v(V) is given by a law of the form :

$$v(V) - v_o = - A.V + B.V^2$$

where v_o denotes the natural convection, unaffected by considered impurity contents, while A strongly depends on impurity content. This formula is quite similar to that giving the heat transferredfrom the heated electrode in similar conditions ([4]).
For higher impurity content the convection inhibition is so strong in the plane of the ring that above a voltage threshold the warm boundary does not rise. The formula is then nearly reduced to its first term and the curve nearly linear (dotted line in fig. 1e). Above the threshold V_t, the heat transfer is carried by the liquid ejected along the wire as observed by Schlieren optics.
Figure 2a shows successive configurations of the standing boundary obtained by increasing the high voltage (for a given impurity content) while figure 2b shows the height of the boundary as a function of V.

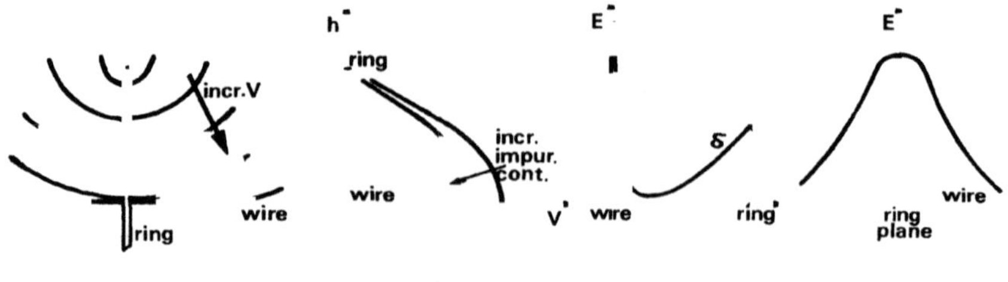

Fig. 2

It appears that impurity has a significant effect in the wire vicinity, whereas in the ring vicinity, the repulsive effect is quasi intrinsic. This shows that warm liquid undergoes repulsive force from the field source. Figure 2c shows the radial field distribution in the plane of the ring. The above repulsion occurs near δ. Figure 2d shows the calculated distribution along the wire. A similar field distribution exists near the ring and might account for the boundary stability with warm liquid rising along the profiles of the curves 2a.
A systematic study of the influence of water content on the electrical and electroconvective properties of C_6H_6 is summarized on figures 3. Fig. 3a, 3b and 3c give, for low (L) and high (H) water content, respectively the increment Φ of heat transfer, the boundary velocity v and the current I flowing between wire and ring (the ring is positive in these experiments). Under low applied voltage, water, like other ionic impurities, impedes convection but at high voltage it favors charge injection hence produces hydrodynamic drag.
Many of the results discussed above are related to combined divergences of the field and temperature. In an attempt to separate out the effects of the field and themperature divergences, we have developed a new experimental set-up consisting of a flat grating made of a single platinum wire stretched between a pair of combs, and a brass knife-edge, set along the light beam, which can be connected to the high voltage source. This two-dimensional configuration is more sensitive than the previous one, and easier to handle mathematically. However, the field diverges only near the kinfe-edge.
We shall show how a comparison of results obtained with these different configurations can improve our understanding of the effect of ionic or dipolar motions on the electroconvective processes in insulating liquids. Potential applications will also be discussed briefly.

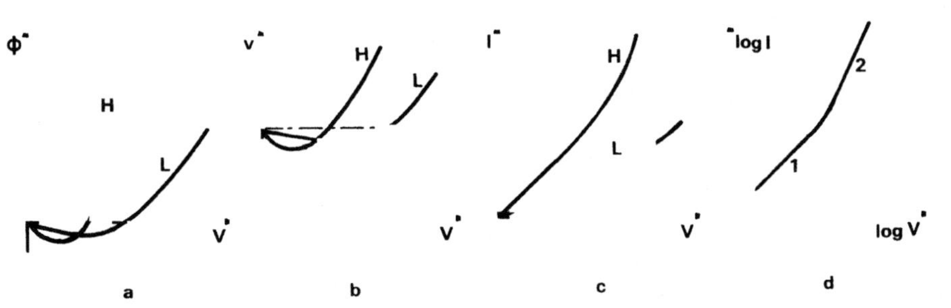

Fig.3

FIELD DISTRIBUTIONS IN CHLOROBIPHENYLS UNDER DIRECT VOLTAGES

J.D. Cross* and M. Eto**

1. Introduction

The field distribution in an interelectrode gap is obviously of fundamental importance in determining the electric strength of an electrode configuration. It is common practice to simply solve Laplace's equation for the configuration and assume that the result obtained represents the true field distribution. While this is quite acceptable in the case of vacuum, gas, or some solids it is not acceptable in the case of liquids of commercial purity. Polar liquid insulants such as Aroclors(R), and nonpolar hydrocarbons containing water as an impurity, have a relatively high ion content. The ionic content is increased when the liquids are subjected to strong electric fields between metal electrodes. The movement of the ions in the electric field produces liquid motion and therefore the charge distribution is controlled by electrohydrodynamics and the controlling equations are not easy to handle. The electric field distribution is more complex than a simple examination of electrode geometry would suggest. As computation of the fields is difficult it is natural to turn to experiment. Measurement of the potential distribution in liquids has been attempted by means of electrostatic probes (1), (2), but it is difficult to avoid distortion of the field by the probes. For polar liquids the Kerr electro optic effect offers an attractive alternative. This technique has been widely used to observe field distributions in nitrobenzene and chlorobiphenyls (Aroclors). Previous studies with the Kerr technique in this laboratory (3), (4), have shown that the field between two parallel plates 5 mm apart in Aroclor 1242 is not uniform. The field distribution could be represented by $E = E_A e^{KX}$ where E_A is the field at the anode, X is the distance the anode and K is a constant. The observation of the Kerr fringes in that work was by long exposure (∿ 1 sec) photography and by visual observation which would not reveal the effect of the rapid eddies that exist in such a liquid under high electric stress. This paper reports an extension of the earlier work in which the electrode spacing was varied from 2 to 8 mm and the low speed fringe observations were augmented by high speed photographs using a Q-switched laser. The effect of an insulating spacer between the electrodes, parallel to the field is also reported.

2. Experimental Procedure

The test cell consisted of a cylindrical stainless steel chamber mounted horizontally with glass windows at either end. Parallel

* Department of Electrical Engineering, University of Waterloo.†
** On leave from Hachinohe Technical College, Hachinohe, Japan.
(R) Registered service mark of the Monsanto Company.

† Waterloo, Ontario, Canada

This work is supported by F.N.R.S. Furthermore we acknowledge support of N.A.T.O. which has made possible continuous cooperation between the authors.

(1) Coelho R., Vanderschueren H. "Influence de la pureté d'un liquide isolant sur la convection thermique en présence d'un champ électrique" C.R. Acad. Sc. Paris, t. 274 (1972), 1259-62 série B.

(2) Vanderschueren H. "Application de l'électroconvection et de la strioscopie à la détermination rapide de la pureté des liquides diélectriques". Thèse - Université de Liège - Coll. Publ. Fac. Sc. Appl. N° 47 - 1974.

(3) Senftleben H., Schnabel P. "Der Einfluss von Raumladungen in hochisolierenden Flüssigkeiten auf den Wärmeübergang unter Wirkung elektrischer Felder". E. Physik 170 (1962) 82-92.

(4) Care J., Swan D. "Some Transient Phenomena in Heat Transfer resulting from electric Stress" Brit. J. Appl. Phys. (1963) 263-266.

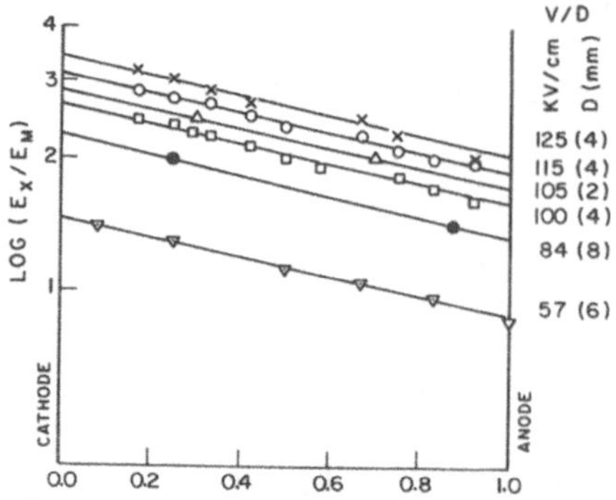

Fig. 1

Field distribution in gaps of 2,4,6, and 8 mm. E/E_m against X/D.

plane electrodes of aluminum 16 cm long by 6 cm wide with 0.5 cm radius edges were mounted horizontally in the chamber. The Aroclor was dried by passing it through molecular seives and filtered (0.8 μm absolute) for several days before degassing and filling the chamber. Water content of the liquid was determined by the Karl-Fischer test (A.S.T.M. D1533). The cell was mounted between crossed polaroids and illuminated with parallel monochromatic light, either from an incandescent lamp and filter or from a Q-switched ruby laser.
The Kerr constant of Aroclor is relatively small and therefore with the electrode length used no more than five fringes could be produced before breakdown. To permit determination of the field at several points in the gap, even at potential differences at which only one fringe appeared, a Babinet-Soleil compensator was used to produce a known phase difference between E and O rays in the cell and hence to position the fringe pattern in any chosen position. For some observations a p.t.f.e. prism 4 mm x 8 mm in cross section and 14 cm long was placed between the electrodes and alligned with the light path, effectively dividing the inter-electrode space into two parts.

3. Results

3.1 The effect of varying electrode spacing.

Figure 1 shows the measured electric field for gaps of 2,4,6 and 8 mm in Aroclor with water content of 10 p.p.m. The field is shown as a ratio of E to E_m plotted on a logarithmic scale against position in the gap. E_m is the characteristic Kerr field of a particular geometry given by $E_m = 1/\sqrt{2\ell B}$ where B is the Kerr coefficient the liquid and ℓ is the path length. For the test

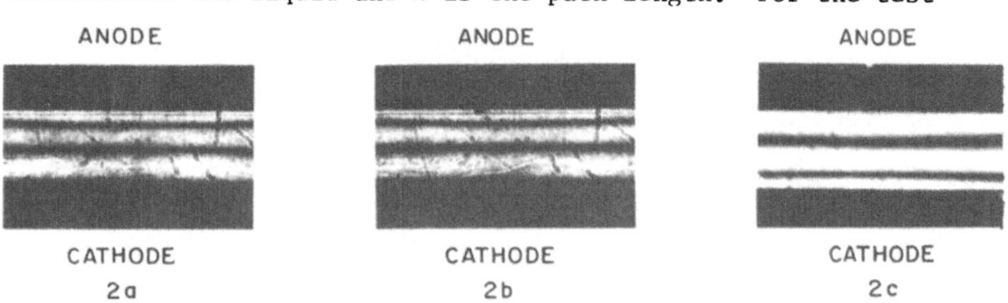

Fig. 2.a. Low speed photograph of Kerr fringes with 42 kV across 4 mm.
Fig. 2.b, 2.c. High speed (<1μs) photograph with 42 kV across 4mm.

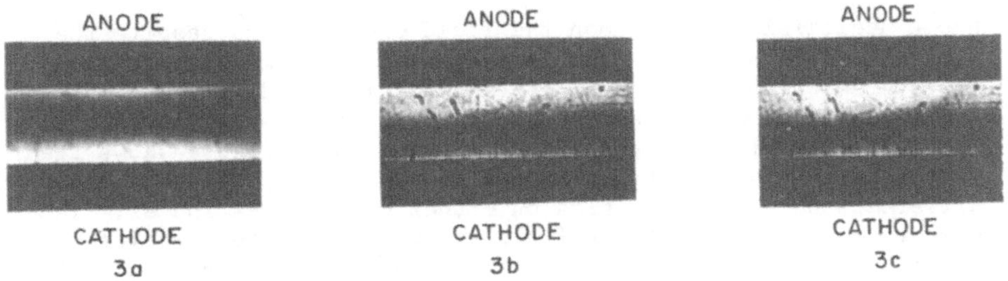

Fig. 3.a. Low speed photograph of Kerr fringes with 28 kV across 4 mm.
Fig.3.b.,3.c.High speed (<1μs) photograph with 28 kV across 4 mm.

cell used $E_m \simeq 48.5$ kV/cm. It can be seen that the exponential field distribution previously reported for 5 mm gaps holds from 2 mm to 8 mm with a form $E = E_A\, e^{kX/D}$ where E_A is the anode field k is a constant, D is the gap spacing and X is the distance from the anode. The exponential distribution is apparently a characteristic of a parallel plate gap in Aroclor rather than a particular distribution for a 5 mm gap. It should however be pointed out that the degree of non-linearity is relatively small and a plot of $E^{1/2}$ against X/D also yields a set of parallel straight lines.

3.2 High speed observations

Fig. 2a shows a low speed (0.1 s) photograph of a Kerr fringe pattern with 42 kV across a 4 mm gap. This is close to breakdown stress. Figs. 2b and 2c show pulsed laser photographs of the same system. Fig. 3a,b and c show a similar sequence with 28 kV applied to the gap. These photographs are typical of the observations made. The high speed photographs show a slight shift in fringe position compared with the low speed photographs: this is due to a difference in wavelength between the laser light and the filtered white light. The similarity between a,b,c photographs is obvious. The effect of the turbulence of the liquid motion on the local field values is small. The high speed photographs reveal small disturbances in the even field distribution and these are more noticeable at lower field strengths but even then they do not represent a significant departure from the long term average.

3.3 Effect of spacer

Observations with a spacer placed symetrically in the inter-electrode gap confirmed the previous report (3) that the spacer

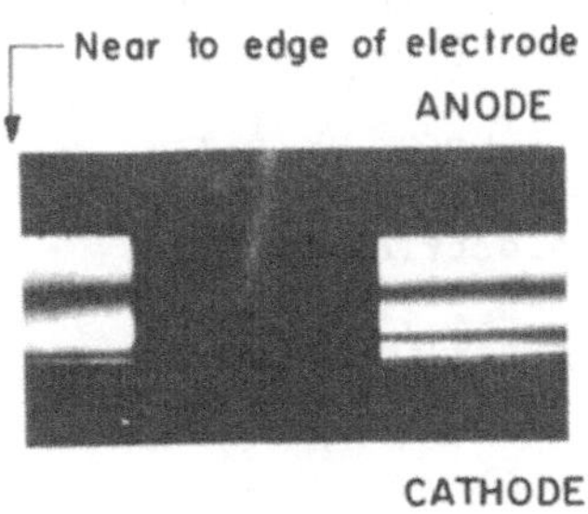

Fig. 4

Kerr fringes in 4 mm gap with spacer asymetrically placed.
42 kV across 4 mm.

has a negligible effect upon the field distribution despite the fact that such a spacer greatly reduces the breakdown voltage of the system. The spacer has been shown (5) to modify the electrohydrodynamic flow and therefore one might expect a difference if the spacer is placed asymetrically in the gap. Fig. 4 shows the pattern for such a system. It can be seen that there is a clear difference between the fields on either side of the spacer. The spacer does not distort the field at its surface but permits the coexistence of two different distributions in the gap. The fringe distribution to the right of the spacer is very close to that for the parallel gap without the spacer and on that side the parallel gap extends for several centimetres. The narrower gap on the left has a more uniform field.

4. Discussion

The observations confirm that the field distribution in Aroclor between parallel plate electrodes is not uniform; It is approximately exponential for a range of electrode spacings. The field at the cathode is some 70% higher than that at the anode. This is consistent with an excess of positive ions probably H^+ created at the anode by a reaction between water and the aluminum electrode. High speed photographs show that the instantaneous field distribution is practically identical to the time average. The scale of the fine structure of charge and velocity is of the order of 0.2 mm for 42 kV across 4 mm and 0.5 mm for 28 kV across 4 mm. Felici (6) has proposed that this characteristic length λ should be approximately given by $\lambda = \frac{\eta t}{\rho}$ where η is the viscosity ρ the density and t the transit time. Flow measurements (5) for similar gaps give t = 4 ms for 42 kV and 10 ms for 28 kV. $\eta/\rho \simeq 37$ therefore $\lambda \simeq 0.15$ mm at 42 kV and $\lambda \simeq 0.36$ mm at 28 kV which agrees well with our observations. The dependence of the field distribution on flow is demonstrated by the effect of the spacer in an asymetric position. Electrostatics would indicate that no such horizontal discontinuity in field should exist yet the reported discontinuity is consistently reproducible. It is clear that designs of liquid insulated apparatus must take into account both the distortion of the field by space charge and the liquid flow patterns that control the charge distribution.

1) A.S. Bloor and M.J. Morant, Probe and charge transit studies of the potential distribution in dielectric liquids. Colloque No. 179, C.N.R.S., Grenoble, 1968.
2) E.O. Forster, Electric conduction in Liquid hydrocarbons. J. Chem. Phys. 37, 1021, 1962.
3) E.A. Cherney and J.D. Cross, Space Charge Effects in Chlorobiphenyls. IEEE Trans. on Elec. Insul., EI-8, No. 1, 10-16 1973.
4) E.A. Cherney and J.D. Cross, Electric Field Distortion at Solid Liquid Dielectric Interfaces, IEEE Trans. Elec. Insul. E.I.-9, No. 2, 37-42.
5) E.A. Cherney and J.D. Cross, Electrical Breakdown at Solid-Liquid Interfaces, IEEE Winter Power Meeting, New York, 1975.
6) N.J. Felici, D.C. Conduction in Liquid Dielectrics Direct Current 2 No. 4, 147-165, 1972.

STUDY OF A THREE ELECTRODE CELL FOR THE RECORDING OF THE TRANSIENT PRE-BREAKDOWN PHENOMENA IN LIQUID DIELECTRICS.

K. Piotrowski• and R. Eyben••

To study the transient pre-breakdown phenomena under a step-voltage in insulating liquid dielectrics (non saturated hydrocarbons), we have used a three electrode cell. The purpose of such a cell was to protect the apparatus against the destruction caused by a possible electrical breakdown of the investigated dielectric at high applied fields ($E > 10^7$ V/m). Figure 1 shows the

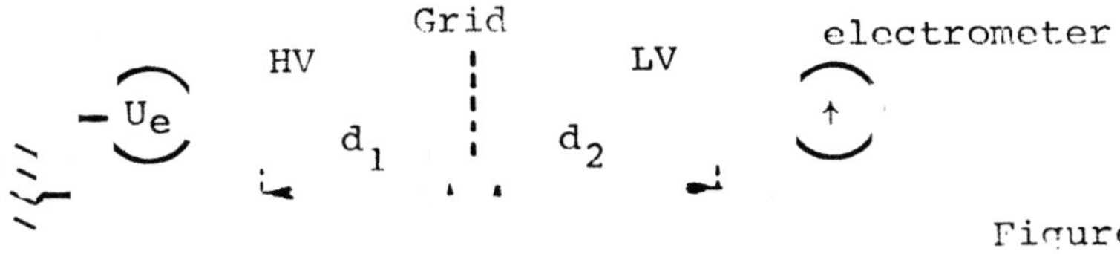

Figure 1.

idealized experimental circuit. The part of the dielectric subjected to pre-breakdown tests is situated between the HV electrode and the grid (HV area). The LV area of the experimental circuit consists of the grounded grid, the LV electrode and the measuring apparatus (electrometer or oscilloscope). The LV circuit is itself the measuring circuit. No external voltage is applied between the grid and LV electrode.

The operating principle of the measuring cell is the following : the application of the step voltage between the HV electrode and the grid sets the charge carriers into motion. The carriers coming close to the grid penetrate into the LV area where they give rise to a current (signal)(fig.2), the duration and shape of which can provide information about : 1) the velocity of the positive or negative carriers, 2) the formation of possible homocharges, 3) the production of partial discharges at the grid level.

The general trend of experimental curves I = f(t) is presented in figures 2a and 2b. During the interval of time τ_f, the current is practically constant. Then, its value increases linearly up to its maximum at the time τ_{tr} or τ'_{tr} .

We have investigated a simplified physical model of

• "Maître de Conférences" at the Applied Science Department of the University of Liège - Dielectric Materials.

•• Junior Assistant at the University of Liège

Fig. 2a

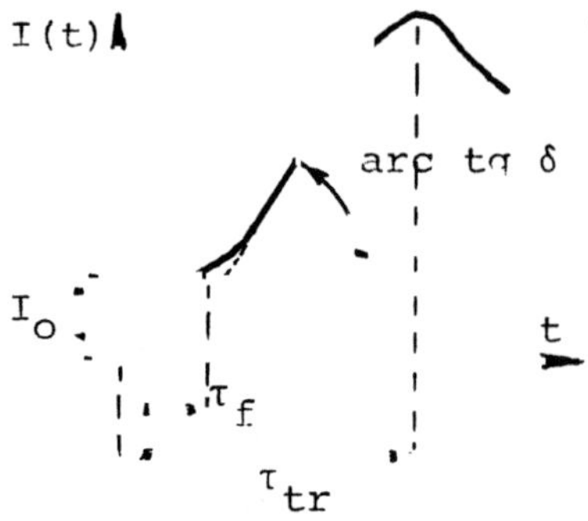

$$\tau_{tr} < \frac{d_2}{\mu E}; \quad \mu = \frac{d_1^2}{\tau_{tr}.U_e}$$

Fig. 2b

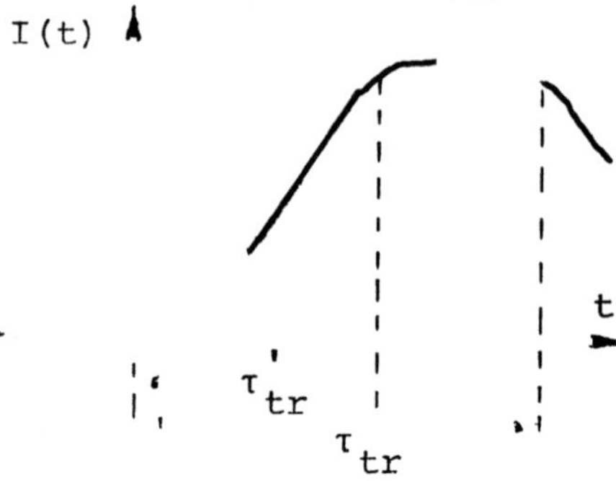

$$\tau_{tr} > \frac{d_2}{E}; \quad \mu \simeq \frac{d_2 d_1}{\tau'_{tr}.U_e}$$

such a cell. The results of this study and the discussion concerning the measurements of the mobility of the charge carriers are presented in this paper. The full treatment of this problem will be presented later on.

Theoretical considerations

During the analytical treatment of this model, we have taken account of : 1) the continuity of the current through the grid, 2) the formation of a space charge close to the wires of the grid. The model has been discussed at the scale of the grid mesh. The following hypothesis are used : 1) proportionality between velocity and electrical field, 2) the outflow of the charges takes place through thin beams of diffusion current from the areas of charge concentration at the surface of the grid, 3) the diffusion current in the LV circuit is calculated assuming the electrodes are plane and parallel, 4) a correcting factor "a" is introduced to take into account the lateral diffusion of the charges.
The model leads to the following prediction for the LV current :

$$I(t) = I_o \qquad \text{if} \qquad 0 < t < \tau_f \tag{1}$$

and $$I(t) = I_o + \delta_{th}(t - \tau_f) \text{ if } \tau_f < t < \tau_{tr} (\text{or } \tau'_{tr}) \tag{2}$$

with $$\delta_{th} = \frac{S}{a.d_2} (e\ \phi_o).(\mu.E)^2 \tag{3}$$

and $$I_o = 2\ \sigma_o.S.K.E \tag{4}$$

δ_{th} is the theoretical slope of the linear part of the current I(t)

$$\tau_f = \frac{\Pi r}{2\mu E} \tag{5}$$

Table I (Samples) 23°C; d_1 = 2 mm, d_2 = 4 mm

Samples	$\sigma_o 10^{11}$	$\phi_o 10^{-17}$	$\mu^+ . 10^{10}$	$\mu^- . 10^{10}$
	S/m	m^{-3}	m^2 $(Vs)^{-1}$	
1 Oil 30 % RH	0,18	0,24	2,5 0,2	- 0,2
2 Oil + VC 0,004 mol	4,5	3,2	5,1	3,8
3 Oil + VC 0,02 mol	14,6	10,0	5,3	3,3

VC - Solution of versatate of Cu in the transf. oil.

Table II (Sample 2)

$E.10^{-5}$	$I_o(A)$		$\tau_f(s)$	
$V.m^{-1}$	exp.	theor.	exp.	theor.
2,5	$2,7.10^{-11}$	$5,2.10^{-11}$	∿ 2	2,4
5,0	6.10^{-11}	10.10^{-11}	1 ∿ 1,5	1,2
25,0	$8,7.10^{-10}$	$5,2.10^{-10}$	0,4	0,3

Table III $\delta_{th} . \delta_{exp}^{-1} = a$ Θ = 23°C

Sample	$E(V\ m^{-1}).10^{-5}$				
	5	25	50	125	Average
1	10	9,0	6,5	12	9,0
2	18	16	16		18
3	20	19	16		

The time τ_f is the space charge time formation. In our model it is equivalent to the transit time of a charge along the 1/4 of the circumference of a wire of the grid at the μ.E velocity.
The two first equations mean that the current is constant during the first part of the phenomenon and then increases linearly with the time.

Experimental verification of the model (examples)

A.- Table II shows good agreement between the theoretical and experimental times τ_f. The same is true for the values of initial current I_o. B.- Table III shows that the correcting factor "a" is independent of E and of the initial carrier concentration ϕ_o(samples 2 & 3). It seems depend on the nature of the carriers (sample 1 and 2 or 3). This factor is very little influenced by the mesh grid dimensions. C.- We have stated that the experimental slope δ_{exp} is actually found proportional to ϕ_o. D.- We have found two types of experimental curves for I(t)(fig.2a and 2b). In the terms of our model their slopes could depend on whether $\tau_{tr} < d_2(\mu E)^{-1}$ or $\tau_{tr} > d_2(\mu E)^{-1}$. The experimental results confirm these conclusions. E.- The value of mobility deduced from the experimental value for time τ_{tr} (fig.2a) agrees with the values found in other sources (Coelho, Gémant) with other methods but for similar physical conditions. The calculations described in the present paper should be considered as a preliminary approach to the theory of our three electrode cell. A plausible explanation of the charge transfer through the grid bas been given, giving theoretical results in good agreement with the experimental measurements. However, the exact mechanism of this processus of charge transfer is not entirely elucidated and one must still study the role played by other phenomena, including electrophoresis.

Measurements of the mobility

The mobilities have been deduced from the time τ_{tr} (fig. 2a)(Table I). For the fields in the range $1,25.10^7 > E > 10^5$ V m^{-1} (Θ = 23°C) our results shows that the mobility is practically independent of the electrical field and of the charge concentration ϕ_o. As concerns pure oil, we find 2 maxima in the curve I(t)(2 types of positive charges). At temperatures higher than 30°C, important modifications of the shape of the curves I(t) are observed (apparent diminution of the mobility with increasing field).

Symbols

$\sigma_o = \phi_o.e.(\mu^+ + \mu^-)$ specific conductivity of the sample.

$E = U_e.d_1^{-1}$

S = surface of an electrode.

2r = diameter of the grid wire.

K = experimental ratio of the LV field to the HV field.

CORONA AND THE MOLECULAR STRUCTURE OF INSULATING LIQUIDS

C.W. Smith* and J.H. Calderwood*

Introduction

Corona is but one of the many phenomena associated with the electrical breakdown of insulating liquids; the experimental conditions determine which of the phenomena will be important (Sharbaugh 1970). For experimental purposes, it is convenient to restrict the spatial extent of the prebreakdown phenomena by using a point-plane electrode system with an adequate gap length to avoid breakdown at the voltages used. In dielectric liquids, the corona discharges are of the order of tens to hundreds of microns in size and although occuring as brief flashes of light (Smith and Calderwood 1968), they are confined within a well defined boundary as can be seen from photographs taken with exposures of several minutes. Shammas et al (1974) found that the radius of curvature of this boundary is proportional to the applied voltage for coronas at points in liquid dielectrics; Sorapipatana et al (1975) have extended the measurements to coronas at points in gases and vapours. Mirza et al (1972) found that the assumption of Paschen breakdown within a small bubble or cavity before breakdown did not lead to the observed variation of breakdown strength with the ambient pressure; neither did it lead to the observed variations of corona onset voltage with ambient pressure. Singh et al (1972) found the radius of the bubbles forming the "spray" associated with corona in dielectric liquids to be about one micron. These bubbles are too small to have cooled from a plasma region 10 μm to 100 μm in size at 10^4 K, a temperature estimated from the corona spectrum (Smith, Kao and Calderwood 1967).

Corona and Molecular Structure

To overcome the difficulties in interpreting the results of the measurements on coronas in liquid dielectrics, measurements were made of the corona radius for a range of voltages and pressures and for a number of gases and vapours, in particular for helium gas and the vapours of n-hexane and n-pentane. (Sorapipatana et al 1975).
Theoretical electron excitation cross sections are available for helium for the appropriate electron energy range, and for those transitions which were observed in the relaxation radiation. This enabled the mean free paths for the electrons to be derived theoretically. It was found that the largest values of the electron excitation cross section occurred for the 2^3S-2^3P transition, while the strongest emission from coronas in helium was at 1.083 μm (1.15 eV), the emission from the 2^3P-2^3S transition.

*Department of Electrical Engineering, University of Salford, Salford M54WT, Lancs, England.

It was then assumed (Adamczewski 1957, Lewis 1960) that the energy gained by an electron from the applied field while travelling a distance of one mean free path λ was greater than, or equal to the excitation energy eV' for this transition within the corona region. The boundary of the corona was thus defined as a surface of constant field E given by

$$E = V'/\lambda$$

This gave correctly the measured value of the field at the boundary of the corona at a point cathode in helium gas within the experimental accuracy; the field at the boundary of the anode corona was found to be twice this calculated value.
Although the fields at the corona boundary in helium gas corresponded to an infra-red excitation, corona emission is also observed in the visible region. It is thus necessary to assume that once the electrons are capable of inelastic collisions they can retain and accumulate any energy in excess of eV' at each collision, until after a few mean free paths they become able to excite higher energy transitions. Some justification for this assumption is found from similarities between d.c. and radio frequency discharges. The elastic collision process results in no change of electron energy because the elastic collision cross section is independent of the electron energy and because of the large ratio of masses involved; but it does produce a large reduction in the drift velocity in an applied field. The inelastic scattering process results in most of the energy loss but has little effect on the drift velocity; the electron either loses all its energy, or the excitation energy corresponding to the first excitation level (Holstein 1946).

Fields measured at the boundaries of coronas at point electrodes in a number of dielectric liquids are listed in table 1 together with some comparative values for the breakdown stresses.
These fields were used in conjunction with the above equation to select appropriate excitation energies and mean free paths. Appropriate excitation energies were found in the infra-red spectra of the liquids concerned (Bentley et al 1968, DMS 1972). Appropriate mean free paths were found to be multiples of the axial projection of the C-C bond distance (Adamczewski 1957), except for cyclohexane where the appropriate mean free path was the mean diameter of the ring. These quantities are also listed in table 1.

A corona region may also be regarded as the source of a transient, wave-like thermal disturbance (Carslaw and Jaeger 1959) which propagates with a velocity

$$v \doteq \sqrt{2K\omega}$$

and whose amplitude is attenuated by a factor ext $-(x\sqrt{\omega/2K})$ after travelling a distance x, where the diffusivity $K = k/\rho\sigma$, k being the thermal conductivity, ρ the density and σ the specific heat of the dielectric. The angular frequency ω corresponds to the temperature fluctuations of the thermal disturbance. Schlieren photography shows prebreakdown phenomena which are not geometri-

Table 1

Liquid	Breakdown Stress V/m	Point Electrode	Field at corona Boundary V/m	Infrared Absorption µm	Mean Free Path nm
n-hexane	$1.56x10^8$	cathode	$7.2x10^8$	13.9	0.123
			$0.8x10^8$	25	5x0.123
		anode	$4.1x10^8$	25	0.123
			$0.8x10^8$	25	5x0.123
n-heptane	$1.66x10^8$	cathode	$7.2x10^8$	13.9	0.123
			$0.8x10^8$	20.5	6x0.123
			$1.6x10^8$	20.5	3x0.123
n-pentane	$1.44x10^8$	anode	$1.2x10^8$	21.3	4x0.123
			$5.2x10^8$	21.3	0.123
cyclo-hexane		cathode	$5.3x10^8$	9.6-9.9	0.24
			$2.7x10^8$	19.1	0.24

cally coincident with the corona and Thomas (1974) has shown that Schlieren photography can detect small thermal changes in the dielectric under these conditions.

It may be significant that the regions shown in Schlieren photographs, the corona regions and the breakdown channel are all comparable in size to the wavelengths of the vibrational modes of the dielectric molecules assumed for table 1.

At radii of the order of 1 µm within a corona region at a very sharp point, the electric fields will be of the order of a hundred times those listed in table 1 and will be sufficient for field ionization to occur. Careful examination of corona photographs shows a very bright region close to the point electrode. The bubbles of the "spray" phenomenon are of radius 0.7 µm to 1 µm and measurements of the prebreakdown transient voltage and current waveforms gave 3 µm as the radius of such a conducting region at the electrode (Smith and Calderwood 1968).

Conclusion

Present results suggest that the sharply defined boundary to the corona region is a surface of constant electric field and that it represents the demarcation surface between regions in which elastic and inelastic collisions occur. It is possible to choose a combination of multiples of the axial projection of the C-C bond distance and energies corresponding to infra-red absorption modes and calculate the field at the corona boundary.

References

Adamczewski, I., 1957, Zesztyty Naukowe Politechniki Gdanskiej, 3, 3.

Bentley, F.F., Smithson, L.D. and Rozek, A.L., 1968, Infrared Spectra and Characteristic Frequencies 700-300 cm^{-1}, (London: Wiley).
DMS, 1972, Working Atlas of Infrared Spectroscopy, (London: Butterworths).
Holstein, T., 1946, Phys. Rev., 70, 367.
Lewis, T.J. 1960, J. Electrochem. Soc., 107, 185.
Mirza, J.S., Smith, C.W. and Calderwood, J.H., 1972, Proc. 4th Int. Conf. on Conduction and Breakdown in Dielectric Liquids, 198, (Dublin: Typografia Hiberniae)
Shammas, N., Smith, C.W. and Calderwood, J.H., 1974, J. Phys. D: Appl. Phys. 7, 2587.
Sharbaugh, A.H., 1970, Report No. 70-C-191, (Schenectady: General Electric).
Singh, B., Smith, C.W. and Calderwood, J.H., 1972, Proc. 4th Int. Conf. on Conduction and Breakdown in Dielectric Liquids, 202, (Dublin: Typografia Hiberniae).
Smith, C.W. and Calderwood, J.H., 1968, Proc. Conf. Phén. Cond. Liq. Isol., 161, (Paris:CNRS).
Smith, C.W., Kao, K.C. and Calderwood, J.H., 1967, Ann. Rep. 1966 Conf.Electrical Insulation and Dielectric Phenomena, Publ. No. 1484, 45, (Washington DC: NAS/NRC).
Sorapipatana, C., Smith, C.W. and Calderwood, J.H., J. Phys. D: Appl. Phys., 8 (to be published).
Thomas, W.R.L., 1974, Proc. 43rd Conf. Electrical Insulation and Dielectric Phenomena, (Washington D.C.: NAS/NRC) (To be published).

BEHAVIOUR OF A WATER DROP IN PRE-BREAKDOWN PROCESSES IN INSULATING OIL

J. Fleszyński*

Introduction

In an electric field of a strength higher than the critical one, a water drop immersed in a liquid dielectric becomes unstable. The critical field strength is dependent on the radius of the water globule and its surface tension as well as the permittivity of the medium (Garton and Krasucki, Taylor, Panchenkov and Cabek) .

This paper deals with the results of phenomenological studies of the effects of different factors on the course of dynamic processes taking place after the water drop has reached instability.

Method and scope of investigation

The tests were carried out in fresh transformer oit at a vertical electrode arrangement. Drops of distilled water introduced into the oil(surface tension of a drop in oil 0.025 N/m) were placed on the bottom plate electrode. The phenomena occurring in the electric field were registered with a high-speed Pentazet 16 film camera (300-3000 frames/s). The voltage of the breakdown of the oil was measured with an oscillograph.

The effects of the following factors were investigated:

a) rate of constant polarity voltage rise (1-50 kV/s and 1.100 kV/ms)

b) water drop diameter changed by decimal parts of millimeter (water drop diameter measuring accuracy 0.01 mm)

c) configuration of the electric field (electrodes: plane - plane, sphere - plane, point - plane)

Most investigations were carried out at a 5 mm electrode gap.

Results

Uniform electric field (electrodes plane - plane)

At low rates of voltage rise (1-50 kV/s) the drop introduced on to the bottom plate electrode becomes detached from it and drifts across the electrode gap. The charged water globule heads for the electrode of opposite polarity, "bounces" off it and drifts in the opposite direction. In further stages of the phenomenon, the water globule elongates in the direction of the field, assuming shapes approximating that of a prolate spheroid. A characteristic of the drop reaching instability is formation of sharp points at the ends of the ellipsoid where every now and then minute droplets are ejected. Under another unstable condition which is reached at a slightly higher field strength, the drop rapidly elongates. With the drop having an adequate initial radius longer than the critical one, a conductive water channel is formed bridging the electrodes. Fig. 1 shows the dependence of the critical radius, r_{cr} , of the water globule on the electrode spacing gap at three rates of voltage rise 1, 10 and 50 kV/s. At a

*Technical University of Wroclaw, pl. Grunwaldzki 13, Wrocklaw, Poland.

given electrode gap the r_{cr} value diminishes with increasing rate of voltage rise.

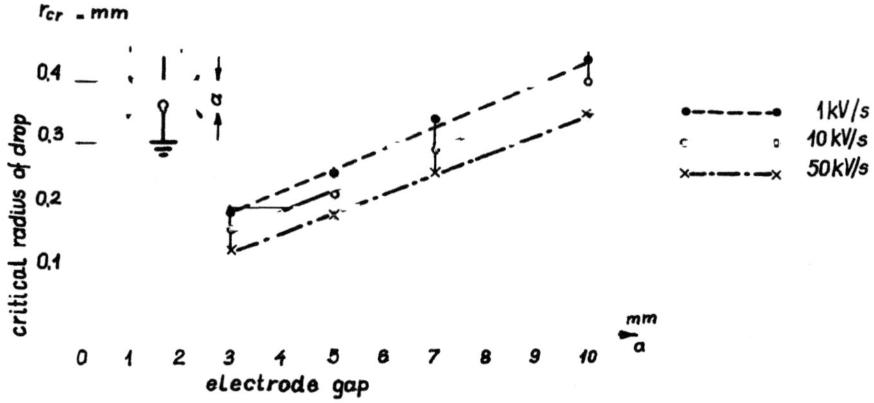

Fig. 1

Drops with radii shorter than r_{cr} do not bridge electrodes. The elongated water channel disintegrates and water-in-oil emulsion is formed which becomes more fine-grained with increasing field strength. The droplets of the water emulsion are dispersed in the oil and expelled from the electrode gap. At a rate of voltage rise of 1 kV/s the breakdown of oil usually occurred after the water emulsion had been completely expelled from the electrode gap. The breakdown voltage of oil is then relatively high -exceeding 50 kV (Fig. 2 - curve 1). At the investigated high rates of voltage rise (1-100 kV/ms), the breakdown voltage of oil was determined by the dynamics of elongation of the drop. The rate of elongation of a drop only slightly increases with increasing rate of voltage rise; it is of the order of 1mm/ms. Therefore the dependence of the breakdown voltage on the rate of voltage rise at 1-100 kV/ms

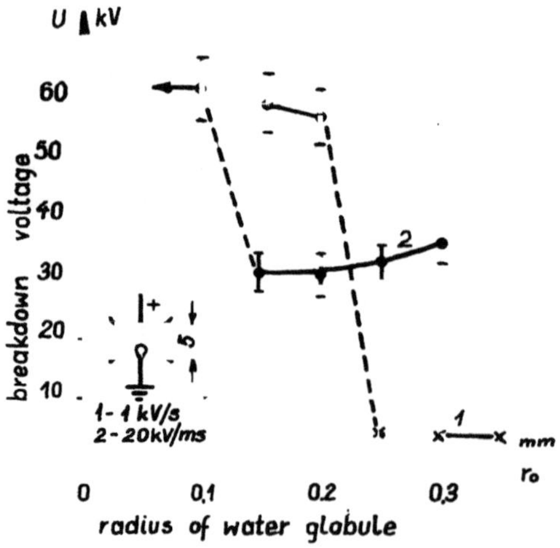

Fig. 2

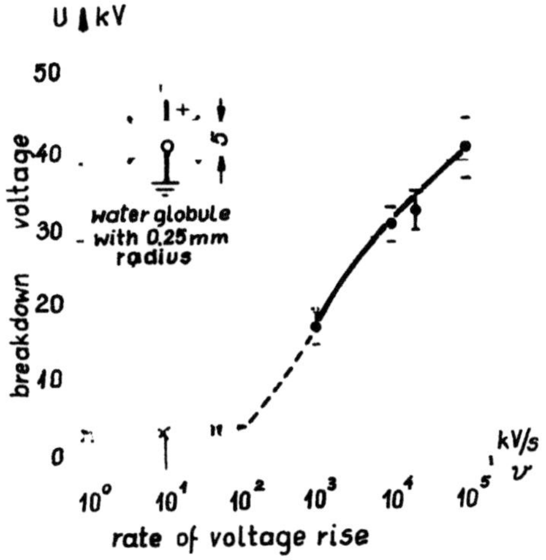

Fig. 3

× water channel bridging of electrodes
• breakdown of oil initiated by an elongated drop
o breakdown of oil after drop disintegration

is of a sharply rising character (Fig. 3). At high rates of voltage rise smaller droplets elongate more rapidly and can diminish the strength of oil in a greater degree than larger drops.

Elongating drops initiate electric discharges only when their initial radius is greater than their critical radius (Fig. 2 - curve 2). The critical radius of a water globule decreases as the rate of voltage rise increaes.

Noteworthy is the path of the electric discharge initiated by an elongated drop. At high rates of voltage rise, the discharge often by-passes it.

Nonuniform electric field

At voltage rise rates of from 1 kV/s to 50 kV/s, there occur in the oil whirlpools and turbulence. The area of strong whirlpools and turbulence at a point-plane electrode arrangement has the shape of a cone the top of which is at the terminal and of the point electrode. A water drop in this area caught in a swift rotational motion disintegrates without coming near the point electrode. Drops which are outside the area of strong whirls of oil move along the field line and once near the point electrode they elongate and disintegrate. As the voltage increases, the atomized droplets are expelled from the electrode gap. At the moment of the breakdown of the oil, no dispersion water is to be seen between the electrodes. At the investigated high rates of voltage rise (1-50 kV/ms), the behaviour of a water drop under an unstable condition was similar to that of one in a uniform field. At voltage actions of millisecond duration, there still occur no whirlpools in the oil, and the process of elongation of the drop continues

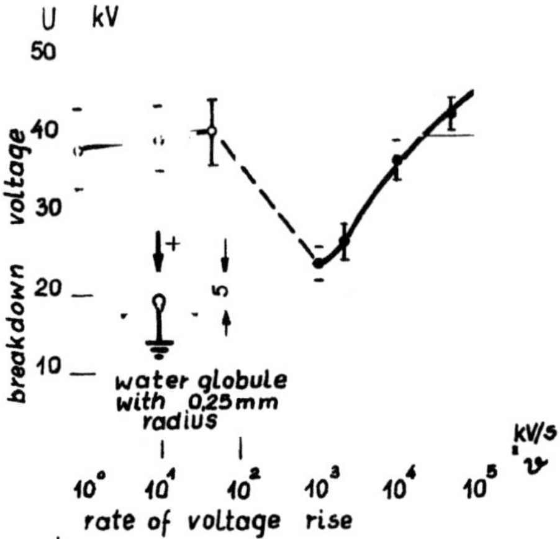

Fig. 4

• breakdown of oil initiated by an elongated drop
o breakdown of oil after drop disintegration

indisturbed. Just like in a uniform field, the breakdown voltage increases with increasing rate of voltage rise (Fig. 4). In a sphere - plane electrode arrangement, water drops were placed in different places with a different field line curvature. A water drop in a place with a strong curvature assumes on reaching instability complicated shapes and does not elongate to such a degree as in a uniform field. Shearing stresses soon bring about the disruption of the forming water channel.

Conclusions

The role and behaviour of a water drop in pre-breakdown processes in oil are dependent on the rate of voltage rise, the size of the drop and the electric field configuration. Depending on the above-mentioned factors, a water drop elongating under an unstable condition either initiates a discharge or disintegrates. In the latter case no water bridges are formed in the oil.

References

1. Garton, C.G., and Krasucki, Z., Proc.R.Soc.A, 280, p. 211, 1964
2. Panchenkov, G., and Cabek, L., Zh.Fiz.Khim., 24, p. 1249, 1968
3. Panchenkov, G., and Cabek, L., Zh.Fiz.Khim., 24, p. 2300, 1968
4. Panchenkov, G., and Cabek, L., Kolloid Z., 31, p. 704, 1969
5. Taylor, G., Proc.R.Soc.A, 280, p. 383, 1964

THE EFFECT OF SF_6 ON PARTIAL DISCHARGE IN TRANSFORMER OIL UNDER NON-UNIFORM FIELD

H. Yamashita*, H. Amano* and T. Mori*

Introduction

In the present paper the effect of SF_6 on the partial discharge in transformer oil under D.C. and A.C. field is explained. The effect of SF_6 on the partial discharge in the oil may be divided into two effects; the one the effect of pressure presented by gaseous SF_6 and the other the electrical negativity of the dissolved one. To make clear these effects, the effect of oil surface pressure and electrical negativity are investigated separately as follows; that is, pressure effect is replaced by Ar gas here and the electrical negativity effect is represented by dissolved iodine.

Experimental

The test cell, made of glass, has the capacity of 400 cc. The electrode system used is needle to plane, and the gap length is 1 mm. Test liquid is transformer oil and is filtered through glass sintered filter, and is introduced into the test cell. The upper gaseous phase of the cell is evacuated to the pressure lower than 10^{-3} torr. Then SF_6 or Ar gas is introduced into the cell to a appointed pressure. In the case of SF_6 let it alone until it becomes equilibriun state.

D.C. field

Fig. 1 shows the characteristics of starting voltage of partial discharge (Vco) and breakdown voltage (Vs) vs. enclosed pressure of Ar. Both Vco and Vs increase as increasing of the enclosed pressure of Ar and their increasing rates are almost same. The frequency of partial discharge pulse decreases and the charge amount of it increases as the pressure on the oil surface increases. From the result that the partial discharge has pressure dependency, it seems that the partial discharge is attributed to the discharge in tiny bubbles generated in the oil and the breakdown of the gap occures when the bubbles have grown up and bridged over the gap. These bubbles may be produced by electro-chemical dissociation of the oil in the high field region between the gap.

Fig. 2 shows the Vco and Vs characteristics of the oil containing iodine. In this case, there is no effect of pressure and only the effect of electrical negativity of iodine may presented. Both Vco and Vs increases as the contents of iodine increases, but, on the contrary, they decreases when the content exceeds a certain value. An addition of iodine seems to increase negative charges, which change the field intensity of the high field region between the

*Faculty of Enineering, Keio University,
832 Hiyoshi-cho, Kohoku-ku, Yokohama, Japan

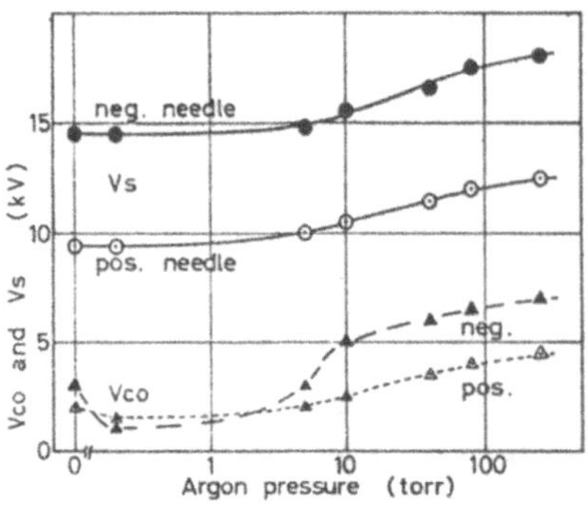

Fig.1 Characteristics of starting voltage of partial discharge (Vco) and break-down voltage (Vs) vs. enclosed pressure of Ar.

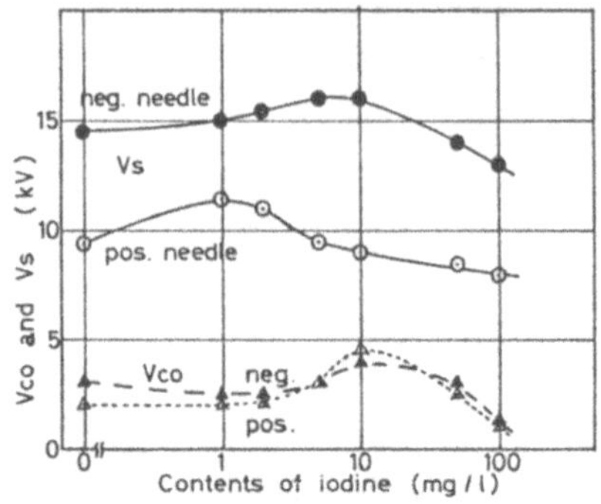

Fig.2 Characteristics of starting voltage of partial discharge (Vco) and break-down voltage (Vs) vs. contents of iodine.

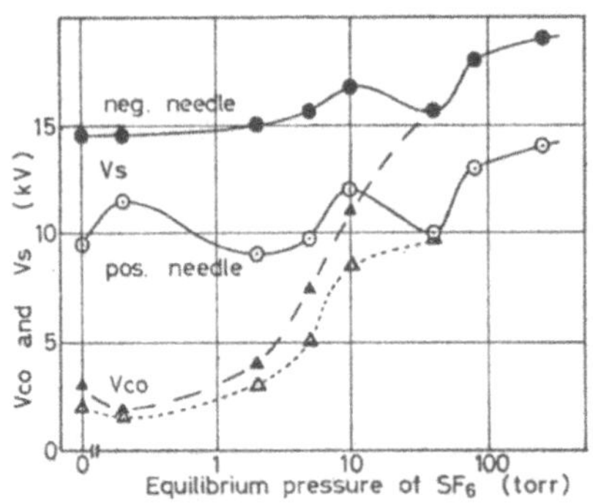

Fig.3 Characteristics of starting voltage of partial discharge (Vco) and break-down voltage (Vs) vs. equilibrium pressure of SF_6.

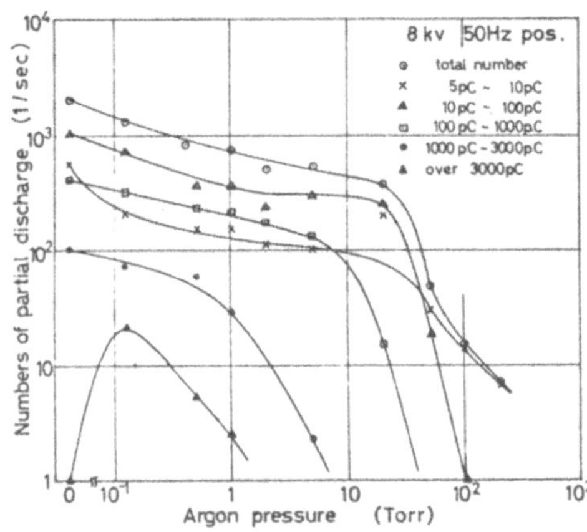

Fig.4 Characteristics of the frequency of partial discharge vs. enclosed pressure of Ar.

electrodes where the partial discharge mainly occurs. In this case the frequency of partial discharge changes, but the charge amount of a pulse does not change at all, and that seems to support the consideration described above.

Fig. 3 shows the Vco and Vs characteristics of the oil dissolved SF_6. The increasing of equilibrium pressure indicates not only increasing oil surface pressure but dissolving quantity. In this case Vco increases rapidly as increasing the equilibrium pressure of SF_6, and coincides with Vs at the equilibrium pressure over 40 torr; that is, the dissolved SF_6 arrests partial discharge and breakdown of the gap occurs directly without any partial dis-charges. As the dissolved SF_6 may also involved in bubbles gener-

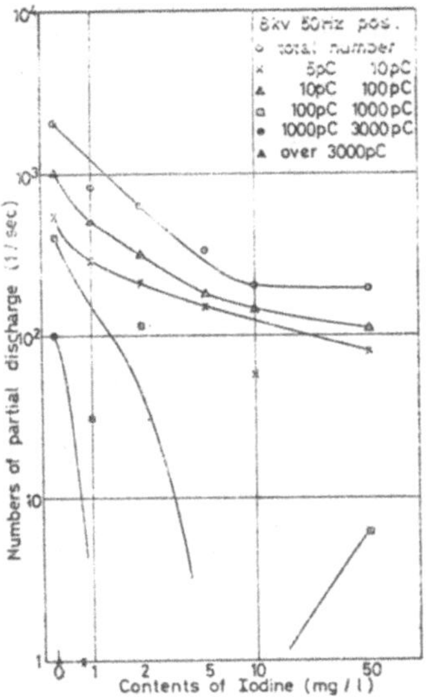

Fig.5 Characteristics of the frequency of partial discharge vs. contents of iodine.

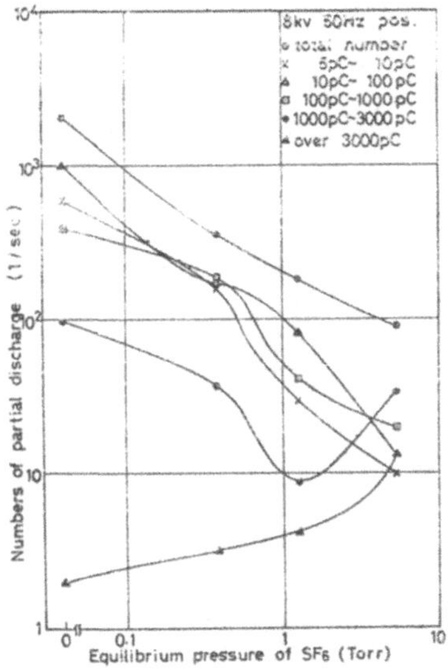

Fig.6 Characteristics of the frequency of partial discharge vs. equilibrium pressure of SF_6. No partial discharge can be detected over 10 torr.

ated in the oil, it can be considered that the breakdown strength of bubble itself increases as the equilibrium pressure increases. This may be supported by the fact that the charge amount of a partial discharge pulse increases rapidly as the equilibrium pressure of SF_6 increases and suddenly the partial discharge disappears in the oil containing SF_6.

A.C. field

Fig. 4 shows the characteristics of the frequency of partial discharge vs. enclosed pressure of Ar. The frequency of partial discharge decreases as increasing of the enclosed pressure of Ar, and it shows the similar tendency to the result under D.C. field. The charge amount of partial discharge exists from smaller one to larger one, and at first larger parts disappear as increasing of the enclosed pressure of Ar. It shows different tendency from the result under D.C. field which shows that at first the smaller charge amount of partial discharge and then the larger one disappear as increasing of the enclosed pressure of Ar. Under A.C. field, the generating and growing process of the bubble is different from the case under D.C. field, so it seems to be discharged without growing up.

Fig. 5 shows the characteristics of partial discharge with containing iodine. As increasing the contents of iodine, the frequency of partial discharge decreases and the charge amount of it decreases, but, on the contrary, the latter increases when it exceeds a certain value. This result is different from that under

D.C. field, and the effect of dissolved iodine under A.C. field is more distinguished than that of under D.C. field. In the case of D.C. field, the effect of iodine seems to affect the inner field as negative space charge, but in the case of A.C. field, the effect of that seems to affect the suppression of partial discharge by electron attachment itself.

Fig. 6 shows the characteristics of partial discharge in the oil dissolved SF_6. As increasing the equilibrium pressure of SF_6 the frequency of partial discharge decreases and the charge amount of larger one increases. And the partial discharge disappears when the equilibrium pressure of SF_6 exceeds 10 torr. In the oil, dissolved SF_6 arrests partial discharge and breakdown occures directly under both D.C. and A.C. fields. The effect of SF_6 on the partial discharge in the oil may be divived into two effects; the one the wffect of pressure presented by gaseous SF_6 and the other the electrical negativity of the dissolved one. So, these two effects of SF_6 may be replaced by Ar gas as gaseous effect and dissolved iodine as electrical negativity. The result in the case of the oil which contains iodine and is compressed by Ar (I_2+Ar) shows that partial discharge disappears when the dissolving quantity of iodine is 50 mg/l and enclosed pressure of Ar is 200 torr. This difference of the effects between SF_6 and I_2+Ar seems to be as follows; SF_6 is a gas and may involve within tiny bubbles generated in the oil, while iodine is a solid and is little contained in the bubble, so the breakdown strength of bubble in the oil containing SF_6 may increase.

Conclusion

The partial discharge in highly degassed transformer oil with dissolved SF_6 is strongly arrested under both D.C. and A.C. fields. Breakdown occurs directly at relatively low equilibrium pressure of SF_6 without any partial discharge.
The partial discharge in transformer oil seems to discharge of tiny bubbles generated by electro-chemical dissociation of transformer oil molecule in high field region of transformer oil.

References

1. A. Nosseir and R. Hawley; Brit. J. Appl. Phys., 1965, vol.16
2. A. Nosseir and I. Megahed; J. Phys. D: Appl. Phys., 1970, vol.3
3. I. Megahed and H. Tropper; J. Phys. D: Appl. Phys., 1971, vol.4
4. E. J. Murphey; J. Phys. D: Appl. Phys., 1970, vol.3
5. H. Yamashita and T. Mori; Trans. IEEJ, 1973, vol.93-A, No.9

BREAKDOWN TIME LAG AND TIME OF FLIGHT MEASUREMENT IN LIQUID DIELECTRICS

K. Arii*, K. Hayashi**, I. Kitani* and Y. Inuishi**

Introduction

Time lag of breakdown provides important knowledge on the mechanism of dielectric breakdown of liquids[1),2)]. Generally observed time lag can be divided into two parts, formative and statistical time lags. From formative time lag t_f, the time required for the nucleation and propagation of breakdown will be obtained. In some cases, the mobility of the high energy electron which causes breakdown can be determined. Statistical time lag t_s, on the other hand, indicates the mechanism of the initial electron supply and breakdown probality. In this paper, time lag measurement in nanosecond range at higher overvoltages (1.5~3 MV/cm) on silicone oil and liquid argon will be presented in connection with the mechanism of breakdown. If situation allows the comparison with time of flight measurement[3)] of carriers will be made.

Experimental set up

Fletcher type single shot square[1),4)] wave pulse generator with 100 ns duration and 1 ns rise time was constructed with coaxial delay line pulse forming network, pressurerized triggered gap, capacitive divider and sample holder. Needle to plane (divergent field) and sphere to plane (nearly uniform field) electrode systems were used. Commercial sewing needles (tip radius a few μm) and steel ball with 2 mm diameter, and phosphorus bronze disk were used. The electrode separation was adjustable to 1 μm by micrometer head directly connected to the upper electrode. Sample liquids were (i) dimethyl silicone oil of commercial grade with viscosity of 10^2, 10^3 and 10^4 cs respectively, and (ii) liquid argon. These liquids were poured into electrode vessel, being followed by the evacuation to 10^{-4} torr for degassing. For each condition, about 20 single shot pulses were applied by triggered gap. Sample liquids were exchanged for each shot in order to avoid the effect of contamination due to previous breakdown. Observed time lag t were obtained by Tektronix 7904 oscilloscope for each shot. Electrodes were exchanged or polished after each set of experiments to avoid the effect of the damage.

Experimental Results and Discussion

1. Silicone oil

(a) Nearly uniform field (sphere to plane electrodes). Fig.1 shows Laue plot of the observed time lag t vs. percentage of not breakdown

*Faculty of Engineering, Ehime University, Japan ; **Faculty of Engineering, Osaka University, Japan

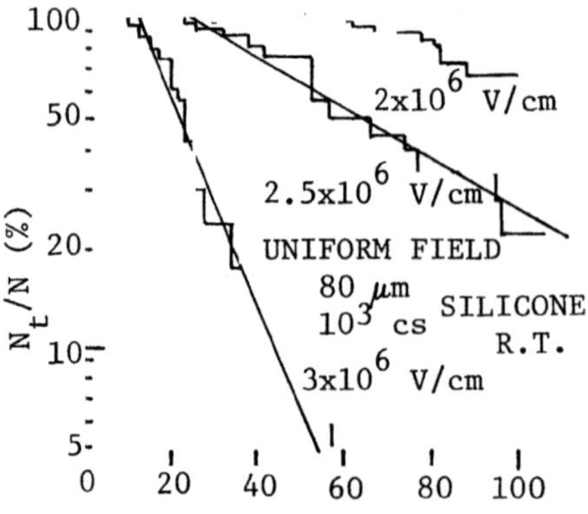

Fig.1 TIME LAG (nanosec) t

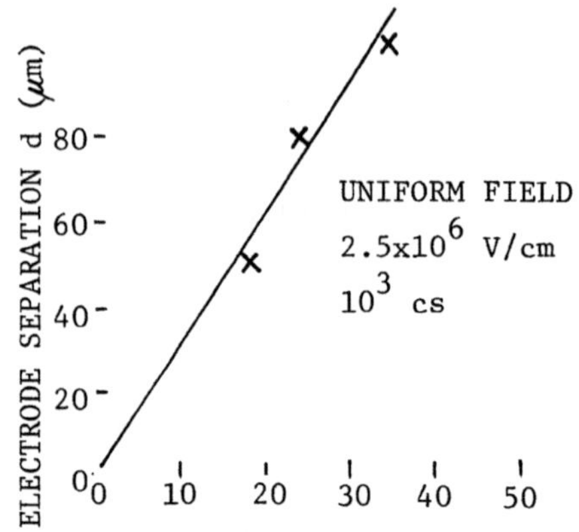

Fig.2 FORMATIVE TIME t_f (ns)

TABLE 1

viscosity (cs)	thickness (μm)	field strength (10^6 V/cm)	formative time (nanosec)	mobility ($cm^2/V\cdot sec$)
		2.0	33.6	0.12
	80	2.5	23.2	0.14
10^3		3.0	14.0	0.19
	50	2.5	17.7	0.11
	100	2.5	34.7	0.12
		2.5	58.8	0.054
10^4	80	3.0	35.2	0.076
		3.5	18.2	0.13

untill time t or N_t/N at various fields in 10^3 cs silicone oil with 80 μm gap length. As well known the experimental relation (Laue)

$$N_t/N = 100 \exp\{-(t-t_f)/t_s\} \quad (1)$$

holds where t_f and t_s are formative and statistical time lags respectively. As shown in Fig.2 formative time lag t_f from equation (1) increases nearly in proportion to the electrode distance d under the same field, indicating the propagation time of breakdown across the gap. Propagation velocity of the breakdown d/t_f thus obtained increases with increasing field as shown in Fig.3. Assuming the single avalanche mechanism of breakdown, the mobility of the free electron responsible to breakdown can be estimated as table 1 from the relation $t_f = d/\mu E$. It should be noted that apparent mobility thus obtained increases with increasing field as shown by dotted line in Fig.3. The magnitude of the mobilities 0.05~0.2 $cm^2/V\cdot sec$ is the range of the free electron mobilities in liquids obtained by Schmidt[7], suggesting the importance of avalanche and or[8] streamer mechanism of the breakdown at these higher overvoltages (1.5~3.5 MV/cm). This, however, does not deny at all the validity of the bubble mechanism[5] of breakdown at lower overvoltages (<1.5 MV/cm), since much longer time lags (ca. 100 times) were observed in these ranges[6]. Statistical time lag t_s decreases much faster than t_f with increasing field as shown in Fig.4. It should be noted that t_f increase to some extent with increasing viscosity as shown in Fig.3. Although the nature of this effect is unknown, next mechanism can be speculated (i) suppression of electron inject-

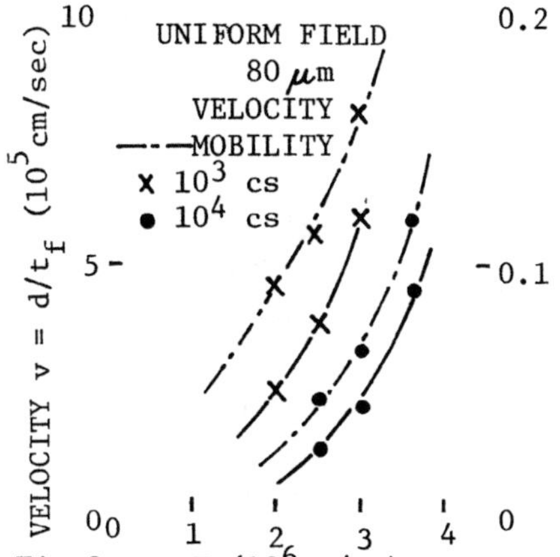

Fig.3 E (10^6 V/cm)

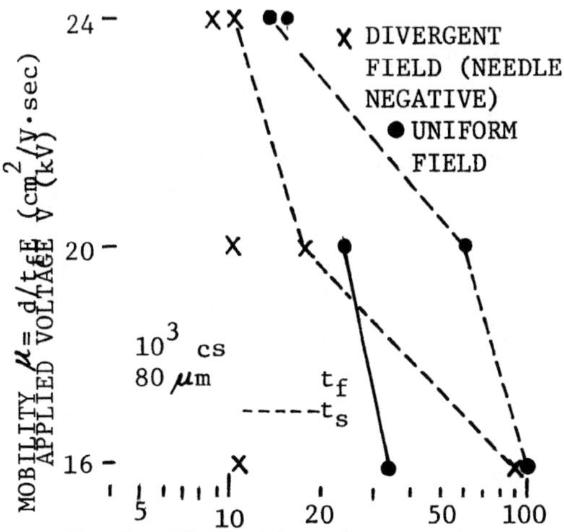

Fig.4 TIME LAG t (nanosec)

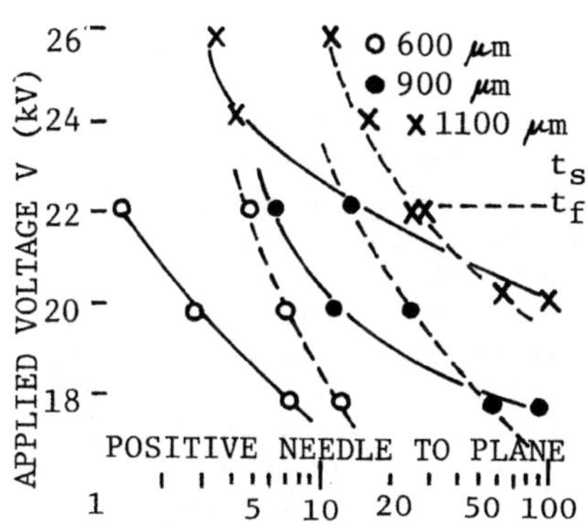

Fig.5 TIME LAG t_s, t_f (ns)

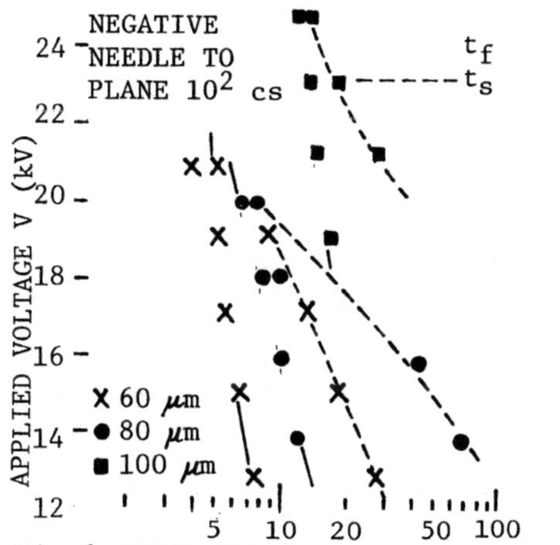

Fig.6 TIME LAG (nanosec) t

ion from cathode asperities by the ionic space charge of trapped electrons, (ii) cathodic field enhancement due to small displacement of positive ions produced by the avalanche, (iii) increase in energy required for bubble formation and propagation at higher viscosity.

(b) Divergent field (needle to plane). Both t_f, t_s in the case of divergent field are smaller than in uniform field under the same applied voltage and gap length as shown in Fig.4. The remarkable polarity effect exists in the relation between t_f, t_s vs. applied voltage as seen in Figs.5 and 6. Namely t_f and t_s in positive needle are much smaller than those in negative needle under the same condition. Average propagation velocity obtained from t_f in positive needle is ca. ten times faster than that in negative needle and in uniform field under the same average applied field as shown in Fig.7. The propagation velocity in positive needle is nearly independent

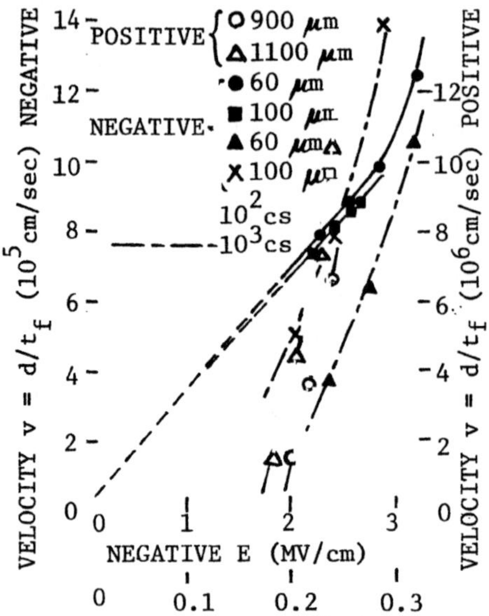

Fig.7

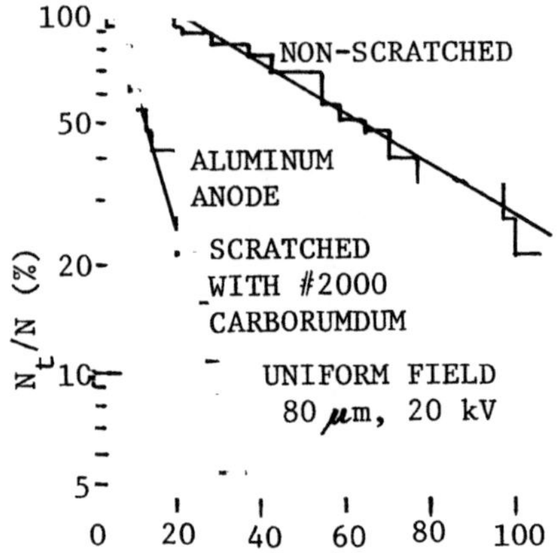

Fig.8

of gap length under the same average field, being insensitive to the viscosity. Propagation velocity in negative needle is much similar to the case of the uniform field under the same applied field, being sensitive to the viscosity. t_s in positive needle decreases much drastically than that in negative needle with increasing applied voltage as shown in Fig.5. These experimental facts suggest that the breakdown in positive needle is established by the positive streamer propagation from positive needle to plane electrode with the velocity of ca. 10^6 cm/sec and that breakdown in negative needle is much similar to the case of uniform field possibly due to cathodic field (needle) shielding by the space charge of trapped electron injected from cathode. In negative needle and uniform field the breakdown may require the propagation of electron avalanche as the prelude to the positive streamer[8]. These speculations were also supported by the effect of mechanical scratching of the anode. As shown in Fig.8, the polishing of aluminum anode with carborundum powder #2000 decreases t_s and t_f very effectively.

2. Liquid argon

Time lags in liquid argon were also measured. Formative time lag gives rather saturated electron velocity around sound velocity (10^5 cm/sec) at higher overvoltages (1~3 MV/cm) in spite of the much higher mobility in the low field region. Time of flight measurement by electron pulse beam gives $\mu = 40$ $cm^2/V\cdot sec$ possibly due to impurities. From Hecht curve the life time of free electron was estimated to be ca. 40 μsec. However, since the experiments on liquid argon are in still premature stage in comparison with the case of silicone oil at this moment, nothing conclusive can be said.

References

1) K. Arii, I. Kitani and Y. Inuishi : I.E.E.Japan, 93-A, No.8, 313 (1973)
2) P. Felsenthal : J.Appl.Phys., 37, 3713 (1966)
3) K. Hayashi, K. Yoshino and Y. Inuishi : Japan.J.Appl.Phys., 14, No.1 39 (1975)
4) R.C. Fletcher ; Phys.Rev., 76, 1501 (1949)
5) A.H. Sharbaugh and P.K. Watson ; Nature 184, 2006 (1959)
6) R.W. Crowe ; J.Appl.Phys., 27, 156 (1956)
7) W.F. Schmidt ; "ELECTRON MIGRATION IN LIQUIDS AND GLASSES (A REVIEW)"(1974)
8) W.R.L. Thomas and E.O. Forster ; 1973 Annual Report, Conf. Elect. Insulation and Dielectric Phenomena, NAS, Washington, D.C. (1974) ; ibid. (1975)

INVESTIGATION OF TIME-LAG CHARACTERISTICS FOR A TRIGGERED SPARK GAP IN A LIQUID DIELECTRIC

J.L.Maksiejewski and J.H.Calderwood*

1. Introduction

In a previous publication (1) we have given a brief account of an investigation into the voltage characteristics of a trigatron spark gap in hexane. We found that triggered breakdowns occurred at main gap voltages which were less than half the normal untriggered breakdown voltage. However, no information was given concerning the time characteristics of the trigatron, and we are now supplementing the results of the previous work by reporting such additional data.

2. Experimental Arrangements

In extending the study of the liquid trigatron, we used the same type of trigatron as in our previous work. The trigatron tested consisted of a spherical and a hemispherical electrode, each of ¾ inch (19.05 mm) diameter and made of stainless steel. A 0.035 inch (0.89 mm) diameter tungsten rod mounted flush with the main electrode formed the trigger electrode, the trigger discharge gap being approximately 0.005 inch (0.127 mm). The trigatron was mounted with its axis vertical in the test cell, which was filled with n-hexane of commercial grade. The main gap settings were measured by means of a micrometer head incorporated in the test cell. The trigatron characteristics were determined with direct voltage of both polarities. The supply unit had a maximum voltage of 100 kV. The circuit arrangements are shown in Fig 1. The hemispherical electrode was permanently earthed, and the main gap voltage V was measured by means of a microammeter in series with a high-voltage resistor.

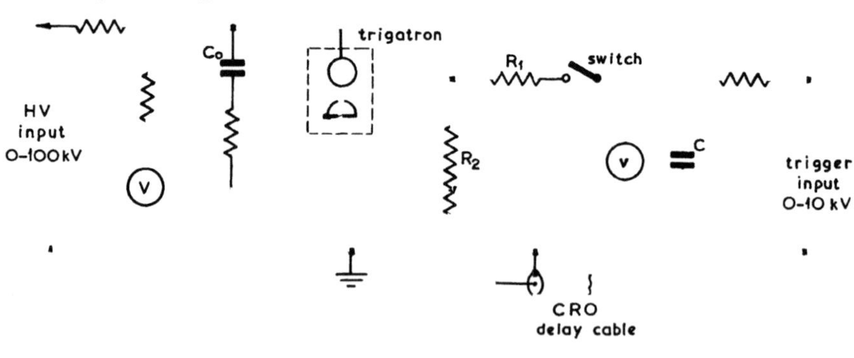

Fig. 1 Circuit arrangement. $C = 0.04\mu F$, $R_1 = 0$, $R_2 = 1 M\Omega$

* Department of Electrical Engineering, University of Salford, Salford, M5 4WT

The trigger-gap discharge was initiated by a voltage impulse of positive or negative polarity from a separate source consisting of a capacitor C charged from a d.c. supply having a maximum output of 10 kV. The voltage v on C could be applied to the trigger electrode by closing a remotely controlled switch. Time-lag, measured as the time interval between the start of the triggering discharge and the beginning of the main trigatron gap discharge, was recorded by an oscilloscope and an associated camera.
The trigger spark was recorded on the oscilloscope by means of a resistive voltage divider. The breakdown of the main gap was also recorded through the same delay circuit utilizing the sudden collapse of voltage V. The collapse of V was recorded by means of the small high-voltage capacitor C_0 and a resistive voltage divider having a common low-voltage arm with the first voltage divider. An aerial was used for tripping the time sweep of the oscilloscope.

3. Experimental Results

Curves were obtained relating the time-lag T with voltage V for a constant trigatron gap spacing, for various polarity combinations. The results for a gap length of 2 mm with opposite polarities of main and triggering gaps are given in Fig 2. We have found· previously that for such polarity combinations the greatest reduction in the breakdown voltage of the main gap occurs. The main gap voltage V in Fig 2 is expressed as a percentage of the

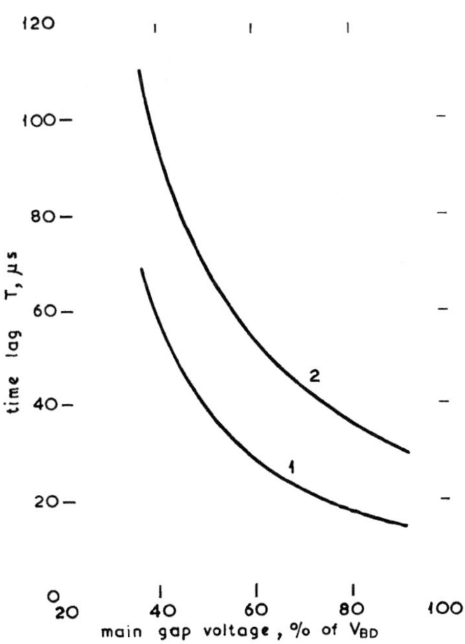

Fig. 2 Time-lag characteristics for the triggered spark gap in n-hexane. Gap length = 2mm.

curve 1 : V-negative , v-positive
curve 2 : V-positive , v-negative

d.c. breakdown voltage V_{BD} for the trigatron spark gap in the absence of a trigger pulse. For each point plotted, 25 time-lags were recorded. The triggering pulses had a rise time of about 0.05 µs. A value of trigger voltage was selected which ensured consistent breakdown of the trigger gap and provided a time delay for the trigger gap of negligible duration. The trigger gap in n-hexane broke down at 3.1 kV; tests were made with a 5kV trigger pulse.

Fig 2 shows that T was dependent both on the polarity and magnitude of the main gap voltage V. With this polarity negative, the time-lags were much shorter. For both polarities of main gap voltage, a rapid increase in T took place as V was decreased.

4. Conclusions

The results show that much shorter time-lags correspond to the case when polarity of the main gap voltage is negative than when it is positive. This seems especially significant, as in our previous investigations we found a greater reduction in the breakdown voltage of the main gap with negative than with positive voltage. Work is in progress which it is hoped will elucidate the mechanism of the breakdown processes in the liquid dielectric triggered spark gap.

(1) J.L.Maksiejewski and J.H.Calderwood, A triggered spark gap in liquid dielectrics, Nature, 220, 5170, 905, (1968).

CATHODE FALL LIKE DISCHARGE IN DIELECTRIC LIQUIDS

H. Kishida, T. Sato and Y. Toriyama*

Introduction

Studies of the breakdown mechanism in liquids have been performed by many investigators. Recently, new experimental apparatus, i.e., the laser-schlieren method[1], the gas bubble chamber[2] and the Kerr effect[3] have been used in this field. These above methods are available to get the spatial development of discharge. However it is very difficult to get the charge distribution between the electrodes just a partial discharge has been brought about. The Dust-figure method[4],[5] has a merit to distinguish the polarity and distribution of charges by discharge. It is important to know a fine structure of the discharge channel on studying the breakdown mechanism in liquid.

The present authors have been studying the surface discharge in gases by using the Dust-figure method. The breakdown in inhomogeneous fields becomes polarity dependent. Positive and negative discharge figures reveal in detail these polarity effect. The negative point electrode ejects electrons. As the impact ionization proceeds, the electron avalanches leave the positive space charges behind, and they steepen the field in front of the cathode. This is called cathode fall[6]. Observing the charges deposited on the insulating surface, it was found that the positive charge produced by back discharge (reverse discharge) existed near the cathode. This phenomenon means that there are negative charges around the cathode. Toriyama[7] suggested that the steep field in the cathode fall gives rise to the secondary electron emission, consequently the avalanche in which electrons are copious occurs along the previous positive column, and the discharge channel is filled with a high conducting mixture of electrons and ions. Recently, this phenomena have been ensured by Murooka[8], who employed the Dust-figure method by the nsec pulse technique.

The aim of this investigation is to examine the possibility of a cathod fall like discharge in liquids. The mechanism of the discharge depends on the level and the waveform of applied voltage. A sharply defined discharge figures in liquids are obtained by applying electrophotographic toners. The light pulses emitted from the point electrode has also been observed by using a photomultiplier.

Experimental procedure

The experimental apparatus is shown in Figure 1. The test cell made of perspex is enclosed in the dark box. A resin plate is inserted between a plane electrode and a point electrode to which an impulse voltage is applied. Electrode spacing can be controlled by

*Department of Electrical Engineering, Musashi Institute of Technology, Tamagawa, Tokyo, Japan

micrometer and the gap can be viewed from opposite sides through quartz windows. For the light pulse studies, a EMI 6256B photomultiplier tube is used with a peak response between 200nm and 400nm. The overall sensitivity is about 100 A/L.
The impulse generator can produce 400/3300 μsec impulse wave, that is available for the phenomena to take place. The corona current and light pulses are recorded on a double-beam oscilloscope. Each test is performed on degassed transformer oil samples. The oil is passed through a 1 μm porosity sintered glass filter being admitted to the test cell. The image of deposit charges on an insulating plate is developed by the following procedure; the positive toner is used to indicate the negativecharge and the negative one to the positive, respectively. An electrified plate is immersed in the liquid developer with a toner. After then, the plate is softly washed in petrol to remove surplus toner.

Results

Figure 2 shows the relation between applied voltage and size of surface discharge when the switching impulses are supplied. The size is the diameter of dust figure. This graph represents that the extension of negative discharge is larger than that of positive, under high stress conditions. On the other hand, when the standard impulse waves are applied, the growth of negative discharge is smaller than that of the positive over the whole range of applied voltages. On applying a negative switching impulse, tree-like sites are spread out over the insulating surface, which are the results of negative discharges having occurred on the front of the voltage. When the voltage is high, copious negative charges are superimposed on some stems of the channels. These are due to the secondary discharge process having taken place at the cathode. On the contrary, there is no negative charge in the stem of a negative site obtained by the standard impulse voltage.
Figure 3 shows the relationship between applied voltage and the charge quantity of surface discharge. In this case, it can be seen that the charge quantity of negative discnarge is more than that of the positive. This graph shows the same characteristics as Figure 2. Therefore, the charge quantity of surface discharges corresponds to the growth of them.
Figure 4 shows typical examples of light outputs from the tip of the point electrode. With a negative point, Trichel like light pulses appear about 5-10 μsec after the first pulse, of which the rise time is approximately 2-5 μsec. THe magnitudes and the frequency of occurrence increase with applied stress. With a positive point, these pulses do not appear.

Discussion and Conclusion

According to the study of Lichtenberg figure in air[9], it was turned out that a Lichtenberg figure displayed on a photographic film represents the light emitted by re-combination of positive ions. The whole character of discharge,therefore,can not be observed by a Lichtenberg figure. A dust figure developed on a dielectric surface represents a sharply defined structure of discharge, and is similar to a photograph taken by the schlieren method[1],

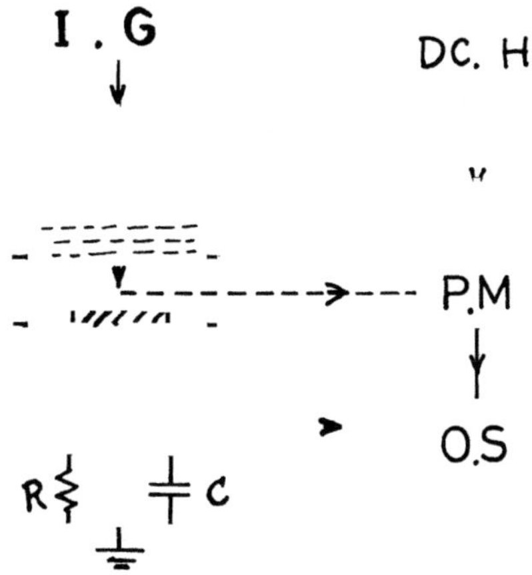

Fig. 1 Schematic diagram of test apparatus.

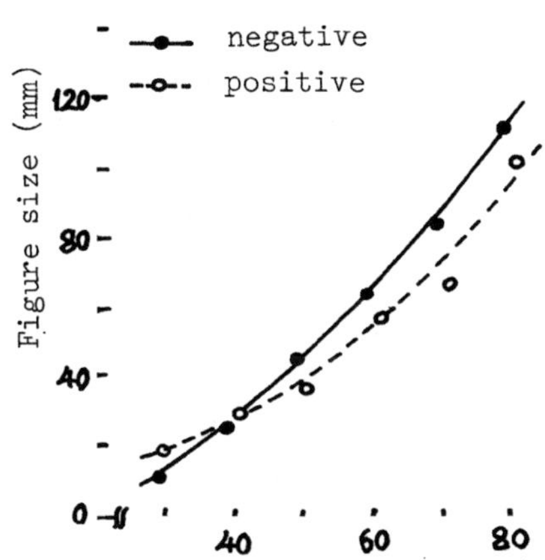

Fig. 2 Extension of surface discharge.

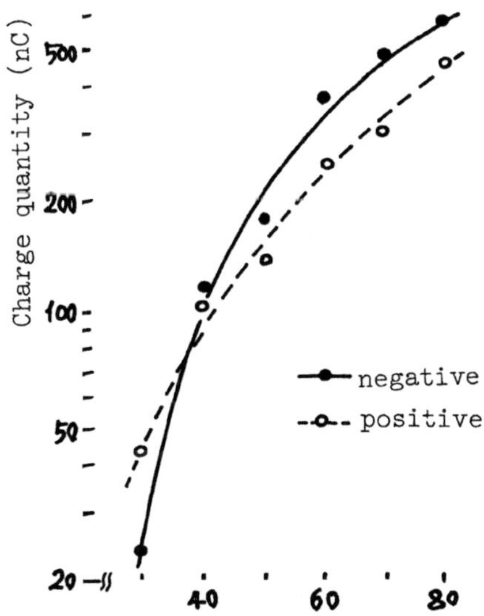

Fig. 3 Charge quantity of surface discharge.

(a) negative point -50kv

(b) positive point +50kv

(Gap. 1.5mm , 10 μsec/div)

Fig. 4 Typical records of light pulses in transformer oil.

[10]. From the point of view on resolution power, the grain size of the toner gives the limit.
The results shows that there is a clear polarity effect on the shape and the size of figures. The case of positive point, since the positive space charges intensify the field at the tip of discharge, the ionization occurs at the tip only of the streamer. The case of negative point, when a partial discharge initiates from avalanches, if the positive charge density near the negative point increases, then the field strength at the cathode is made strong. Consequently, the electron emission can be occurred, then the secondary discharge may be produced. The results can be interpreted to show that the time lag from the primary discharge to the occurrence of the cathode fall like discharge is about 5-10 μsec. This time lag is about 50 nsec in air[8]. The difference between liquid and gas is due to the mobility and the distribution of charges.
In conclusion, one may say that the negative discharge in liquids submitted to high and lasting stresses is intensified by secondary emission processes due to the strong field by the cathode fall.

Acknowledgments

Thanks are to Matsunaga Science Foundation. The authors also wish to thank Y. Yoshida and S. Nogami for their help in carrying out the experiments.

Reference

[1] Morikawa E 1972 IEE. Jap. 92-A 27
[2] Murooka Y and Toriyama Y 1974 J. Appl. Phys. 45 4 1707
[3] Riddiford C L and Jerrerd H G 1970 J.Phys. D 3 1314
[4] Toriyama Y 1931 Phys. Rev. 37 619
[5] Kishida H, Sato T and Toriyama Y 1972 IEE. Jap. 92-A 246
[6] Merrill F H and Von Hippel A 1937 J. Appl. Phys. 10 873
[7] Toriyama Y 1967 IEE. Jap. 87 1626
[8] Koyama S and Murooka Y 1973 IEE. Jap. 93-A 375
[9] Sone M, Toriyama K, Hayakawa K and Toriyama Y 1974 IEE. Jap. 94-A 523
[10] Fießig R 1966 ELEKTRIE 3 99

MEMORY EFFECT IN DIELECTRIC LIQUIDS BEING SUBJECTED TO RECURRING DISCHARGES

B. Bommeli, C. Frei, M. Peter*

Introduction

It is a well known fact that the breakdown strength of a dielectric liquid is lowered when the liquid has been exposed to electrical discharges beforehand. When waiting a while after discharges had occured the dielectric liquid will recover and eventually restore to its original strength, or nearly so. This aspect of breakdown behaviour merits, in our view, close examination, since it is likely to yield new insight into the mechanism of electrical breakdown in liquids. In a previous attempt we had investigated the lowering of the breakdown strength as a function of the time interval between recurrent discharges (ref. 1). The results indicated a decrease of the mean electrical field for discharge ignition being approximately proportional to the logarithm of the frequency of repetition of the discharges. Such a dependence was shown to be in agreement with an avalanche breakdown model when assuming a multiplication of the available charge carriers by a conditioning effect of the dielectric liquid. The conditioning would come via the contamination produced by the preceding sparks, the nature of the agent, e.g. trapping states, being hypothetical. Further work showed some evidence that the observed memory effect of the dielectric has a limited duration. Since these experiments were performed with large electrodes where the discharges could

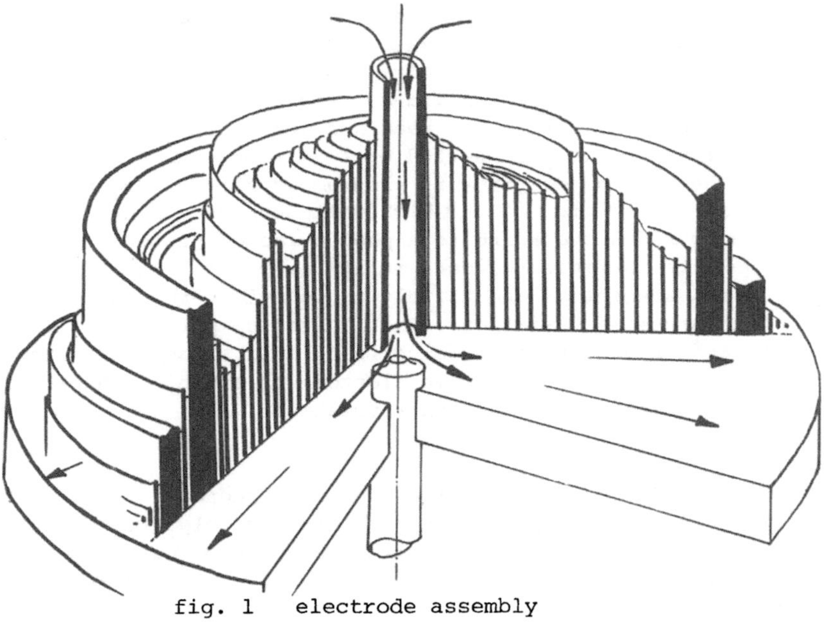

fig. 1 electrode assembly

* Département de Physique de la Matière Condensée, Université de Genève, CH - 1211 Geneva 4.

occur anywhere over the whole electrode surface (ref. 2), a measure of the degree of contamination could not be achieved.
In order to have a better control of the conditioning of the liquid we devised a new geometry, whereby the liquid is first treated by repetitive sparks and immediately thereafter subjected to a measurement of its strength. Our aim is to establish the relationship between the mean breakdown strength $\overline{E}_B$ and the concentration of the supposed contaminating agent c_C. Further, this geometry should enable us to detect lifetime effects and to recognise an influence of the electrode material.

Experiment

The experimental set up (fig. 1) consists of an electrode system having rotational symmetry. The electrodes on the centre axis can be moved independently from the outer electrodes. One of them is a capillary through which the dielectric liquid, a paraffin oil, is introduced. In the gap between these electrodes sparks are generated at the desired repetition frequency by an electrical discharge machine (Charmilles D10). Its servo control actuates the centre electrodes. The contaminated liquid moves then outward, staying within the gap formed by the outer electrodes. The gap is adjusted to a few tens of micrometers and its value measured by means of a inductive transducer. One of these electrodes is composed of annular sections which are electrically insulated between each other. The radii of the concentric electrodes increase near the centre in steps of 0,5 mm, further outword thicker sections are included. Within this part of the electrode system the dielectric liquid is tested and the radial or time of propagation dependent characteristics measured. To this purpose a high voltage high current impulse generator is connected consecutively to the 30 annular sections. The voltage peaks of the rapidly quenched test discharges are fed via an attenuator to a pulse stretcher from which they are read and memorised by a minicomputer, after having been converted by an analogue to digital converter. The on-line computer is programmed firstly to operate the sequential application of the high voltage high current pluses to the electrodes according to a pseudo random pattern, secondly to accumulate the measured data in its memory - a run is made up of 3000 measurements - and thirdly to calculate the mean values and their standard deviations. Finally the computer plots the results as a function of the electrode radii either on a display screen or by means of a graphic plotter. In addition the results may be stored on tape for further treatment.

Results

A typical computer plot of a measurement sequence is shown on fig. 2. A steep rise appears on the immermost electrode sections. Then a voltage maximum occurs followed by a slow but steady decrease with increasing radii. The maximum shows up invariably at the third section, or next to it, and no shift occurs when varying the propagation velocity of the dielectric liquid by a factor of 20. On the other hand, a change of pressure at which the

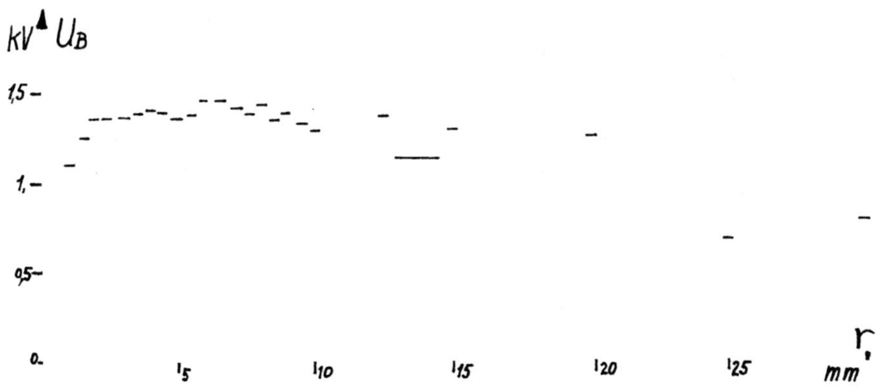

fig. 2 example of breakdown voltage measurement as function of electrode radius, outer electrode gap : 20 µm, centre electrodes : copper positive, steel negative, arc duration : 6 µs, average current : 3A, peak current : 8A, flow rate : $1{,}4 \cdot 10^{-9}$ m^3/s.

paraffin oil is injected, or a modification of the parameters of the sparks at the centre electrode (length, current, repetition frequency) influences the voltage level of the maximum.
When the conditioning of the liquid at the centre electrode is stopped, one may still notice a decrease of the breakdown strength at the outer electrodes, though at a higher voltage level. This effect is dependent upon the rate of the dielectric strength measurements and points therefore to a non negligible influence of the test discharges on the liquid.

Discussion

Since we are unable, for the time being, either to avoid experimentally the instrumental decrease of the breakdown voltage at large radii or to take it into account by a correction term, we have to renounce at drawing a conclusion about the life time of the contamination. It might look as if the breakdown voltage

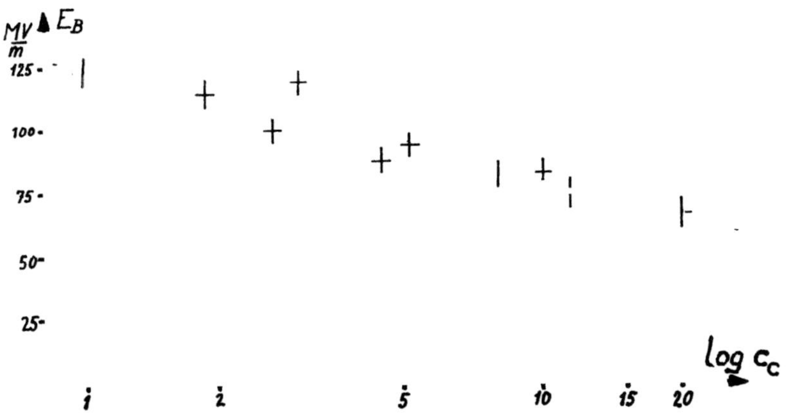

fig. 3 breakdown strength of paraffin oil (Chevron CCL51) as function of concentration of contamination c_c in relative units.

increase in the immediate vicinity of the centre would be related to a decay time constant. However, the independence of the region of increase on the flow rate is incompatible with whatever reasonable assumption of a life time function might be put forward. So one is led either to admit a short-range interaction in order to explain the lowering of breakdown strength when approaching the centre electrode (e.g. due to residual ionization, light emission, shock waves, etc.) or to impute the fact to an incomplete mixing of the contaminating agent with the liquid in the immediate neighborhood of its production.

However, it is possible to extract valuble information from our data : the dependence of breakdown strength on the concentration of contamination. Although not known in absolute units a relative concentration may be assessed. It should be proportional to the repetition frequency of the discharges which condition the dielectric liquid and inversily proportional to the flow rate.

In order to eliminate the dependence of the concentration on the presumed decay the breakdown field strength would have to be extrapolated to $r = 0$. Since any decay constant appears to be long with respect to our time scale (some tenths of a second) we neglect the small difference and refer to the maximum voltage of $U_B(r)$.

In fig. 3 the mean value of the breakdown strength is plotted as a function of the logarithm of the concentration of the contamination. The logarithmic relationship agrees with our breakdown model for recurring discharges (ref. 1). It corroborates in a direct way the dependence of the breakdown strength on the concentration of the contaminating agent which was proposed as a hypothesis at the time of our previous measurements.

Acknowledgments

It is a pleasure to thank the Ateliers des Charmilles SA for having financially supported this work. We would also like to thank M. Rey Mermier and M. Rudolf for their contribution to the experimental effort.

References

1) Frei C., Perréard E., Peter M. : Helv. Phys. Acta 45, 778 (1972)

2) Bommeli B., Frei C. : to be published.

THE ASSIGNMENT OF APPROPRIATE DIELECTRIC STRESSES IN LIQUIDS

G.K.H. Simcox*

Introduction

Liquid dielectrics are in common use in the electrical equipment industry and are often used, in perhaps greater variety, for the insulating media of high voltage pulse generators.
The requirements for these two areas are distinct and different. For the former, long term reliability and economic upkeep are mandatory and, for the latter, the emphasis is on peak performance at low duty cycle. On the one hand the dielectric is stressed continuously by A.C.-D.C. voltages and, on the other, the stresses are of microsecond duration or less, with seconds between stresses. This alone would indicate the use of different stress levels but, for liquids, there are added complications due to contaminants. These seriously reduce the electrical performance of liquids for prolonged A.C.-D.C. stress applications but have little influence upon the short, microsecond, pulse performance.
Because of these differing requirements and characteristics there are interesting contrasts of style and content in the publications of research workers in these respective fields.

Prolonged A.C./D.C. Stress

For economic reasons, mineral oils are most frequently used and the published breakdown data reflect this. However, for this stress-time regeme, when contaminants strongly influence dielectric behavior, the performances of most liquids are similar. Further these performances under A.C. or D.C. stresses are comparable provided that the peak A.C. stress ($\sqrt{2}$xRMS) is equated to the positive D.C. level. For negative D.C. polarities the operating stresses can be 50% higher. Most of the A.C. data for oil are therefore applicable to other liquids under these conditions.

There is a continuing technological effort on mineral oils to preserve their electrical and mechanical properties for prolonged dielectric and cooling duties. Efforts are made to reduce the oxidation rate and to obtain long term compatibility of the dielectric and system components. Generally, this is of less concern for pulse generators.
The basis for much of the A.C. dielectric breakdown data can be found by reference to Weber and Endicott[1]. These workers used automated testing equipment to yield a large number of oil breakdown data between electrodes of varying area. The intent of their experiments was to define safe operating stresses for equipments with large stressed areas. They recognized the statistical nature of breakdown and wished to determine the 'best fit' theoretical

* Director of Engineering, Megapulse, Inc., Bedford, Mass.,U.S.A.

distribution of breakdown and derive from it a relationshp between electrode area and dielectric strength for near uniform fields. Their conclusion was that the breakdown data follow Extreme Value Theory[2] rather than Normal Probability and that the distributions for increasing area are similar distributions shifted successively towards zero volts.
Further, that extreme value theory yields a precise equation for predicting the electrode area effect:

$$VA_1 - VA_2 = 1.80 \ \mathrm{SLog} \frac{A_2}{A_1}$$

where: VA_1 VA_2 are the modal voltages (value of highest frequency). For two areas A_1 and A_2. S is the standard deviation.

In practical terms from their experimental data, this implied that a ten-fold increase in electrode area reduced the dielectric strength of oil by 16%.
For power applications the interest in the form of the breakdown distribution is concentrated upon the probabilities of the lower breakdown values. In this respect an extremal distribution predicts higher probabilities of the lower values than would a normal distribution and would indicate the use of lower design stresses and the presence of a larger area effect.
J.K. Nelson[3] agrees that, under carefully controlled test conditions the data should be extremal but holds that random environmental factors will tend to 'normalize' the distribution.
Both E. Simo[4] and Nelson et.al.[5] have carried out large scale experimental programs for breakdown in oil which emphasize the statistical method for evaluating the dielectric under varying environmental conditions.
Typically dielectric strengths for duty in this time regeme are below 100 Kv/cm for electrode areas in the order of 100 cm^2.

Impulsive Stress

In sharp contrast to the above, high voltage, short pulse generators designed for nuclear simulations and experimental physics programs are required to give peak performance for short periods. On occasions the experimental data are more important than the equipment integrity. Consequently the peak dielectric stress which can be reasonably supported is often of more concern than is the form of the breakdown distribution.
When the pulse energy is stored in the electric field between electrode structures according to:

$$U = \tfrac{1}{2} E_o E_r E^2$$

where Eo = Permittivity of freespace
Er = Relative permittivity of the dielectric
E = Electrical stress

the advantages of high values of Er and E are apparent for the achievement of high energy densities. For this reason liquids

are often chosen for the insulating media and, since the influence of contaminants is small for microsecond stress-times, each liquid has its unique dielectric properties.
When the propagation time of the breakdown mechanism between apposing electrodes is of the same order as the effective stress-time, measurement of the propagation velocities provides a useful comparison of liquids as dielectric media.

J.C. Martin and his group at A.W.R.E. Aldermaston, England, contributed much to the present knowledge in this short pulse regeme. In the 1960's they examined the possibility of a breakdown model based on a streamer mechanism, at least in the time scale of 0.1-1.0 μsec. Herbert[6], one of this group, obtained integral relationships for the mean streamer velocities in several liquids as functions of the amplitude and time duration of the applied voltages. From his data for oil, carbon tetrachloride and glycerine, for both pulse polarities, he suggested that:

$$v = \frac{d}{t_{EFF}} \quad \alpha V^{n};$$

where v = Mean Vel. cm/μsec
d = Gap spacing, cm.
V = Applied volt. MV
η lies between 1.2 and 1.8
t_{EFF} = Effective stress time in μsec.

For a square pulse, t_{EFF} is the pulse duration but, for most cases, the pulse may be described as a ramp, (V=kt), and t_{EFF}, now needs definition. After substitution of the ramp function and integration over the time to breakdown, one obtains:

$$t_{EFF} = \frac{T_{max}}{\eta+1}$$

, where T_{max} is the whole time to breakdown.

t_{EFF} depends on the value given to η and, from latter, larger scale experiments, Martin gave η = 3 as a better value. For this t_{EFF} can be taken as the time from 75% V_{max} to V_{max}. Following from this:

$$Ft^{1/3} \quad \alpha \; d^{-2/3}$$

, where F is the peak stress MV/cm
t is the effective time in μsec.

Martin[7] and his workers carried out extensive experiments with liquid dielectrics, using a wide range of electrode areas, with uniform and non-uniform fields, using fast and relatively slow pulse generators. He summarizes that any electrode finish less than gross roughness is unimportant and that solid impurities and additives have little influence upon pulse breakdown in the 1μsec region.

For uniform fields he has provided the proven relationship:

$$F t^{1/3} A^{1/10} = k;$$ where F and t are as previously defined.

A is electrode area in cm^2

k is a constant for the liquid.

Values given for k are: oil, 0.5; alcohols, 0.5; glycerine, 0.7; castor oil, 0.7; deion. water, 0.3.

As a comparison, for $t_{EFF} \sim 1\mu sec$. In oil, the design stresses for these applications are at least 5 times higher than could be contemplated for A.C. designs of comparable electrode area.

Conclusions

The contrasts in dielectric requirements are as noticeable as are the experimental styles. But, with the increasing requirements for pulse generators with higher duty cycles, it can be expected that these two branches of the technology will tend to merge. For increasing pulse durations and repetition rates, the effects of contaminants will increase and the exploitation of streamer propagation times will be limited.

References

1. Weber, K.H. and Endicott, H.S.; Transactions of AIEE, (June 1956).
2. Gumbel, E.J.; National Bureau of Standards, Applied Math Series 33, (1954).
3. Nelson, J.K.; Proc. IEE. 118, (November 1971).
4. Simo, E.; IEEE. Trans. on Electrical Insulation EI-5, (December 1970).
5. Nelson, J.K. et. al., Proc. IEE 118 , (February 1971).
6. Herbert, H.G.; Dielectric Strength Notes, Note 10, AWRE, (October 1966).
7. Martin, J.C.; Circuit and E.M. System Design Notes, Note 4, AFWL (April 1970).

ON THE MECHANISMS OF DIELECTRIC BREAKDOWN IN LIQUIDS

A. Grinberg and D. M. K. de Grinberg*

Introduction

One of the still unsolved problems in the field of dielectric breakdown of liquids is the origin of bubbles observed by many research workers. Their presence has been invoked as an important ingredient in the extrinsic theories but yet no generally accepted mechanism has been proposed to explain the formation of these bubbles. Among the difficulties that must be overcome in this respect, there is the time required for the formation of a bubble, which must be compatible with the results of experiments performed by using very short voltage pulses or irradiation with lasers of very short pulse width.

A model for bubble formation

According to Mirza et al. (1) micro-breakdown is dependent on the instantaneous power density across a very small region; their calculations give an energy of $\simeq 10^{-5}$ J in a volume of $\simeq 10^{-12} cm^3$, i.e., an energy density of about 10^8 atm. For processes like this, where a high amount of energy is transferred to the medium in times short compared with the relaxation time, Bullough and Gilman (2) have suggested that an internal "explosion" may develop and, as a consequence, large negative pressures are produced in finite volumes. These ideas have been applied by the present authors (3) to explain the electrically induced lattice damage in ionic crystals, and now we would like to show how they can be extended to the case of the breakdown of dielectric liquids.
We shall assume that the energy of the pre-breakdown discharge is absorbed in a cylindrical region of radius a. The duration of this energy transfer is $t \simeq \delta$, where δ is comparable to or less than the time required by an elastic wave to cross the cylinder. As a consequence, the pressure inside this region $(r<a)$ raises its value from zero to P_0 but the volume remains constant during the short time $t \simeq \delta$. Therefore, surface tractions are generated on the internal surface r=a. For $a<r<\infty$ the initial value of the pressure is zero. The transient effects of the pressure pulse were calculated by Bullough and Gilman (2) who showed that the sudden absorption of energy in a cylindrical region of the medium generates negative pressures of extremely high magnitude. In fact, a rarefaction wave will move from $r=a$ towards the centre of the cylindrical region; after some time the pressure in the wave becomes negative and eventually diverges. The negative pressure extends over a substantial fraction of the cylindrical radius. This is equivalent to assuming an "implosion", the inverse of an explosion.
The propagation of the shock wave generated by this implosion would originate, by cavitation, micro-bubbles in the bulk of the

* División de Ciencias Básicas e Ingeniería, Universidad Autónoma Metropolitana, Azcapotzalco, México 16, D. F.

liquid. Other effects of this process will be a high increase of the energy in the medium due to the sudden compression and expansion inside the cylindrical region, and the high values of the shock temperature, both of which may contribute to vaporization and decomposition of the liquid.
We have described the two-dimensional case (cylindrical symmetry); as was pointed out by Bullough and Gilman (2), implosions can also be expected in the three-dimensional case and we shall refer below to experimental situations of both types.
The successive events conducing to macroscopic breakdown of a liquid subjected to an applied external voltage would be, then, the formation of a pre-breakdown avalanche, local absorption of energy causing an implosion, formation of micro-bubbles by cavitation, micro-discharges inside the bubbles and, finally, the formation of a streamer.
Hence, the size and local concentration of micro-bubbles play an important rôle in the development of the process. Both of them will depend, of course, on parameters like temperature, external pressure, viscosity, etc., but the intensity of the pre-breakdown discharge will be the main controlling parameter. This suggests that the statistical time-lag would be related to the probability of formation of a pre-breakdown avalanche intense enough as to originate a critical concentration of micro-bubbles of a critical size. On the other hand, the formative time-lag would be related to the time required for the formation of a streamer in the bulk of the liquid.
Let us discuss now the case of dielectric breakdown induced in liquids by laser irradiation. The main experimental observations reported by several investigators (4-10) working with water, benzene, carbon tetrachloride and other organic liquids, refer to dielectric breakdown, shock waves related to it and not to stimulated Brillouin scattering and with associated pressures higher than 200 kbar, light emission and small bubbles along the beam path.
The initial step in the process leading to dielectric breakdown would be the absorption of the laser beam energy. The characteristics of this absorption will be different according to the presence or not of particles in suspension, or when dealing with dye solutions where multi-photon absorption is the predominating absorption mechanism.
When unfiltered water or an unfiltered solution of benzanthracene in benzene was tested with a Q-spoiled laser, spherical pressure transients propagating out from each point of breakdown were detected (6, 7). When a well filtered sample of the dye solution was used, such points of breakdown were absent and a relatively diffuse pressure transient was observed. Carome et al. (7) deduced from such observations that a cylindrically symmetrical pressure gradient is produced, either directly or indirectly, as a result of a laser light absorption process.
A change from spherical to cylindrical pattern was also observed by Barnes and Riekhoff (8) using double-distilled and well-filtered water and a Q-switched laser. At the minimun laser power level for breakdown, a spherical pattern was observed; if the laser was operated above this minimun power the spark became approximately cylindrical in shape, the length of the cylinder increasing

along the direction of the laser beam as the laser power was increased.
All these authors (4-10) mention that the intensity of the pressure associated with the shock waves is high; Bell and Landt (6) have calculated that value from the measured average shock velocity. The pressure values are in the order of 10^5 atm, which is high enough to make our model of bubble generation by cavitation applicable also in this case.
The difference between this case of laser-induced breakdown and the case of breakdown under the action of an external voltage would refer to the steps occurring, in the later case, after the formation of the bubbles. In other words, in laser-induced breakdown only local or micro-breakdown will be produced.
It can be suggested here that the statistical time-lag in this case would be related to the statistical starting characteristics of an electron avalanche, in accord with the situation in the conventional case of breakdown under an external voltage.

Intrinsic and extrinsic breakdown

There has been considerable controversy on whether breakdown in liquids is an intrinsic or an extrinsic process, in other words, if the measured breakdown field strength has an intrinsic physical meaning or not. It seems possible to gain some clarification on this point by taking into account the results obtained in the study of liquid breakdown induced by laser irradiation and by comparing them with the observations of laser-induced dielectric breakdown of ionic crystals. This comparison is legitimate since several investigators have pointed out that the breakdown characteristics of liquids are very similar to those of ionic crystals, especially at high rates of rise of the applied field (11), and that the same mechanism, i. e., avalanche ionization, is operating in both types of media (12, 13).
In the case of laser-induced dielectric breakdown of ionic crystals, a reproducible field strength characteristic of the bulk material could be established (13-16) because the effect of impurities and inclusions was eliminated by the small scale of volume elements and time intervals in which the breakdown occurred.
In the case of laser-induced dielectric breakdown of water, a different behaviour was observed when tap water or filtered water was irradiated (6, 7, 10). In the first case several shocks were detected whereas only one or two shocks were observed in filtered water. Furthermore, multiple points of breakdown were observed when small amounts of Al_2O_3 powder were introduced into filtered samples of H_2O that previously had exhibited only simple-point bahaviour (7). Hence, at least in this case of electrode-less dielectric breakdown of liquids an intrinsic value of the breakdown field can be clearly defined and measured, and Roach et al. (9) have calculated it by selecting the five lowest intensity values for which one or two shocks were produced and the five highest values where no shocks were produced and averaging the ten values. The intrinsic optical breakdown thresholds were measured for several ionic crystals (14, 17); these values are closely related to the corresponding d. c. dielectric strength, a fact that was used to conclude that the same mechanism is operative in both types of

experiments. In the case of liquids, the threshold value was determined for 2-propanol and other eight organic liquids not mentioned in the paper (9), and no comparison was made with values determined by more conventional techniques. It is mentioned, however, that the high values determined are in agreement with the general trend of increase of the breakdown field strength as the time of voltage application is shortened.

References

1. J. S. Mirza, C. W. Smith and J. H. Calderwood, 4th. Inter. Conf. on Conduction and Breakdown in Dielectric Liquids, Dublin (1972), 198.
2. R. Bullough and J. J. Gilman, J. Appl. Phys. 37 (1966) 2283.
3. A. Grinberg and D. M. K. de Grinberg, J. Appl. Phys. 45 (1974) 2007.
4. R. G. Brewer and K. E. Rieckhoff, Phys. Rev. Lett. 13 (1964) 334.
5. E. F. Carome, C. E. Moeller and N. A. Clark, J. Acoust. Soc. Am. 40 (1966) 1462.
6. C. E. Bell and J. A. Landt, Appl. Phys. Lett. 10 (1967) 46.
7. E. F. Carome, E. M. Carreira and C. J. Prochaska, Appl. Phys. Lett. 11 (1967) 64.
8. P. A. Barnes and K. E. Rieckhoff, Appl. Phys. Lett. 13 (1968) 282.
9. J. F. Roach, W. Zagieboylo and J. M. Davies, Proc. IEEE. 57 (1969) 1693.
10. M. P. Felix and A. T. Ellis, Appl. Phys. Lett. 19 (1971) 484.
11. K. C. Kao and J. P. C. McMacth, IEEE Trans. EI-5 (1970) 64.
12. E. Yablonovitch and N. Bloembergen, Phys. Rev. Lett. 29 (1972) 907.
13. N. Bloembergen, I. E. E. E. Q-E10 (1974) 375.
14. E. Yablonovitch, Appl. Phys. Lett., 19 (1971) 495.
15. D. W. Fradin, E. Yablonovitch and M. Bass, Appl. Opt. 12 (1973) 700.
16. D. W. Fradin and M. Bass, Appl. Phys. Lett. 22 (1973) 157.
17. D. W. Fradin and M. Bass, Appl. Phys. Lett., 22 (1973) 206.

BREAKDOWN OF OIL USING ONE BARE AND ONE ELECTRODE COATED WITH AN INSULATING FILM

H. Bertein*

Introduction

The present work deals with the discharges which occur when a voltage is applied between a bare electrode and a coated one immersed in transformer oil;similar numerous works concerning discharges taking place in gas instead of oil,have been made by the author during the last ten years, they used a technique consisting of powdering the insulating coating in order to reveal its charges .From the distribution, sizes and shapes of the charged sites,it has been possible to analyse in detail the features and the mechanisms of partial discharges in gases(1). In the course of the experiments made on discharges in oil, remarkable similarities have appeared with the phenomena observed in gases;therefore the results of the studies of discharges in gases have provided many information which have helped us to carry out the study of discharges in oil.

We must emphasize that all the phenomena described hereafter have been observed too, with other conditions of pressure and voltage in many dielectric liquids as different from one another as liq.N_2 and dibutyl sebacate.

Experimental Procedures

Two sphere plane electrode systems are used. For both,the lower plane electrode is covered with an insulating film of thickness t, of relative permittivity ε and of very high resistivity, the upper spheric electrode is bare.For the first system called here "contact electrode system",the sphere, 5 cm diameter, is in contact with the film;for the second system called "spaced electrode system" a portion of sphere, 20 cm diameter, is separated from the film by a spacing obtained with plexiglass block as shown in fig.1. Both electrodes are immersed in a filtered and dried transformer oil of resistivity $5.10^{14}\Omega\cdot$cm.

Observations of the phenomena during the voltage application are made;observations of the "powder patterns" i.e. of the charges carried by the discharges through the oil and trapped on the film surface,by materializing them on powdering are made too after the voltage application .With air as medium, the surface of the film is powdered with a two coloured mixture which reveals the charged sites with distinguishing one polarity from the other;with oil, a two layer insulating plate is used, the upper one being a film of polyethylene constituting the bottom of the liquid container, the lower one being a thin plexiglass plate . After the voltage application, the former is pulled off, during the separation, the air ionizes at places facing the trapped charges, then an identical distribution of charges with a marginally lower density is deposited in the surface of the plexiglass plate and is materialized on powdering.

* Laboratoire Central des Industries Electriques -
92 Fontenay aux Roses - France

Observations

Alternating Voltage . When the increasing voltage U (crest value) applied to the contact electrode system reaches an onset value called "Inception Voltage U_{im}", discharges are seen near the axis, they coincide with the sudden appearance of numerous gas bubbles. If U is maintained at U_{im}, the phenomenon extends over a circular region which rapidly enlarges, then stabilizes, some of the longest discharges appearing in the form of sparks creeping on the film (later on, these latter discharges will be called "Complex Discharges" and the other ones "Simple Discharges") If U is lowered now, the discharges keep on to occur, then they cease at a voltage U_{em} widely lower than U_{im} , U_{em} is called "Extinction Voltage" If the voltage is raised above U_{im} the film breaks down at a voltage U_o . Viale et al.(2) ,using an oscillographic technique, have observed such phenomena with paper stacks . Sometimes the very first discharge breaks down the film (U_{im}=U_o), then the phenomena are not observable . On condition that the film breakdown does not result from the very first discharge, U_o is observed to depend only on the film material and thickness (3) irrespectively of the pressure and treatment of the oil. In order to render the effect the most easily observable, U_o-U_{im} must be large. For a given oil, a high value of U_o can be reached by using a film capable of withstanding the discharges, thus polyethylene is far more suitable than mylar or glass;moreover, the film must be thick enough , say at least 100 µm. For a given film, a low value of U_{im} can be reached by using a non degassed oil at a reduced pressure; for a film of a given t/ε value, and with non degassed or partially degassed oil, U_{im} and U_{em} are found independent from the film material. In figure 2, curves are shown of U_{im} and U_{em} versus t/ε .

The spaced electrode system provides the value U_i and U_e of the inception voltage and of the extinction one corresponding to a parti-

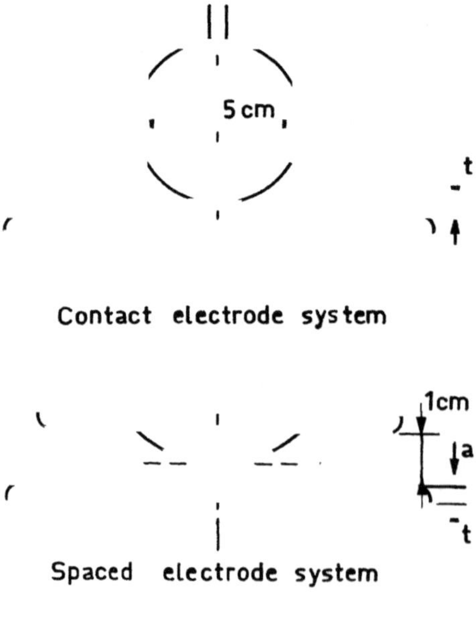

Fig 1

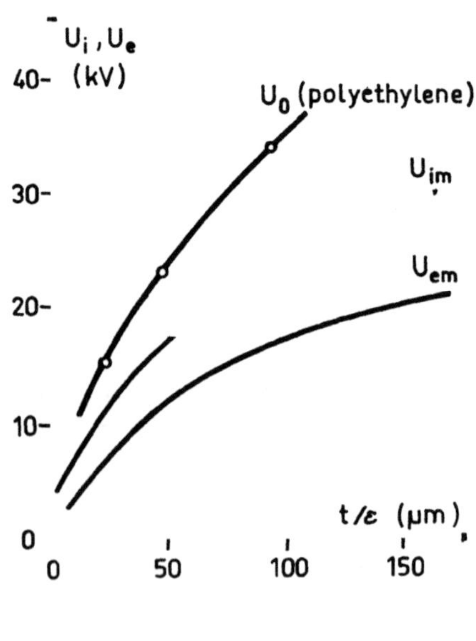

Fig 2

cular value a of the spacing; many comparative measurements of U_i made with a = 1 mm have shown that the corresponding voltage across the gap $U_a=U_i/(1+t\varepsilon'/a\varepsilon)$ (ε'being the relative permittivity of the fluid) has the same value, whatever the film material, as the breakdown voltage of oil when measured without film, i.e. approximatively 65 kV with a thoroughly degassed oil . U_e/U_i is found nearly the same for non degassed and partially degassed oil, its value being about 0.7 for values of a ranging from 0.5 to 2 mm . Later on, the ratio $a\varepsilon/t\varepsilon'$ will be designated by the symbol η .

Powder patterns have been performed in oil .Sites of simple discharges appear in the form of small radial zones, they exist alone just above the inception voltage, elsewhere they coexist with sites of complex discharges which assume the form of trees becoming predominant when U and η increase ; the positive trees appear at values of U and η significantly smaller than negative ones, and the former are longer and thicker than the latter.

Similar observations of discharges have been previously made in air where the onset voltage is far lower than in oil . The powder patterns are far larger in air than in oil, though their general features are similar, except that in air positive trees are identical in size as negative ones and appear at the same values of U and η Figure 3 schematizes powder patterns in oil and in air .

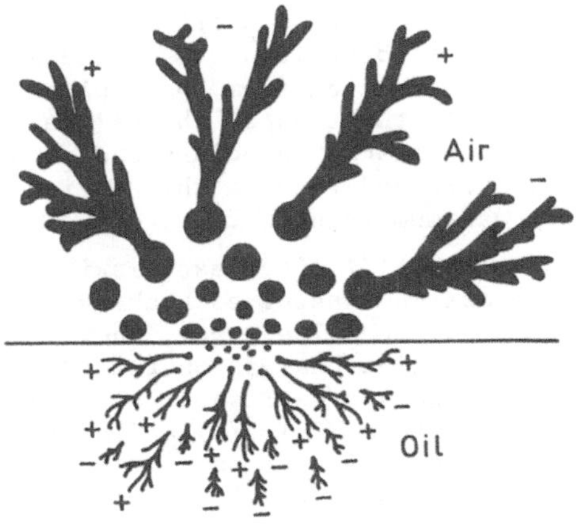

Fig 3

Moreover, at the inception voltage, the values of η above which trees appear is of the order of 100 in air, in oil it approximates to 5 .

Direct Voltage . With the contact electrode system, discharges and bubbles are observed during the rise and during the decrease of voltage, and a discharge breaks down the film at a raising voltage of about 1.5 U_o (2).With an oil of a lower resistivity and with a low rate of voltage rise, most of the applied voltage appears across the film, then no discharge is seen prior to the film breakdown which occurs at a voltage corresponding to its Intrinsic Dielectric

Strength, i.e. about 70 kV for films of 100 μm of polyethylene or mylar; Zaky et al (4) have found the same result. As to the powder patterns, they show the same types of charged sites as with alternating voltage.

Similar observations have been previously made in air, the inception voltage is then far lower than in oil.

Mechanism of Partial Discharges Distribution

The oversimplified following picture results of numerous observations and measurements of simple partial discharges in gases . In the course of works on oil, it has appeared that this picture is suitable too . As to the complex partial discharges, it fits too provided that one considers that the remanent field strength is far smaller than the value g_r definited below which fits only with simple discharges.

Distribution of Discharges in a Uniform Field . Let an increasing direct voltage of instantaneous value u be applied between two plane parallel electrodes, one of them being coated. When u reaches $U_a(1+1/\eta)$, the field strength reaches the breakdown value $g=U_a/a$ in the gap, then a discharge is initiated, it carries across the gap a charge which is trapped close to its impact on the film. This charge creates a field of the opposite direction which counteracts the applied one, so that the discharge extinguishes when the resulting field at this place is reduced to its critical "remanent value" $g_r=U_r/a$; this phenomenon lasts about 10^{-9} s. Afterwards, u continuing to increase, the field resulting from the applied field and from the non uniform one due to the trapped charge reaches the value g at another place, then a new discharge occurs and so on, several discharges deposit charges here and there ; this series of discharges ceases when u reaches its maximum value U .

With an alternating voltage,(5) after each maximum U, as the resulting field presents a maximum facing each charged site, a new series of discharges occurs at each alternance, they tend to deposit alternatively positive and negative charges on the same sites. When lowering U, the number of discharges decreases, the last one extinguishes when the sum of the voltage $U/(1+1/\eta)$ and of the half voltage drop $0.5(U_a-U_r)$ in front of the last charged site has decreased below U_a . Thus, the extinction voltage value is $U_e=0.5(U_a+U_r)(1+1/\eta)$ Figure 4 shows the general aspect of the curves $U_i=f_{(a)}$ and $U_e=f_{(a)}$

Distribution of Discharges in a Non Uniform Gap with Alternating Voltage . The phenomenon is somewhat more complicated here :the very first discharge takes place at the minimum U_{im}, a_{im} of the curve $U_{im}=f_{(a}$ suppose that U is maintained at U_{im}, the trapped charge creates in its neighbourhood a field which, after the first maximum of voltage, enhances the applied one, then at the second alternance several new sites corresponding to values of a sligthly different from a_{im} are charged . After the next maximum of voltage, the field due to these charges enhances the applied field and so on, the region occupied by the charged sites extends progressively during several alternances until its frontiers reach the level of the roots a' and a'' of eq.$U_e=f_{(a)}$ corresponding to the particular value U_i, where they stabilize;thereafter, whatever the variations of U, the frontiers of the region occupied by the charged sites correspond definitively to eq.$U_e=f_{(a)}$. In

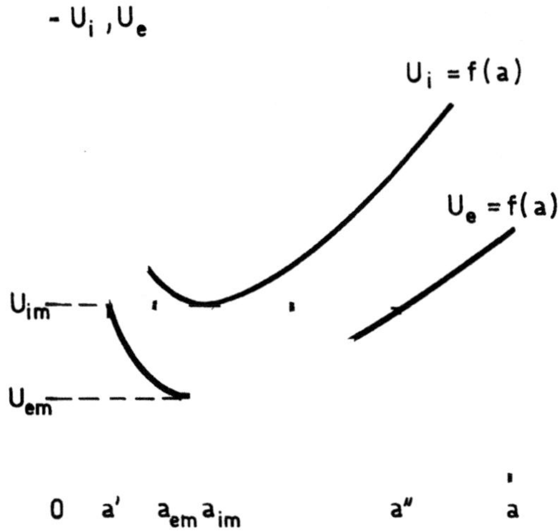

Fig 4

particular, as U decreases, this region narrows and the very last discharge occurs at the minimum U_{em},a_{em} of eq.U_e=f(a). Remind that the variations of U_{im} and U_{em} against t/ε have experimentally been plotted in figure 2.

Mechanism of a Partial Discharge

The charged particles carried by the discharge and trapped close to its impact on the film create in the fluid a field counteractiong the applied one, thus modifying the discharge evolution as early as the first avalanche has crossed the gap. The higher η,the lower this field. According to a and η values, the conducting channel crosses the gap conformably to the Streamer mechanism, the Townsend one or an intermediate one (6) so the role of the cathode material decreases with raising a and η. Since our valuations of U_a made with a=1 mm have shown no wall material dependence, the channel is here a streamer.

Once the streamer has crossed the gap, a radial field is created , then the depositing charge is spread out by tangential discharges on an increasing area centred on the impact. There are the charged particles let by the tangential discharges on their path which draw the sites which will be materialized on powdering; their shapes are explained in the following manner;when an increasing voltage is applied to a rod, corona first takes place around it and if the voltage exceeds a critical value, corona is replaced by a spark (6).Here the streamer plays the role of the rod and the charge depositing on the impact supplies to it an impulse voltage which directly depends on η and on U .If η is lower than a critical value η_o, the voltage supplied to the rod during the charge deposition remains lower than the sparking value, then corona occurs alone and a radial site is drawn;if η exceeds η_o, this voltage reaches its sparking value, then corona ceases and is replaced by a spark, thus the site drawn in that way is constituted of a radial site developing into a tree the importance of which increases with η and U.

According to the works previously carried out in air at pressures ranging frmm 0.1 to 1 atmosphere, the increase in pressure causes the size of the site to decrease, as to the polarity, it has effect neither on the tree length nor on η_0 value . On the other hand, Howell (7) has shown that in a rod gap with high pressure air, the sparking voltage is widely lower when the rod is positive than when it is negative. Now, as it has been seen, the charged sites are far smaller in oil than in air, as to η_0 it is far smaller in oil than in air too, proving that the charge density on the sites is far higher in the former case, at last η_0 is far smaller for positive trees than for negative ones, the positive trees being longer and thicker than the negative ones. All these features strongly give evidence that the discharges take place within a very high pressure gas space, that agrees with calculations of Krasucki (8) and measurements of Smith and Calderwood (9).

Conclusion

In order to prevent the development of complete interelectrode breakdown, numerous investigators have used pointed electrode techniques, so they have obtained a partial breakdown which has allowed them to examine the prebreakdown events. For the same purpose, the present author has used a (nearly) uniform gap bounded by an insulating film, then events corresponding to two very different stages of the breakdown evolution are provided. It must be emphasized that here the oil gap breakdown is complete, and since the pre-streamer stage occurs in a uniform field, this pre-streamer stage is nearly the same as that of a spark occurring between bare parallel electrodes.

References

1 Bertein H. 1973 J.Phys. D. Vol 6 1910
2 Viale F.,Wind G., Fallou M., Leroy G., Wendling J. 1968 rap. 12.12 CIGRE
3 Bertein H. to be published
4 Zaky A.A., Zein El Dine M.E. and Hawley R. 1965 Br. J.Ap.Phys. Vol.16 437
5 Bertein H. 1966 Rev. Gen. Elec. t.75 n° 2 165
6 Meek J.M. and Craggs J.D. 1953 Electrical Breakdown of Gases Oxford Clarendon Press
7 Howell A.H. 1939 Trans. Amer. Inst. Engrs. 58, 193
8 Krasucki Z. 1966 Proc. Roy. Soc. 294 393
9 Smith C.W. and Calderwood J.H. 1968 Proc.Conf. Liq. Isol. CRNS 161

ESTIMATION PROBLEMS IN THE STEP-FUNCTION BREAKDOWN TEST

J.E. Brignell*

Introduction

The purpose of this paper is to bring the attention of the experimentalist certain useful facts and formulae relating to the interpretation of breakdown time-lags. Furthermore, it is pointed out that the traditional Laue plot method yields a rather poor estimation of the time-lag components, which, in view of its laborious nature, makes it rather pointless.

The Laue plot method is a rather crude example of a method of *order statistics*, and while it is not suggested that the experimenter should burden himself with the mathematical background to this subject, it is strongly recommended that he should take note of some of its results. In particular, formulae are available for Best Linear Unbiassed Estimations (BLUE) which not only give a result which is most accurate in a certain sense, but also take account of 'censoring'; that is the loss of all results lying outside certain pre-defined bounds. Censoring of time-lag results occurs because the right hand tail of the distribution is unbounded and will not fit into the finite range of a practical time-measuring instrument.

The time-lag distribution

The step-function test is preferred for its high information yield in comparison with the impulse test which gives information in the binary form characteristic of a Bernoulli trial[1]. Indeed, given accurate estimation of the step-function test parameters and their stress dependence it is possible to predict the results obtainable from any form of impulse.

The breakdown time-lag appears to have two components - the formative time-lag, T_f, and the statistical time-lag, T_s. The latter is the most stress-dependent phenomena in breakdown studies (it approximately halves for every 0.1 MV/cm increase in stress) and its value compared with the mean and standard deviation of T_f determines which of the four characteristic manifestations of the general distribution of time-lags is observed. These are shown in functional form in table 1 and as curves in Fig. 1. The symbols have the following meaning.

U( )	unit step function
μ	conditional failure rate = $(\bar{T}_s)^{-1}$
T	mean formative time
σ^o	standard deviation of formative time

* Department of Electrical and Electronic Engineering, The City University, St. John Street, London EC1V 4PB, England.

Table I

Distribution Type	Conditions T_f	σ_f	Density function , $f(T)$
a. One parameter exponential	$<<T_s$	$<<T_s$	$U(T)\mu\ exp\left[-\mu T\right]$
b. Two parameter exponential	$\sim T_s$	$<<T_s$	$U(T-T_o)\ \mu exp\left[-\mu(T-T_o)\right]$
c. Three parameter exponential	$\sim T_s$	$\sim T_s$	$\mu exp\left[(\sigma^2\mu^2/2)-\mu(T-T_o)\right].\frac{1}{2}\left\{1+erf\left[\left\{(T-T_o)/\sqrt{2}\sigma\right\}-(\sigma\mu/\sqrt{2})\right]\right\}$
d. Normal	$>>T_s$	$>>T_s$	$\left\{1/\sigma\sqrt{(2\pi)}\right\}\ exp\left[-(T-T_o)^2/2\sigma^2\right]$

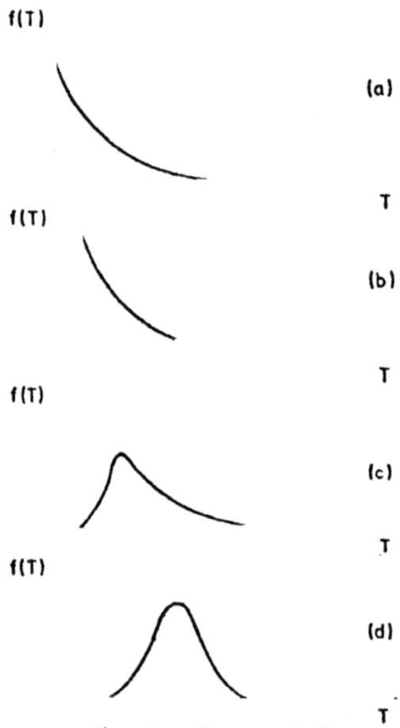

Fig. 1 Shapes of distributions in table 1

Distributions (a) and (b) apply to time lags of more than a microsecond, the range occupied by most of the published work, to which the Laue plot has been applied. Here a summary of a theoretical and simulation study of some estimators is given; details are available in a separate report[2].

Some estimators

Five different estimators have been studied - two practical, two from order statistics and one new proposal for a simple alternative. For comparison, it is convenient at this point to change to the symbolism adopted in the statistical literature for the two parameter exponential distribution((b) in table 1, with (a) as a special case);

$$f(x) = \frac{1}{\sigma} \exp\left|-(x - \theta)/\sigma\right|.U(x - \theta)$$

Note that σ is now the standard deviation parameter for the exponential distribution. The methods studied were:

(i) The BLUE formulae, which for a complete sample are

$$\sigma^* = n(\bar{x} - x_1)/(n - 1), \qquad \theta^* = (nx_1 - \bar{x})/(n - 1)$$

here σ^* and θ^* are estimates of known variance[2,3]. $\bar{x}$ is the mean and x_1 the smallest value.

(ii) A straight line fitted using the estimate of the expected values for the ordered normalised exponential distribution rather than L_A or L_N. These estimates are

$$E\{U_i\} = \sum_{j=1}^{i} \frac{1}{n - j + 1}$$

(iii) The conventional Laue plot reduced to a more rigorous form. This is done by noting that the independent variable (the Laue estimate of the ith ranked standardised variate $L_A(i) = -\ell n(1 - i/n)$ which has the same values for a given size of experiment) is plotted vertically, so a horizontal least squares fit is performed.

(iv) The Laue plot modified according to a suggestion by Weibull advocated by Gumbel[3] and used by Beddow[4]. Here to avoid the loss of the nth point L_A is replaced by

$$L_B(i) = -\ \ell n|1 - i/(n + 1)|.$$

These were all studied by simulation with some analysis[2], and as a result two very simple estimators for σ and θ were proposed and tested.

$$\theta^* = x_1$$

$$\sigma^* = \bar{x} - x_1$$

2 The Laue estimates L_A and L_B plotted against the ideal exponential expectations. Also shown are some simulated ranked samples for n = 20 averaged over 100 repetitions.

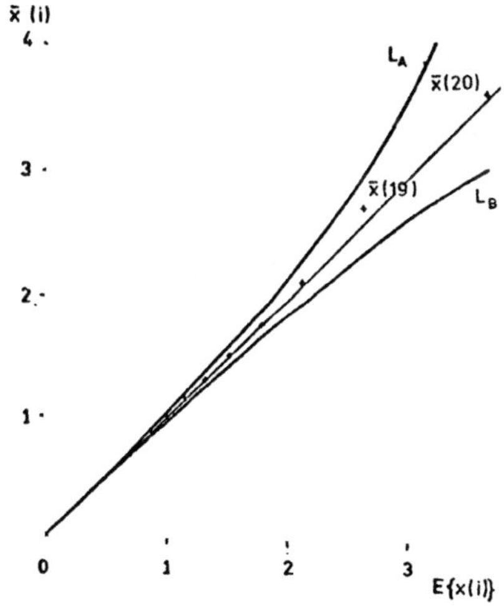

Summary of Results

The Laue estimators for the expected values deviate from the ideal in opposite directions. This bias is shown in Fig. 2 where $L_A(i)$, $E\{U_i\}$ and $L_B(i)$ are plotted against $E\{U_i\}$. Also in this figure are plotted some average values from the simulation study in which groups of samples were taken in blocks of n = 20 and ranked.

Owing to lack of space the main results[2] will be summarised verbally. A plot of the r.m.s. error of the various estimates as a function of n shows, as would be expected, that the BLUE is the best. All the others are considerably worse for θ^* (which corresponds to T_f) and marginally so for σ (which corresponds to $\bar{T}_s$). The worst method of all is the modified Laue plot using L_B, e.g. at n = 20 it produces four times the BLUE error in θ^* and $1\frac{1}{2}$ times in σ^*.

When, for a constant n = 20, the errors are plotted against the number of results censored from the right, for the first four points the two Laue plot methods actually improve! This is because of the deviation shown in Fig. 2 couple with the large variance of the higher order statistics. Indeed, apart from the

BLUE, all the estimates for θ improve continuously with censoring and are more or less equivalent to each other.

Distribution plots for the various estimates confirm that the scatter of all of them is much greater than for the BLUE, especially for θ. Furthermore it is seen that the smallest value recorded is a very good (though biassed) estimator of θ, and is insensitive to censoring. Without censoring, the mean minus the smallest value is a good estimator for σ, but this measure is sensitive to censoring.

Discussion

An important point which is often not appreciated by experimenters is that once a group of experimental results is placed in ascending order, as is done for all these estimators, they cease to be statistically independent. The common practice of connecting Laue plot points by a staircase before visually fitting a line should be abandoned, as it is optimistic in this context (conversely for fitting the points between confidence limits it is pessimistic and therefore acceptable).

The most important result from these studies is that the formative and statistical time-lags can be estimated accurately from the average time lag $\bar{T}$ and the smallest value recorded T_1 by

$$\mu = (\bar{T} - T_1)^{-1}, T_f = T_1$$

This enables the estimate to be made quickly during the course of an experiment. Only in the case of severe censoring (10% or more of the larger data) is it necessary to use the full BLUE formula[2,5].

If a plot is needed, as a test for the exponential distribution for example, then it would be better to use $E\{U_i\}$ in (ii) above rather than L_A or L_B as the independent variable.

References

1. Lewis, T.J.: 'The statistical basis of impulse testing', Proc. I.E.E., 1958, 105 C, p27.

2. Brignell, J.E. 'Estimation problems in the step-function breakdown test' Research Report No. E75/1, The City University, London.

3. Gumbel, E.J.: 'Statistics of Extremes' (Columbia University Press, 1966).

4. Beddow, A.J.: Ph.D. Thesis, University of London, 1968.

5. Sarharn, A.E., and Greenberg, B.G. (Editors): 'Contribution to order statistics' (Wiley, 1962).

BREAKDOWN AND INFRA-BREAKDOWN PHENOMENA AT LOW RATES OF STRESS APPLICATION

B.R. Mears and J.E. Brignell*

Introduction

In recent years a considerable degree of understanding of the statistical (if not physical) nature of electric breakdown in liquids has been achieved, though this has mostly been concentrated in the region of short breakdown times under constant stress. Of greater technological importance, however, is the question of 'direct-stress' breakdown, which involves inherent conceptual difficulties over the inextricable relationship between the fundamental variables of stress and time. It has been established that there is a small dependance of ultimate breakdown voltage on the rate of rise of applied voltage, but while it is possible to develop a simple statistical model to explain the dilution of the potential effects of availability of time for breakdown necessary for such a small dependence[1], the testing of such a model is not an easy task.

A matter of even greater technological importance is the question of whether there is possibly some relationship existing between any of the observable infra-breakdown phenomena and the probability of breakdown of the sample at a higher stress. The establishment of such a relationship might lead to a non-destructive test for insulation structures containing dielectric liquids.

The experimental study of such questions if fraught with difficulties. The breakdown phenomenon itself is subject to wide statistical scatter, and there are problems of control of experimental variables, the length of breakdown sequence necessary to estimate statistics, the stationarity of such sequences, etc.

One key to the easing of such problems is rigorous, automatic and versatile control of the laboratory high-voltage source. Such control has been developed in the authors' laboratory in three stages. Firstly[2] the high-voltage generator was controlled by an electronic integrating circuit, which allowed the application of ramps of varying rates of rise and the 'holding' of the voltage at a constant level. The inadequacy of such a technique was exposed when the need was found to extend the range of working to lower rates of rise and longer hold-times; here the inevitable effects of drift and leakage interfere. For this reason a purely digital method was developed[3] which overcame these problems, and at the expense of 'discretisation' of the output voltage removed the lower limit on speed of operation. Nevertheless, there were still reasons for wishing to extend the

* Department of Electrical and Electronic Engineering, The City University, St. John Street, London, EC1V 4PB, England.

versatility and automation of the high-voltage supply. It is necessary to change the ramp rate by orders of magnitude to investigate its effect; so a sequence long enough to produce reasonable statistics may have a duration measured in days. Furthermore, the combination of breakdown tests with other infra-breakdown tests, such as pulse-height-analysis or transient waveform analysis[4], implies the need for complicated manoevres over this sort of length of time. For these reasons it was considered worthwhile to develop a computer controlled high-voltage supply and this paper describes some of the preliminary work on liquids which is made possible by such a development. The working of simultaneous experiments by these techniques is not without problems and at the time of writing it is only possible to perform transient analysis with breakdown, though pulse height analysis is being added.

Apparatus

The computer controlled high-voltage supply has been discussed in general terms elsewhere[5], and with its special problems of safety, etc., it proves to be an excellent didactic example of laboratory on-line computing technique. Control is exercised through a specially designed hardware-software package, which not only makes it easy for the programmer to prescribe patterns of operation, but also prevents him from implementing certain dangerously erroneous sequences of actions. The occurrence of a breakdown is registered through an amplifier connected to a resistor is series with the test-cell and is used to cause a computer interrupt, which resets the voltage to zero ready for the next trial and records the maximum voltage achieved. The computer is not, however, fast enough to limit the breakdown damage so it is backed up by a fast hydrogen thyratron by-pass circuit[2]. The design of the test-cell is not critical, and for this work various cells have been used, depending on the subsidiary experiments to be made.

Early measurements

With this equipment available, a lot of the necessary preliminary statistical work became possible. Initial questions to be answered were: What is the true nature of the breakdown distribution under completely controlled conditions and with a very slowly applied stress? How dependent upon the rate of rise is the breakdown voltage at low application rates? How do variations of the technique of voltage application affect the breakdown probability? Are there any features of the stationarity or auto-correlation of breakdown sequence which throw light upon some of the traditionally held views such as the existence of a 'conditioning' process?

Work in Progress

Perhaps the work most suitable for illustration in the present format is that concerned with the breakdown distribution, and the two figures included in this paper are examples of histograms automatically plotted by the computer. Both of these examples would have been very difficult to obtain by traditional manual methods.

The first is concerned with the actual form of the distribution. On theoretical grounds[6] this ought to be a certain shape associated with weak-link phenomena, the chief characteristic of this shape being a pronounced negative skew. The computer allows a sufficient number of highly-controlled measurements for the shape to be accurately determined. Fig. 1 shows a histogram from five hundred measurements of the breakdown voltage of air-saturated n-hexane between 1 cm dia. spherical stainless-steel electrodes at a gap of 70 μm. With this number of results the classical weak-link (or smallest value) distribution is quite clearly obtained. The mean strength of this sample was 0.825 MV/cm.

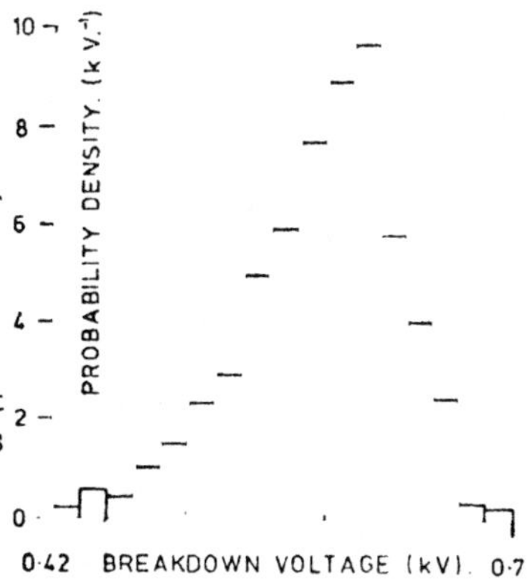

FIG 1 HISTOGRAM FROM 500 BREAKDOWN MEASUREMENTS

Fig. 2a shows the reproduction by computer of an earlier experiment[1] to test whether the breakdown voltage is affected by the rate of rise of applied voltage. Histograms for two rates of rise are shown: 1 kV s^{-1} and (dotted) 100 Vs^{-1}. The results are very similar to those obtained previously, and owing to the large scatter inherent in the breakdown phenomenon, it is not easy to demonstrate the statistical significance of the difference. The computer, however, allows us to increase the ratio of rates-of-rise by a further order. In Fig. 2b the rates are 1 kV s^{-1} and (dotted) 10 V s^{-1}. The effect is now quite apparent. The gap was 100 μm and the mean strengths were 1.062 MV/cm (fast ramp) and 0.86 MV/cm (slow ramp).

It will be noted that there is a loss of the smallest-value shape in these two experiments, which is due to the experimental technique of alternating the ramp rates for successive groups of ten breakdowns to obviate spurious effects of non-stationarity[1]. It will be noted that the duration of this experiment (35 hours) makes it difficult to perform without the aid of a computer.

Discussion

Even above its convenience, the great advantage of the computer in this sort of measurement is its objectivity. Demonstration of the rate-of-rise effect is difficult for a ratio of only 10:1 as the effect is smaller than the underlying phenomenon. The tolerance of the computer to extended experimental runs has permitted an increase of the ratio to 100:1, which allows the effect to be demonstrated with clarity. Similarly, the repeated occurrence of the characteristic negatively skewed distribution is easily shown with the very large numbers of measurements obtainable by computer, and automatic plotting takes much of the labour out of the process.

Tests have shown that one of the experimental fields in which human intervention can introduce subtle distortions is breakdown measurement[5]. A great fault of some of the earlier manual work was the failure to identify the shape of the distribution or the great degree of scatter inherent in breakdown. Knowledge that the process is a weak-link one leads to a number of further results such as the logarithmic area effect[6]. The rate-of-rise effect is probably the key to this earlier failure, as there is a strong tendency in manual measurements for the rate-of-rise to fall as the expected breakdown level is approached.

The main problem in the next stage of this work, the attempt to correlate infra breakdown phenomena with ultimate breakdown strength, is to decide which feature of the phenomena to select for correlation It would be over-optimistic to anticipate an early result, if any.

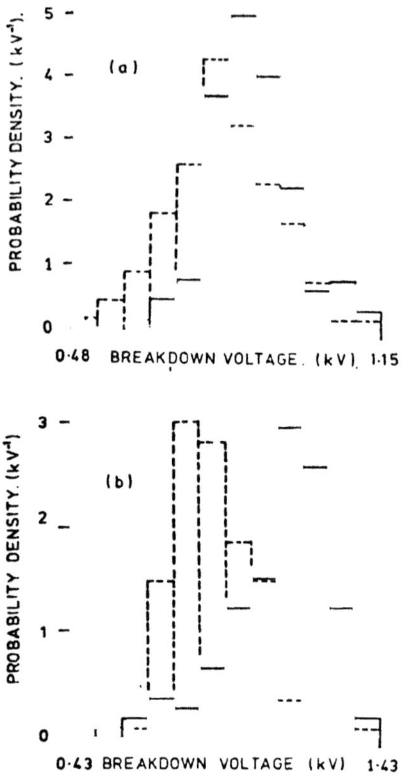

FIG 2. THE RATE OF RISE EFFCT
(a) 1kV S^{-1} AND (DOTTED) 100 V S^{-1}
(b) 1kV. S^{-1} AND (DOTTED) 10 V.S^{-1}

Acknowledgment

We wish to thank the U.K. Science Research Council for supporting this work with an equipment grant and the maintenance of one of us (BRM).

References

1. Brignell, J.E., 1966, Proc., I.E.E., 113, 1683.

2. Brignell, J.E., 1963, J. Sci. Inst. 40, 576-8.

3. Reichert, A.R., Rhodes, G.M., Brignell, J.E., 1972, J. Phys. E., 5, 587.

4. Buffam, C.J., Brignell, J.E. This conference.

5. Brignell, J.E., Rhodes, G.M., 'Laboratory On-line Computing', Intertext, London, 1975.

6. Gumbell, E.J., 'Statistics of Extremes', (Columbia University Press, 1966).

IONIC MOBILITY MEASUREMENT IN DIELECTRIC LIQUIDS BY THE METHOD OF THE FULL SPACE CHARGE LIMITATION

D. Goodstein, A. Savoia and F. Scaramuzzi*

Measurements of ionic mobilities have provided a rich source of information about fluids, and particularly about the superfluid state in liquid helium, during the past fifteen years. A considerable number of techniques have been deviced for these measurements, most of them designed to extract the time of flight of the ions' across a region of constan electric field. An alternative method, first presented by Scaramuzzi[1], makes use of the fact that, under full space charge limited conditions, the current - voltage characteristic of a diode depends simply on the zero field mobility of the ions[2]. The results of this technique have been shown to be in good agreement with those of time of flight measurements. Recently, an application has arisen which requires far more sensitive measurements of changes in mobility than had ever been made before by any technique. For this purpose, the full space charge (FSC) method has been adapted to make differential measurements which can detect changes of one part in 10^4 or less in the mobility [3].

The apparatus for FSC measurements consists basically of a diode made up of two flat parallel disc electrodes immersed in the medium whose mobility is to be measured. One electrode, the source, is impregnated with Po^{210}, an α emitter which ionizes the medium, creating a thin (typically 0.2 mm) layer of intense plasma.

A voltage is applied between the electrodes to extract from the plasma the ions of the desired sign. The current reaching the second electrode, the collector, is measured by an electrometer. At low applied voltage, the current extracted from the plasma is limited by the distribution of space charge in the medium, because the local, screened electric field at the plasma is equal to zero. Under these conditions, Poisson's equation is easily integrated to give

$$i = a\mu V^2$$

where i is the collected current, a is a calculable geometric constant, μ is the mobility and V is applied voltage.

To obtain μ from this apparatus, one first measures i as a function of V. A plot is then made of i versus V^2. The linearity of the results confirms that the FSC condition is satisfied, and the slope of the line is proportional to μ. A typical series of results of this kind is shown in Fig. 1.

* Laboratori Nazionali del CNEN, Frascati (Roma), Italy and California Institute of Technology, Pasadena, California 91109

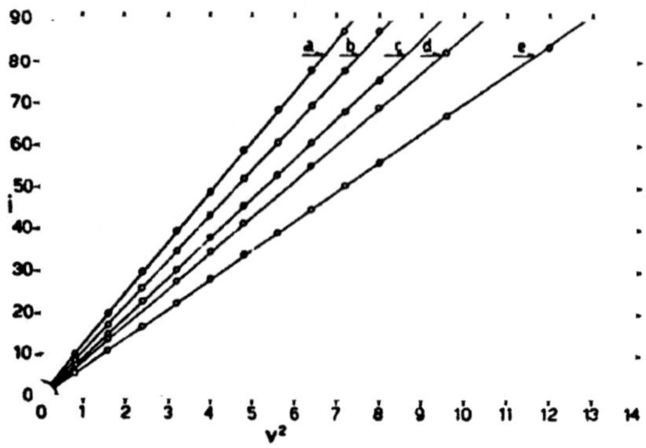

Fig. 1 Plots of i versus V in FSC conditions, for different temperatures at standard vapor pressure:

a - T ≐ 2.003 K, μ = 8.697 x 10^{-2} cm^2/V s
b - T = 2.048 K, μ = 7.797 x 10^{-2} cm^2/V s
c - T = 2.095 K, μ = 6.827 x 10^{-2} cm^2/V s
d - T = 2.124 K, μ = 6.207 x 10^{-2} cm^2/V s
e - T = 2.172 K, μ = 5.049 x 10^{-2} cm^2/V s

Compared to other techniques, the principal advantage of the FSC method is the simplicity of the apparatus. On the other hand, the absolute value of μ depends sensitively on details of the geometry, and requires precise measurements of small currents (typically 10^{-12} to 10^{-13} amp). The method is most effective when at least one value of μ is known so that all of the constants may be normalized, and changes in μ measured. It is also more effective than other techniques when λ is small.

Recently, controversy has arisen over whether the mobility of ions in liquid helium has an infinite slope in passing through the superfluid phase transition. The issue could be resolved only be measuring the rate of change of μ within 10^{-3} K of the transition. Measurements with the required sensitivity had never been made before, but the FSC method has been successfully adapted to this task.

At a given pressure and temperature of the medium, a voltage in the FSC region is applied to the diode, and the resulting current collected. Current of opposite sign is also fed into the electrometer and adjusted in such a way that the electrometer reads zero. The sensitivity of the electrometer may then be increased until it is limited by the internal noise of the device ($\sim 5.10^{-17}$ amp in our case). The mobility is then changed by changing, say, the temperature of the medium. The current from the external source is kept constant, and the voltage across the diode is varied to bring the electrometer reading back to zero.

The change in mobility is then proportional to the change in V^{-2}.

Results of the measurements at the superfluid transition are shown in Fig. 2. Relative changes in mobility, $\Delta\mu/\mu$ close to 10^{-4} are reported. The ultimate sensitivity of the apparatus has been improved since those data were taken, principally by regulating the temperature of the input resistor of the electrometer. That improvement had the effect of reducing significantly slow drifts in the calibration of the apparatus that had previously occurred during the course of a days' experiment.

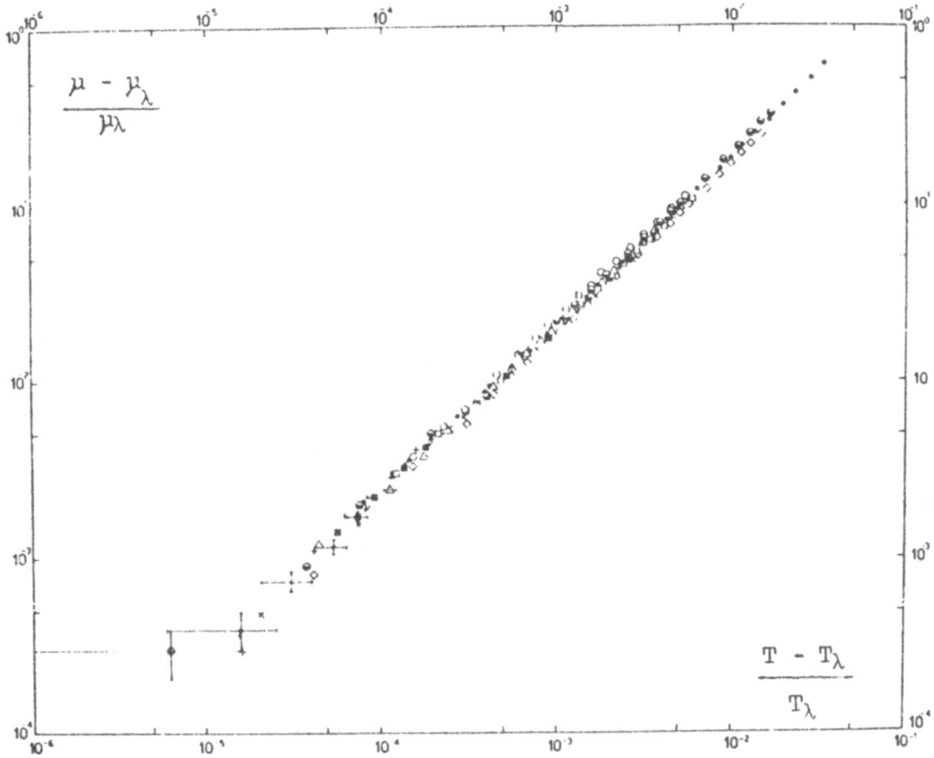

Fig. 2 Plot of $\log_{10}((\mu-\mu_\lambda)/\mu_\lambda)$ for positive ions as a function of $\log_{10}((T-T_\lambda)/T_\lambda)$. μ_λ is the value of the mobility at the λ transition. The errors shown are due to the uncertainty in the determination of the λ parameters. The various symbols refer to data taken in different runs.

The data in Fig. 2 are all taken in superfluid helium. We have had less success on the normal liquid side of the transition, where our sensitivity is no better than $\sim 10^{-3}$ in $\Delta\mu/\mu$.
The difficulty may be due to heating from the radioactive source, which must be rather intense for the FSC method, or it may be due to some relaxation phe-

nomenon that proceeds more slowly in normal liquids than in the superfluid. We shall continue to investigate this point.

References

1. F. Scaramuzzi, Internal Report LNF 63/79 CHEN-Roma, 1963
2. H.F. Ivey, Adv. Electron. Phys. 6, 137, 1954
3. D. Goodstein, A. Savoia and F. Scaramuzzi, Phys. Rev. 9, 2151, 1974

CORONA LIGHT AND CORONA PULSE IN LIQUID NITROGEN

Yuzo Takahashi* and Toshiaki Sekino*

1. Introduction

One of the authors (Y. T.) has already reported electrical characteristics of corona discharges in liquid nitrogen (1) (2). Not many papers have been so far published on corona light in liquefied gases (3). In the present investigation, light and pulses of coronas in liquid nitrogen are observed.

2. Experimental Details

A point-plane electrode system was used in d.c. and a.c. (50 Hz) fields. The point electrode was a steel needle 3 μm in tip radius, and the plane a brass disc 80 mm in diameter. The point was placed above the plane. The gap spacing was 30 mm wide. Experiments were carried out in a glass Dewar jar. The liquid nitrogen was boiling (at 77 K at atmospheric pressure), and probably became contaminated by ice particles and liquid oxygen condensed from air in the jar during the course of experiments. The light emitted in the gap was observed through a window of the jar and measured with a photomultiplier 1P21.

3. Results

3. 1. Appearance of the corona light

A faint circular light appeared in the gap just in front of the point tip, which could be discerned with a dark adapted eye (2). The diameter of the lighting volume seemed to be of the order of millimeters. The volume and intensity increased as the applied voltage was raised. At high applied voltage, the light sometimes flickered, and 'twigs' of light of several millimeters in length and tiny strong light spots appeared (1). Color of the light seemed to be blue with the naked eye.

3. 2. Discharge current and photocurrent

In d.c. corona, discharge current i_d and photocurrent i_p were larger with the point negative than with the point positive for a given applied voltage, and this is possibly ascribed to the cathodic emission from the point tip. When applied voltage was varied, i_p was approximately in direct proportion to i_d, as shown in figure 1. The plot with the point negative had a slight tendency of saturation. The ratio i_p/i_d was larger with the point positive than with the point negative by approximately 40 %.

*Department of Electrical Engineering, Chuo University,
1-13-27 Kasuga, Bunkyo-ku, Tokyo, Japan

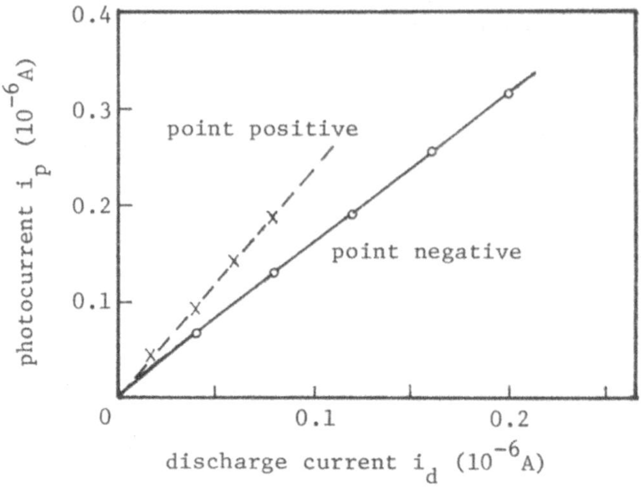

Figure 1. Photocurrent vs discharge current in d.c. corona.

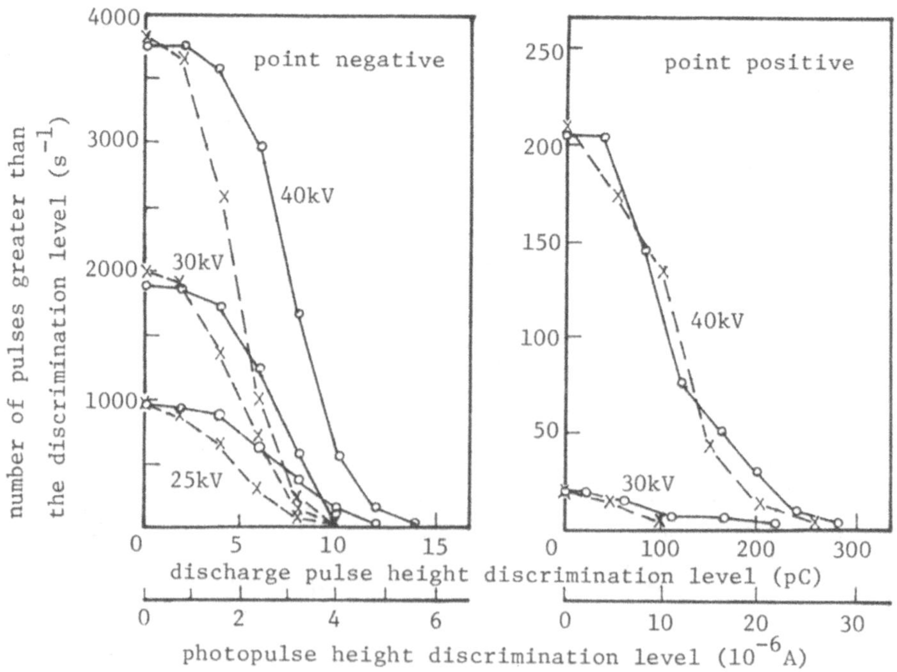

Figure 2. Distribution of discharge pulse height (solid curves) and photopulse height (dotted curves) in d.c. corona. The parameter is applied voltage.

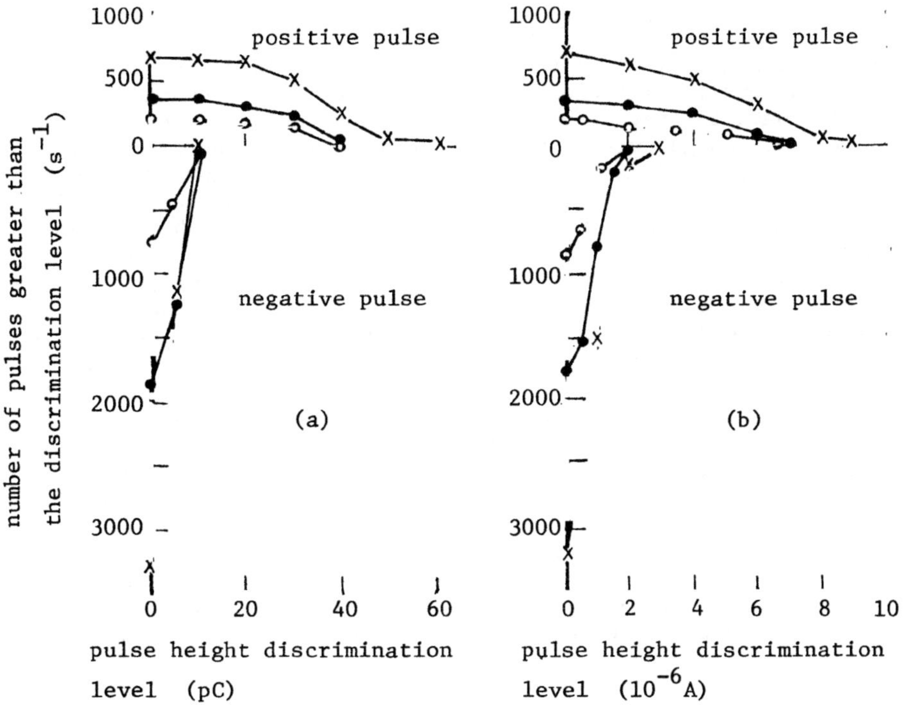

Figure 3. Distribution of (a) discharge pulse height and (b) photopulse height in a.c. corona. o :17kVrms •:20kVrms, x: 25kVrms.

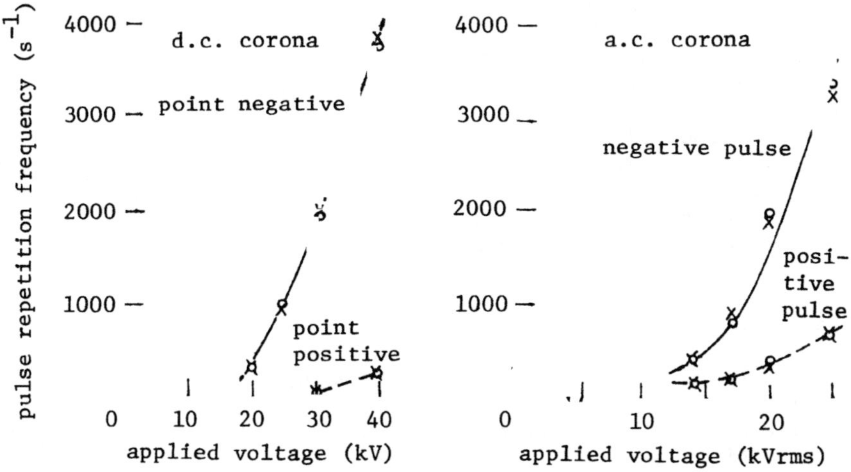

Figure 4. Repetition frequency of discharge pulses (o) and photopulses (x) in d.c. corona and a.c. corona.

3. 3. Discharge pulse and photopulse

The discharge pulse and the photopulse appeared on an oscillogram simultaneously. Therefore the occurrence of a discharge pulse corresponds to that of a photopulse. The distributions of height of the discharge pulses and the photopulses in d.c. corona are shown in figure 2, and those in a.c. corona in figure 3. The distributions of the discharge pulses and the photopulses resembled each other. It seems that the negative pulses and the positive pulses in a.c. corona correspond to the pulses in d.c. corona with the point negative and the point positive, respectively. The negative pulses were more frequent and smaller than the positive pulses. The d.c. corona with the point positive was intermittent, and its pulses were very large. The positive pulses in a.c. corona were smaller in magnitude and greater in number than the pulses in d.c. corona with the point positive. On the other hand, the distributions of the negative pulses in a.c. corona were similar to those in d.c. corona with the point negative. The total number of the pulses per second for a given applied voltage, i.e. the repetition frequency of the pulses, is shown in figure 4. The repetition frequency in d.c. corona with the point negative was more than ten times as great as that with the point positive, while in a.c. corona the frequency of the negative pulses was only fivefold greater than that of the positive pulses.

4. Discussion

The repetition frequency of the corona pulses is dependent upon polarity of the point electrode. When electrons and ions that have been produced by a discharge in front of the point tip are neutralized, a subsequent discharge will occur. Since the mobility of positive ions is small, it takes them possibly long time to go across the gap and be neutralized at the plane electrode in the case of the point positive, while electrons can be neutralized soon at the point tip. With the point negative, on the other hand, it is likely that positive ions are neutralized at the point tip, and electrons drift rapidly toward the plane. Therefore corona pulses will be more frequent with the point negative than with the point positive. In the case of a.c. corona, residual electrons produced by discharges in the preceding negative half cycle of 50 Hz a.c. probably facilitate the neutralization of positive ions in the positive half cycle. Therefore the repetition frequency of the positive pulses in a.c. corona will be greater than that in d.c. corona with the point positive.

References

(1) Takahashi Y 1973 Bull. Fac. Sci. Engng, Chuo Univ. 16 109 — 118

(2) Takahashi Y and Ohtsuka K 1975 J. Phys. D:Appl. Phys. 8 165 — 169

(3) Coelho R and Sibillot P 1970 Rev. Gen. Elect. 79 29 — 39

THE OBSERVATION OF DIELECTRIC BREAKDOWN IN LIQUID NITROGEN USING DISCHARGE FIGURES

H.Mitsui*, Y.Toriyama** and J.H.Calderwood*

1. Introduction

The use of Lichtenberg and dust figures in the study of discharges in some gases and in transformer oil is well known. It has now been found possible to apply these techniques to liquid nitrogen also, and the results are reported in this paper.

2. Test Equipment and Experimental Procedures

The test chamber (Fig 1) consists of a cylindrical glass vessel 40 cm high and 20 cm diameter thermally insulated with styrol foam. One of the stainless steel electrodes is a needle (tip radius 0.5, 1.0 or 5.0 mm); the other is a plane disc 14 cm in diameter, on which is mounted a 3 mm thick ebonite plate.

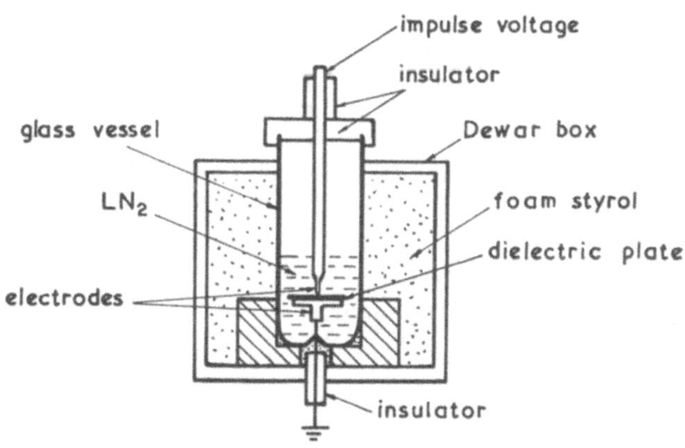

Fig. 1 Test cell and electrodes.

Lichtenberg figures were obtained by placing a photographic sheet on the plate, with the needle in light contact with the sensitive surface, introducing liquid nitrogen and applying an impulse voltage (1 x 40 µs) of either polarity. Dust figures were obtained in a similar manner, but without the presence of the photographic sheet. In this case, the pattern is developed after the discharge by sprinkling on the ebonite surface before it is removed from the liquid nitrogen a mixture of electrified sulphur and red lead. For the study of pressure effects, a special pressure chamber was used.

* Department of Electrical Engineering, University of Salford, Salford, Great Britain
** Department of Electrical Engineering, Masashi Institute of Technology, Tokyo, Japan

3. Results and Discussion

3.1 Figure Patterns

(a) Dust Figures. In general, at lower voltages there is little difference between positive and negative dust figures obtained in liquid nitrogen, though at higher voltages significant differences are observed. In particular above a certain voltage the positive figures become tree-like, and increase in size very rapidly with increase in voltage. The negative figures are more leaf-like, and grow much more slowly with voltage. Typical positive and negative dust figures are shown in Fig 2 and Fig 3 respectively. These figures are different from those obtained for discharges in air, but resemble those obtained for transformer oil.

(b) Lichtenberg Figures. Enlarged microscope photographs of typical negative and positive Lichtenberg figures are shown in Fig 4 and Fig 5 respectively. In general, the negative figure appears to be continuous and mainly along the surface, although it partially breaks into the photosensitive layer: however, the positive figure does not always appear to be continuous. The reason may well be that in the negative case, the electric field driving the electrons away from the needle has a component which attracts them towards the plate and so causes them to enter the photo-sensitive layer of the film; in the positive case, the electrons are attracted towards the needle and there is a field component repelling them upwards from the film into the liquid nitrogen, so causing breaks in the figure. The dust figures, produced in a different way, do not reveal this difference in figure continuity.

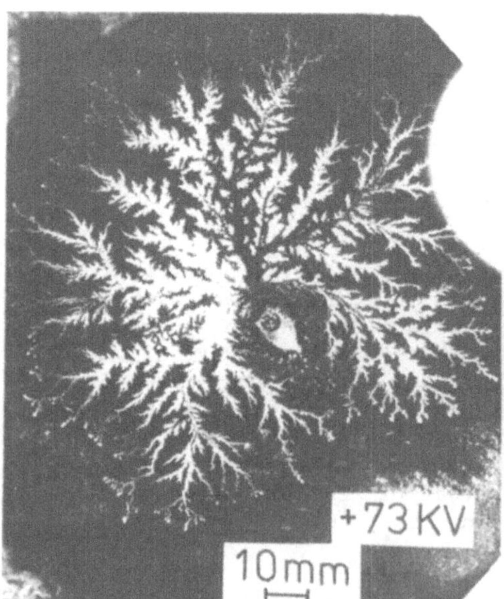

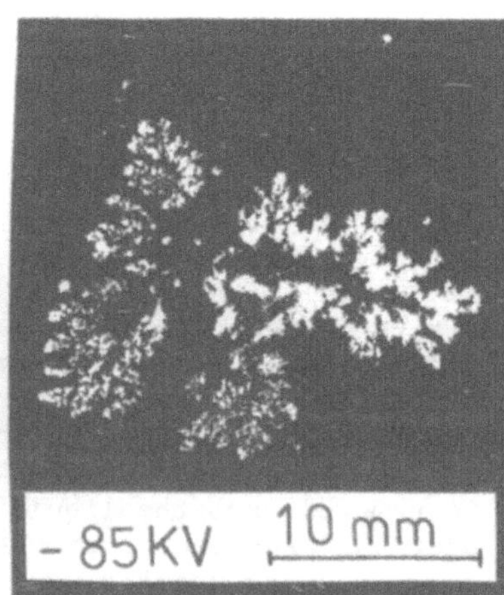

Figs 2 and 3 Positive and negative dust figure in liquid nitrogen

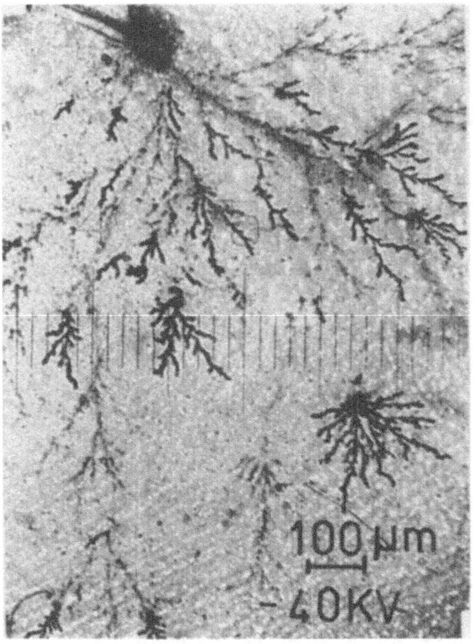

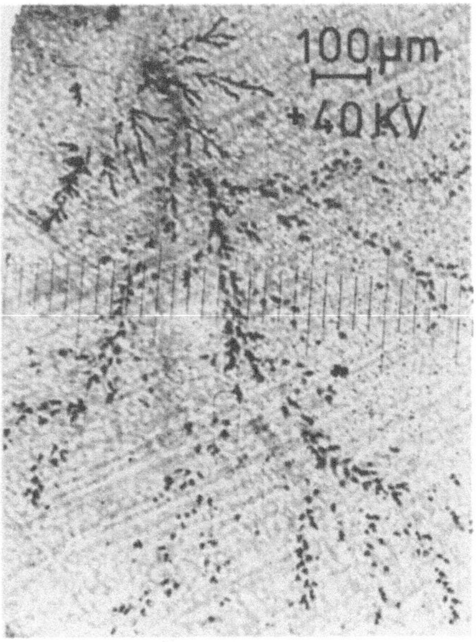

Fig 4 Negative Fig 5 Positive
Enlarged microscope photograph of the Lichtenberg figure in liquid nitrogen

3.2. Relationship between size of dust figure and applied voltage

As already mentioned in section 3.1, the size of the figures increases with increase in applied voltage; for a positive voltage, the rate of increase being suddenly greater when a certain critical voltage is reached, which is also related to the appearance of a tree-like figure. Experimental results for figure size as a function of voltage are shown in Fig 6, from which it can be seen that the critical positive voltage increases as tip radius increases. It is therefore possible that the same critical field is involved in all cases. It is

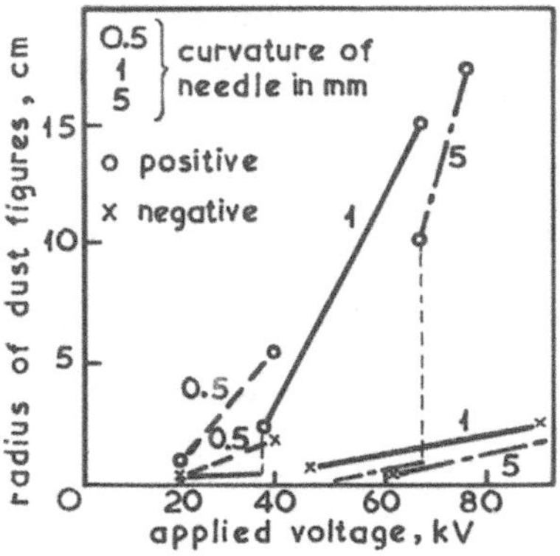

Fig. 6 Applied voltage and radius of dust figures.

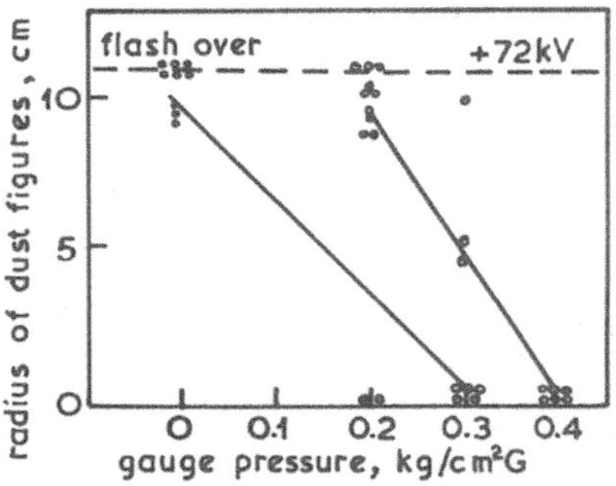

Fig. 7 Pressure effect.

interesting to speculate on the possible physical significance of such a field. Other investigators[1][2] have reported that the pulse strength required to produce a discharge is lower for nitrogen gas at $77^{o}K$ than for liquid nitrogen. It therefore seems likely that the evolution of bubbles would have an important influence on electrical discharge in liquid nitrogen. It may be that the critical field is that required to produce bubbles so vigorously that the nature of the discharge is altered, leading to the appearance of the tree-like figures with a rapid growth.

3.3 Effect of pressure on positive dust figures

It can be seen from Fig 7 that the pressure above the liquid nitrogen affects these figures significantly, the bigger being the pressure the smaller being the figure. The characteristic pattern of the figures is also affected, in that the tree-like shape is difficult to distinguish at a gauge pressure of about 0.4 to 0.5 kg cm^{-2}, and is never observed at 1 kg cm^{-2}. This observation is consistent with the importance to the discharge process of bubble production.
However, it has been reported [3] that at gauge pressures below 4 kg cm^{-2}, Lichtenberg figures do not alter significantly from those obtained at atmospheric pressure. The reason for this may be the difference in the discharge mechanism brought about by the possible burrowing of electrons into the photographic plate, which is of course not present for the dust figure experiments.

4. Conclusion

The experimental observations lend support to the suggestion that in liquid nitrogen, dielectric breakdown takes place in three stages. In the first stage, electronic multiplication occurs due to ionization by collision: in the second stage bubbles are produced when the energy transfer from electrons is sufficient to cause local boiling: the third stage involves the process of producing discharge in bubbles. Evidence for this third stage is given by the mist-like patterns observed in the tree branches of the Lichtenberg figures after flashover took place. The effect, which was only present if flashover occured, was attributed to discharges in the nitrogen gas bubbles.

References

1) K. Horii and Kosaki, Reports of study meeting on Electrical discharge, IEEJ, ED-72-23, 1972
2) T. Miura and Z. Jwata, Reports of study meeting on Insulating materials, IEEJ, IM-71-3, 1971
3) Kosaki and K. Horri, Reports of study meeting on Insulating material, IEEJ, IM-73-9, 1973

SOME ASPECTS OF BREAKDOWN IN LIQUID NITROGEN

L.P. Ó Gallchobhair and T. J. Gallagher*

1. Introduction

At the Dublin Conference on Dielectric Liquids, electrode effects on the strength of liquid nitrogen were reported (Keenan 1972). The results indicated that both the cathode and the anode are important in controlling the breakdown voltage. A tentative explanation was offered, attributing this control to electron capture by electronegative oxygen impurity molecules, along the lines suggested by Swan and Lewis for the anode influence on the d.c. strength of liquid argon (Swan and Lewis 1961). The results in this paper represent an attempt to confirm this explanation. By introducing oxygen as an impurity, it is shown that there is a significant alteration in the strength as the oxygen content is increased. The paper also examines the statistical properties of the breakdown sequences, including the distribution and the scatter of the results.

2. Apparatus and Procedure

The arrangement of the test cell is shown in Fig.1. The $\frac{3}{8}$" diameter hemispherical electrodes were made of copper, steel, brass or aluminium. Parallelism was ensured by

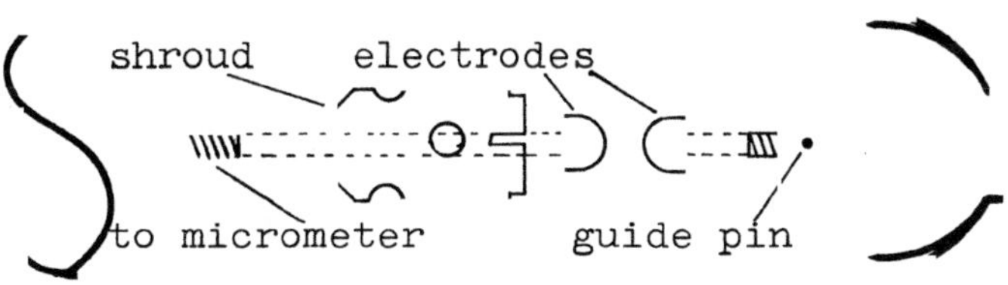

FIG. 1. Test Cell

mounting the top electrode in a PTFE shroud which was guided by two pins in the lower electrode holder. Windows were included in the shroud to prevent trapping of nitrogen inside the shroud and to allow observation of the gap. For the measurements, a constant gap of 80 μm was used which was set using a micrometer. An automatic electrical system, which was similar to that described by Secker and Hilton (1972), was used to control the experiment. Moreover, each breakdown value was recorded on a Digital Equipment Co. PDP-12 minicomputer, which allowed online calculation of the results.

*The authors are with the Department of Electrical and Electronic Engineering, University College, Dublin, Ireland.

Electrode preparation was by high speed buffing as used by Gallagher and Lewis (1964), followed by immersion in n-hexane to prevent oxidation (Hancox 1957). The electrodes were then examined under a metallurgical microscope to ensure a consistent finish, before assembly in the test cell.
The test system was evacuated to 10 μtorr for at least 24 hours before nitrogen gas, which was nominally oxygen free, was liquefied using an external jacket of liquid nitrogen, at atmospheric pressure. For tests requiring the addition of oxygen, known quantities of the gas were introduced into the cell before the nitrogen was liquefied.

3. Results

3.1 Electrode Effects. Table 1 compares the results of all previous investigations on the strength of liquid nitrogen with those obtained in the present work.

Table 1. Electric strength of nitrogen (MVm^{-1}), with different electrode materials, and with added oxygen (brass electrodes) († present work).

Steel	Brass	Cu	Al	Au	Pt	Br/Al	Ref.
100	-	-	-	-	93	-	1
180	-	-	-	150	224	-	2
80.5	-	-	-	-	-	-	3
146	83	-	114	-	-	77	4
162	70	104	-	-	-	-	5
-	100	-	80	-	-	74	†

% O_2	.6	.8	1
Strength	104	131	143

It is obvious that there is a definite electrode influence on the strength. This is true, not only of results obtained by different workers, but within series of measurements in the same laboratory. For instant, the present work yielded a strength of 100 MVm^{-1} and 80 MVm^{-1} for pairs of brass and aluminium electrodes, respectively, whereas Keenan (1972), from this laboratory obtained values of 83 MVm^{-1} and 114 MVm^{-1} for the same metals. The opposite tread in values is probably due to the fact that Keenan used a technique involving polishing with a diamond paste to preapre the electrode surfaces, rather than the buffing method used here. It is worth noting, however, that despite these different procedures the scatter in both sets of measurements was about equal. The coefficient of variation, in this work, varied from 15% to 22% about an average of 17.5%.

3.2 Oxygen Effects. Also included in Table 1 are some preliminary measurements on liquid nitrogen with added amounts of oxygen. There is a consistent rise with increase in oxygen concentration. Furthermore, this rise is accompanied by a change in the distribution of results in a sequence, as can be seen in Fig. 2. In the tests with oxygen added, some very high values of breakdown strength were recorded, which occasionally exceeded 200 MVm^{-1}. However, the scatter was not reduced by the oxygen, the average coefficient of variation for the above tests being in the region of 18%. Since experimental conditions are altered in a liquid-electrode system by each successive discharge, scatter in results is to be expected in any sequence of breakdown measure-

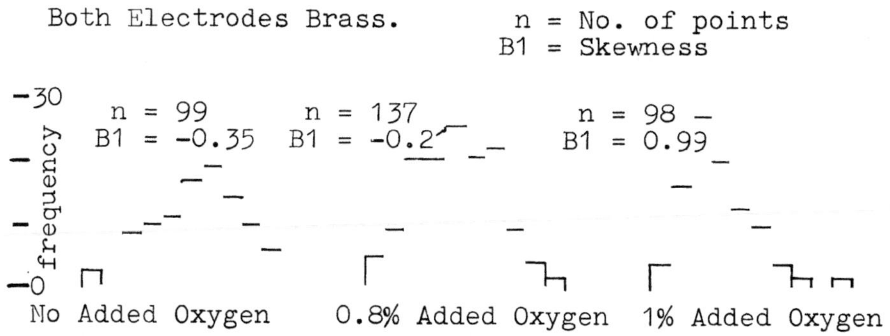

FIG. 2. Distributions of Results (Normalised)

ments. However, large scatter seems to be an inevitable feature of work on liquid nitrogen, whether control of the experiment is manual (Keenan 1972), or automatic, as in this work. At present, a physical explanation for the large scatter is not obvious. However, it may possibly be associated with induced liquid motion, which could create randomly-changing conditions at the electrode-liquid interfaces.

3.3. Miscellaneous Comments. On a few occasions when the gap was closed too tightly during a check on its setting, the breakdown strength was reduced in an anomalous fashion as is shown in Fig. 3. The scatter in results disappeared for some period of time even though the voltage at breakdown continued to rise. It was noted that slight variations in the vapour pressure of the sample affected the results. On occasions when the level of the refrigerant was adjusted, the breakdown voltage appeared to be affected. This may be explained by assuming that the test liquid was near to its boiling point. Jefferies and Mathes (1970) have found a twofold increase in the 60 Hz breakdown voltage of nitrogen by increasing the absolute vapour pressure from 4 to 3 bar in order to prevent boiling.

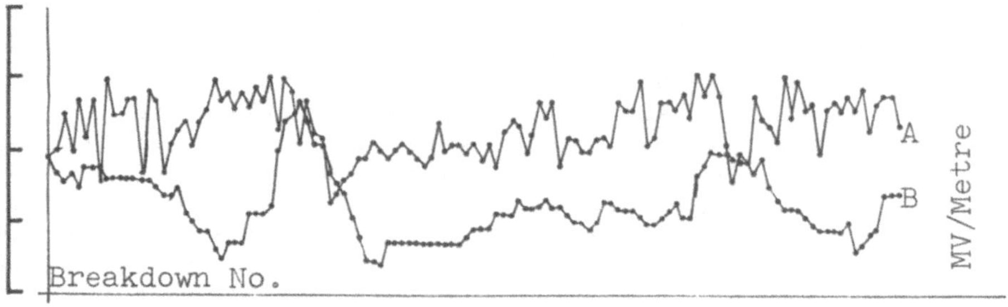

FIG. 3. Graph A Sequence of Breakdown Values
B Local Coefficient of Variation %

4. Conclusions

It is clear from the results here that the value of the breakdown strength of nitrogen can be altered by a change in the electrode metal or by the addition of impurities to the liquid. The effects mentioned above are under further investigation and it is hoped to supplement them by measurement of the conduction current pulses, provided these exist.

5. Acknowledgements

The authors would like to thank Professor J.J.Morrissey, Head of the Electrical and Electronic Engineering Dept., for the facilities provided, and University College, Dublin for monetary support.

References

1. Kronig R., Van der Vooren A., (1942), Physica 9, 139.
2. Swan D.W., Lewis T.J., (1960), J.Electrochemical Soc., 107, 180.
3. Mathes K.N., IEEE Trans., (1967) Vol. EI-2, No.1. April,
4. Keenan T.F., (1972), Conference on Breakdown and Conduction in Dielectric Liquids, Dublin, July 25-27.
5. Kawashima A., (1974). Cryogenics 14, 217.
6. Swan D.W., Lewis, T.J. (1961), Proc. Phys. Soc. LXXV111, 448.
7. Gallagher T.J., Lewis T.J., (1964). Brit.J. Appl. Phys., 15, 491.
8. Hancox, R., 1957, Brit. J. Appl. Phys., 8, 476.
9. Jefferies M.J., Mathes K.N., Sept. (1970), IEEE Trans. EI-5, 3.

STATISTICAL ANALYSIS OF THE DIELECTRIC BREAKDOWN IN LIQUID NITROGEN UNDER UNIFORM FIELD CONDITIONS *

L. Centurioni, G. Molinari, A. Viviani **

1 - Introduction

A study of the statistical behaviour of the breakdown strength of liquified gases seems to be of interest since they can both be used as dielectric liquids and allow the investigation of the behaviour of very simple liquids without giving rise to pollution effects by decomposition products during tests.

In this paper, a statistical analysis of the breakdown strength of liquid nitrogen in the millimeter gap range under uniform field conditions, with various electrode spacings and different uniform field areas, is reported. Tests have been carried out under controlled conditions, and with a new type of uniform field electrodes, which have some advantages over the commonly used ones. Finally, the obtained results are discussed.

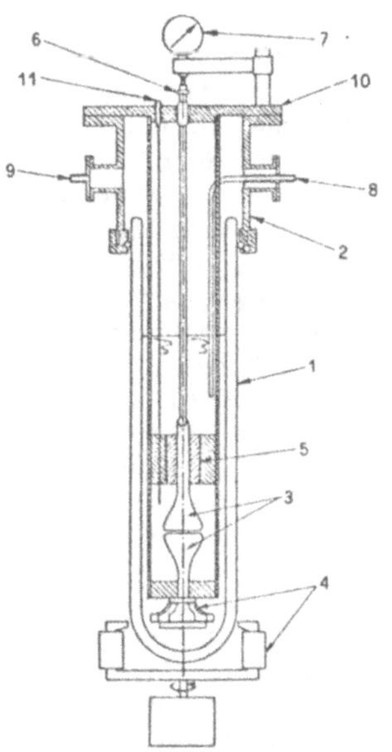

Fig. 1 - Testing apparatus
1, dewar - 2, supporting structure - 3, electrodes - 4, stirring device - 5, micrometric screw - 6, high voltage connection - 7, gap setting control - 8, LN2 inlet - 9, vacuum and bubblers connection - 10, upper flange - 11, sampling connection.

2 - Testing apparatus

The testing apparatus shown in Fig. 1 has been designed to carry out tests with as accurate a control as possible of all the test conditions. Particular care has been used to avoid liquid nitrogen pollution during transfer and tests. To this aim, the apparatus has been vacuum-cleaned up to 10^{-2} torr to remove polluting oxigen and water vapours, while nitrogen vapours exhaust during transfer and tests through a bubbler, preventing polluting air from re-entering. Liquid nitrogen to be tested is stored in an air-tight dewar, and is filtered before entering the test dewar. Gap setting is possible by means of a stainless steel tube connecting an electrode to the outside of the upper flange through the micrometric screw 5 in Fig. 1; the measurement of the gap setting is made by means of a comparator with electric zero. The gap setting has proven to be very stable, irrespective of the liquid nitrogen level in the dewar. To avoid

* Work supported by CNR, Gruppo Macchine Elettriche
** Department of Electrical Engineering, University of Genova, Viale F. Causa 13, 16145 Genova, Italy

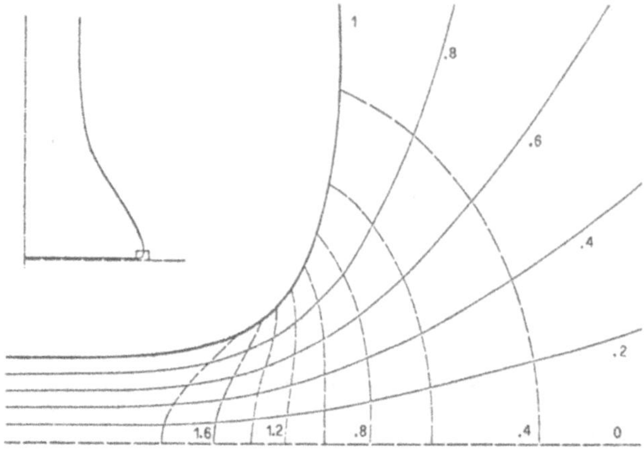

Fig. 2 - Electrodes shape and field patterns in the border area. equipotential lines (V), ——— equifield lines (V/mm)

settling of impurity particles, a stirring device has been provided, which is elettromagnetically driven from outside the dewar. More detailed informations on the testing apparatus can be found in Ref.1.

The uniform field electrodes used for the tests are shown in Fig. 2. The profiles of these new electrodes have been computed by a procedure similar to that used by Rogowsky, applied to a three-dimensional geometry instead of a two-dimensional one[2]; the field patterns generated by the electrodes are also shown in Fig. 2. These new electrodes, having the same uniform field area, present smaller overall dimensions, and, therefore, are more suitable to an area effect analysis in liquified gases, as in this case they reduce the dimensions of the involved dewars. Furthermore, a better definition of the uniform field area is obtained by means of a steeper variation of the electric field outside the uniform field area. By means of this field pattern, a strong reduction of the liquid volume subject to a strong electro-dielectrophoretic force is also obtained; it can be useful to reduce the changes in impurity concentrations along the electrode border.

3 - Experimental results

To perform the tests, boiling nitrogen has been admitted through a stainless steel filter with a mesh 80 μm in side. Some samplings during the tests have shown, on the average, the presence of 80 ppm of argon, 50 ppm of oxigen and 5 ppm of water. The time interval between the tests have been of 1 minute, and a stirring of the liquid has been performed after each test. Tests have been made on three sets of electrodes 30, 50 and 70 mm in uniform field area diameters, corresponding to 706, 1963 and 3848 mm^2 uniform field areas, respectively. For each electrodes set, tests have been performed with .2, .4, .6, .8, 1. mm spacings; for each spacing, a sample set of 160 breakdown voltages have been obtained. The first ten tests of each sample set have been discarded, since they have displayed a notably lower value than the following tests; a tipical diagram of breakdown strength versus the test order is plotted in Fig. 3. The main parameters of the statistical analysis of

TABLE I

Main statistical parameters of the test data

		smaller area					medium area					larger area				
		.2 mm	.4 mm	.6 mm	.8 mm	1. mm	.2 mm	.4 mm	.6 mm	.8 mm	1. mm	.2 mm	.4 mm	.6 mm	.8 mm	1. mm
Mbs*(MV/m)		27.2	23.7	22.6	21.5	20.8	23.9	21.2	19.1	18.6	18.2	22.2	19.6	18.2	17.8	17.2
S (MV/m)		.42	1.23	1.72	1.63	2.05	.40	1.01	.60	.67	.98	.32	.58	.67	.81	1.18
$\sqrt{b_1}$		-.96	-.14	-.17	-.17	-.05	-.88	-1.3	-.57	-.72	-.70	-1.01	-1.0	-.85	-1.0	-.99
b_2		3.64	3.36	3.16	2.96	3.13	3.68	4.32	3.71	4.51	3.23	4.15	5.03	4.44	4.20	3.92
extreme val. distribution	ts	4.10	8.70	11.0	5.30	12.1	4.60	10.2	6.92	18.4	13.8	2.30	3.80	4.60	6.30	7.00
	γ	7	7	7	7	6	7	7	7	7	7	6	8	7	7	6
	p%	77	28	15	62	6	71	18	64	1	6	89	87	70	50	24
normal distribution	ts	21.5	9.00	4.30	8.20	6.80	18.3	38.5	8.40	33.9	21.1	13.3	9.30	7.60	10.7	18.3
	γ	7	7	7	7	6	7	7	7	7	7	6	8	7	7	6
	p%	<.5	26	74	32	44	1	<.5	30	<.5	<.5	4	32	37	16	<.5

Mbs, Mean breakdown strength - S, Standard deviation - $\sqrt{b_1}$, Skewness - b_2, Curtosis - ts, test statistic - p%, probability - γ, degrees of freedom - *, r.m.s. values

the 15 groups of 150 data obtained in this way are listed in Table I. An estimate of the data fitting to an extreme value distribution can be obtained both by the chi-squared tests performed on the data, and by the estimates values of skewness and curtosis. These latter are to be compared with the theoretical skewness and curtosis values (-1.14 and 5.4, extreme value distribution; 0. and 3., normal distribution). Plots of the mean breakdown strength versus the gap setting and the uniform field area under test are given in Figs. 4 and 5, respectively.

4 - Discussion

The analysis of the statistical parameters displayed in Table I shows that an increase in the uniform field area gradually gives rise to a better fit of data to an extreme value distribution. In fact, whereas for IEC spherical electrodes (which can be considered as uniform field electrodes with very limited area) the distribution turns out to be normal in the same gaps [1], a doubtful fitting to the extreme value distribution can be found in the case of electrodes 706 mm^2 in area, and an increasingly better fit is found as area increases.

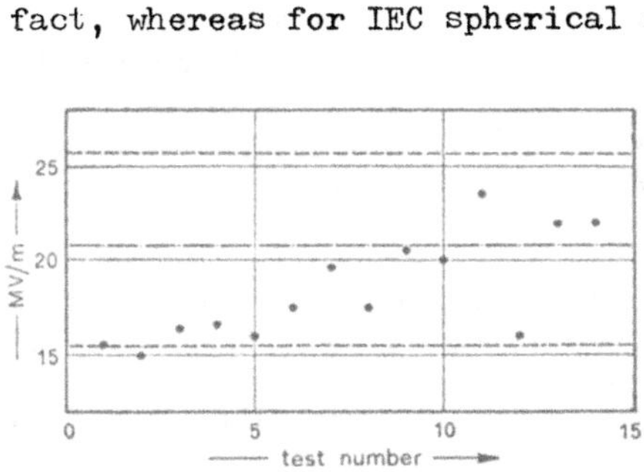

Fig. 3 - Breakdown strength E versus test order for the initial tests of a sample.

—·—, mean value of the test set
-----, minimum and maximum values

As can be seen in Figs. 4 and 5, a remarkable decrease in breakdown strength has been found both as uniform field area and electrode spacing increase, according to the known behaviour of dielectric

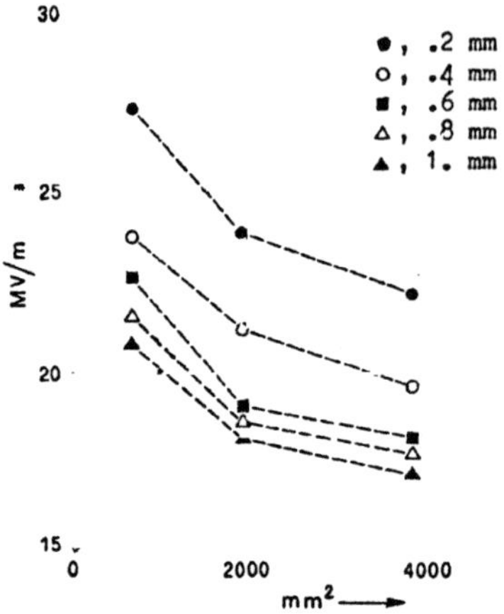

Fig. 4 – Breakdown strength versus uniform field area for the various gap settings.

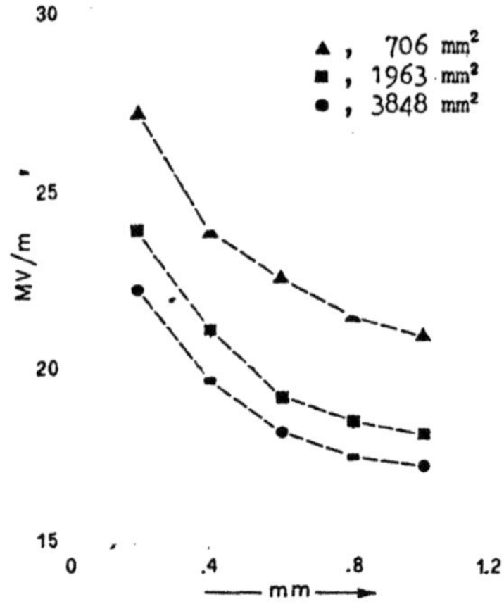

Fig. 5 – Breakdown strength versus gap setting for the various uniform field areas.

liquids commonly investigated. A sligthly better fit to the extreme value distribution has been obtained, for each uniform field area, by the lower spacings with respect to the higher ones, as can be seen in Table I. This phenomenon seems to suggest the presence of two separate area and distance effects rather than a volume effect, since two pairs of electrodes with the same volume of liquid subject to a uniform field could give rise to breakdown strengths following different distributions.

Finally, it can be noted that the stirring time necessary to limit the mean increase in breakdown strength during tests down to values around some volts per test has turn out to be half the one which has been necessary in the case of IEC spherical electrodes [1]. This fact seems to confirm the strong influence of impurities on breakdown phenomena, as well as the very limited electro-dielectrophoretic effect of the new uniform-field electrodes with respect to the spherical ones.

[1] G.Molinari:"An apparatus for breakdown tests in liquid nitrogen under controlled conditions and statistical analysis of the dielectric strength of liquid nitrogen with IEC spherical electrodes" to be published

[2] L.Centurioni, G.Molinari, A.Viviani: "Optimal design of uniform field electrodes for breakdown tests in liquid dielectrics" Report SL-10-74, Dept. of Electrical Engineering, University of Genova, Nov. 74

[3] A.Kawashima:"Electrode area effect on the electric breakdown of liquid nitrogen" Cryogenics, Vol. 14, n° 5, 1974, pp. 217-219

[4] J.K.Nelson, B.Salvage, W.A.Sharpley: "Electric strength of transformer oil for large electrode areas" Proc. IEE, Vol. 118, n° 2, 1971, pp. 388-393

[5] M.J.Jefferies, K.N.Mathes: "Dielectric loss and voltage breakdown in liquid nitrogen and hydrogen" IEEE Trans. on Electr. Insul., Vol. EI-5, n° 3, 1970, pp. 83-91

DIELECTRIC BREAKDOWN OF FLOWING LIQUID NITROGEN UNDER D.C. AND A.C. VOLTAGE *

L. Centurioni, B. Delfino, G. Molinari, A. Viviani **

1 - Introduction

In many electrical apparatuses using superconducting or cryoresistive materials, liquid nitrogen, used as a thermal shield, can also be employed as a dielectric, often under flowing conditions.

On the other hand, the presence of a velocity field in the breakdown area can also be useful to assess the influence of impurities or inhomogeneities on the breakdown process: in fact, if this influence is strong, velocity can notably alter breakdown phenomena. Since liquid nitrogen is a particularly simple dielectric liquid, the study of the dielectric breakdown of liquid nitrogen under flowing conditions seems to be interesting from both an engineering and a physical point of view.

In this paper, results of a.c. and d.c. breakdown tests on flowing and stationary liquid nitrogen are reported. Tests have been performed at various speeds with a test cell previously developed [1,2]. A statistical analysis of the obtained data is also given.

2 - Testing apparatus

A schematic view of the testing apparatus is shown in Fig. 1. Prior to tests, the apparatus is vacuum-cleaned through connection 5 in Fig.1 to ensure minimum pollution of oxygen and ice crystals. Liquid nitrogen is then admitted into the pump dewar 1 and into the thermal stabilisation dewar 2 through the filtered inlets 4. Nitrogen vapours exhaust through the bubbler 6, and backstreaming of oil is stopped by the liquid nitrogen trap 13. Through the stainless steel filter 7 and the thermal stabilisation coil 9, liquid nitrogen to be tested is pushed by the pump 3 into the test cell 10, and then into the pump dewar through the hot-wire flowmeter 11. In order to obtain, at the greatest possible extent, the influence of velocity on breakdown strength without a superimposed pressure effect, a low-lift pump has been selected, and the circuit downstream the test cell has been designed in order to minimize pressure drops. A control of the bubbles absence in the liquid flow can be made both by means of the glass windows 8 and by the analysis of the flowmeter response.

An axial section of the test cell is shown in Fig. 2. The inner electrode is spaced and centered with respect to the outer one by means of three PTFE spacing rings, 3, 4 and 5 in Fig. 2. The upper ring 3 also ensures the separation of the liquid nitrogen under test from the liquid nitrogen of the thermal stabilisation bath,

* Work supported by ENEL under contract CRE 11/1974
** Department of Electrical Engineering, University of Genova, Viale F. Causa 13, 16145 Genova, Italy

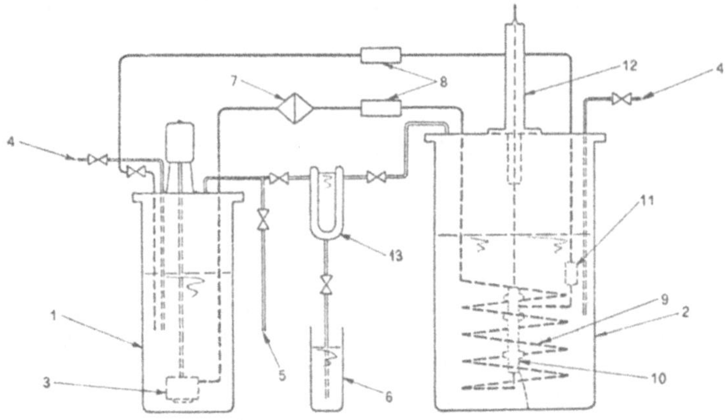

Fig. 1 - Schematic view of the testing apparatus

1, pump dewar - 2, thermal stabilisation dewar - 3, pump - 4, nitrogen inlets - 5, vacuum connection - 6, bubbler - 7, filter - 8, glass windows - 9, thermal stabilisation coil - 10, test cell - 11, flow rate meter - 12, high voltage bushing - 13, LN2 trap.

whereas the lower ring 5 gives rise, through a series of small holes, to a localized pressure drop ensuring uniformity of flow conditions in the test cell. Experimental measurements on a Reynolds model of the test cell have shown that a turbulent flow is present in the test cell at every velocity chosen for the tests reported in this paper. More detailed information on the flow patterns in the test cell can be found in Ref. [2].

3 - Experimental results

Tests have been made on boiling nitrogen under flowing and stationary conditions under a.c. and d.c. voltage of both polarities. To perform such tests, the apparatus has been cleaned up to 10^{-2} torr, and a stainless steel filter with meshes 80μm in side has been inserted in the filter 7 in Fig. 1. Samplings on the liquid nitrogen used for tests have shown, on the average, the presence of 80 ppm of oxygen, 60 ppm of argon and 30 ppm of water. The voltage rise has been obtained by means of an automatic device with a uniform rise rate of 3 kV/s. Tests have been made at three different flow rates, i.e., .1 l/s, .2 l/s and .3 l/s, corresponding to a nitrogen average speed in the test cell of 1.275 m/s, 2.55 m/s and 3.825 m/s, respectively. Furthermore, by removing the spacing ring 3 in Fig. 2, tests under stationary conditions have been performed using the thermal stabilisation dewar as a container of the liquid under test. The

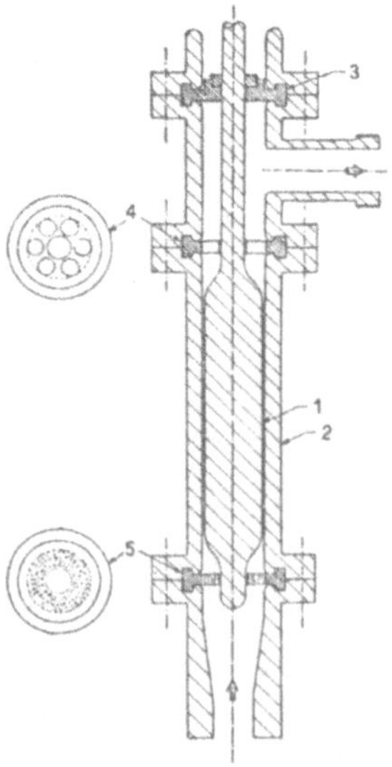

Fig. 2 - Axial section of the test cell.

1, inner electrode - 2, outer electrode - 3,4,5 spacing rings

TABLE I

Main statistical parameters of the test data

	stationary			flowing								
				.1 1/s			.2 1/s			.3 1/s		
	a.c.	d.c.+	d.c.-	a.c.	d.c.+	d.c.-	a.c.	d.c.+	d.c.-	a.c.	d.c.+	d.c.-
Mean break. strength (MV/m)	24.3*	19.5	22.9	27.4*	21.3	25.4	25.6*	19.3	23.7	23.6*	16.0	19.9
Standard deviation (MV/m)	1.3	1.5	1.4	1.9	2.0	1.8	1.7	1.4	1.7	1.9	1.4	1.7
Skewness, $\sqrt{b_1}$	-1.1	-1.3	-1.2	-.06	.38	-.15	.07	.08	-.4	.02	.54	-.02
Curtosis, b_2	4.6	4.3	4.5	3.4	2.7	2.7	3.1	2.5	2.4	2.6	3.2	2.4
extreme val. distribution: test statistic	12.8	6.4	9.3	9.5	21.4	12.2	14.4	12.9	9.1	7.0	20.5	17.3
extreme val. distribution: degrees of freedom	6	5	5	6	6	6	6	6	5	6	6	6
extreme val. distribution: probability %	5	27	10	15	<.5	6	2.5	5	11	42	<.5	.75
normal distribution: test statistic	32.0	33.8	40.9	8.7	7.0	8.8	11.6	3.5	4.6	4.0	13.9	9.8
normal distribution: degrees of freedom	6	5	5	6	6	6	6	6	5	6	6	6
normal distribution: probability %	<.5	<.5	<.5	19	32	19	7.5	74	47	68	3	14

* peak values

time interval between tests has been of 1 minute. The main results of the statistical analysis of the 12 test groups performed under the above stated conditions are shown in Table I. For every group, 110 tests have been worked out; the first ten tests of each group have been discarded to ensure the reaching of a steady state condition. Fig. 3 gives the mean dielectric strength versus the flow rate in the test cell, whereas Figs. 4 and 5 show an example of plottings on normal and extreme value probability papers.

4 - Conclusions

An analysis of the data reported in Table I shows that the turn from extreme value to normal distribution, shifting from stationary to flowing conditions, takes place under both d.c. and a.c. voltages, according to a.c. results previously obtained [1,3]. This distribution shift is shown both by the chi-squared tests and the value of skewness and curtosis, as compared with the theoretical ones (0.0 skewness, 3.0 curtosis, normal distribution; -1.14 skewness, 5.4 curtosis, extreme value distribution). A confirmation of this trend can be also obtained by probability

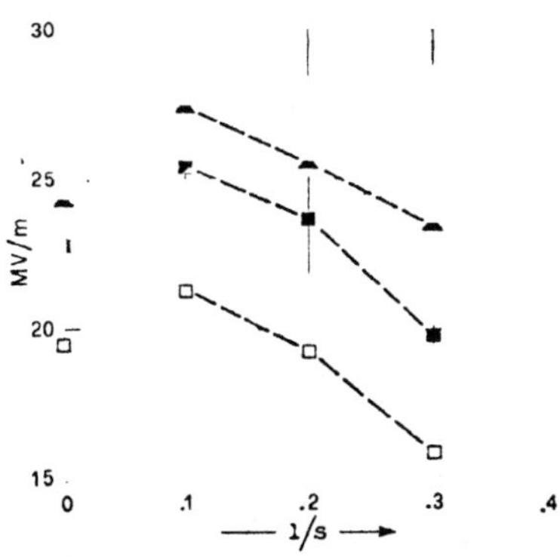

Fig. 3 - Mean breakdown strength versus test flow rate.

▲ , a.c. peak values
□ , d.c. plus polarity
■ , d.c. minus polarity

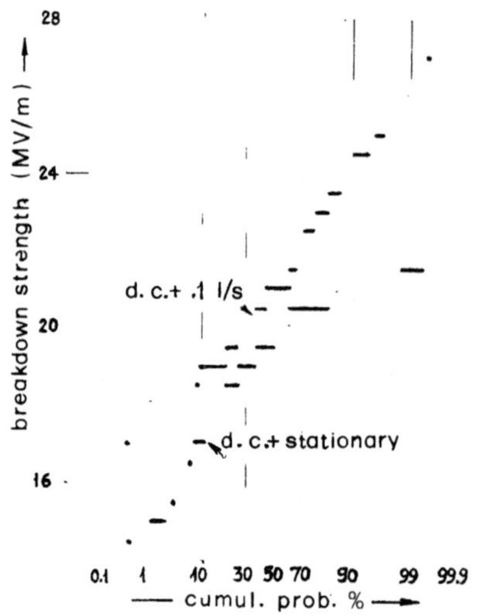

Fig. 4 - Normal probability plottings.

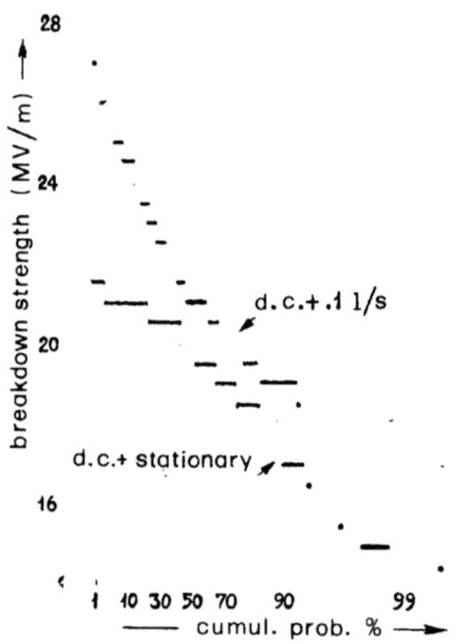

Fig. 5 - Extreme value probability plottings.

plottings, as can be seen in Figs. 4 and 5. Since the distribution shift appears not to be affected by the type and polarity of applied voltage, nor by test velocities, the previously reported hypotheses on the influence of impurities particles and inhomogeneities [1] seem to gain further support.

As far as the mean values of breakdown strength are concerned, one can note from Fig. 3 that an increase in breakdown strength occurs turning from stationary to flowing conditions; however, as speed increases, breakdown strength falls down, reaching values lower than the stationary ones for a speed of 3.825 m/s. This effect seems to be due to the turbolence amount, which rapidly increases as speed does, and gives rise, at the highest speeds, to an easier formation of gas bubbles in the test cell. Therefore, such a process seems to be very sensitive to pressure variations, so that a residual influence of the pressure on the amount of mean breakdown values shifting cannot be excluded.

Finally, it can be observed that under all testing conditions the maximum a.c. breakdown strength was found to be higher than the d.c. one for both polarities, according to the current experience concerning other liquified gases.

[1] L.Centurioni, G.Molinari, A.Viviani: "Experimental contribution to the study of the a.c. breakdown voltage in flowing liquid nitrogen", being published on 'L'Energia Elettrica'

[2] L.Centurioni, G.Molinari, A.Viviani: "Study of the dielectric behaviour of flowing liquid nitrogen", Journée d'etude sur le comportement des diélectriques à basse température, Paris, 14 oct. 1974

[3] J.K.Nelson, B.Salvage, W.A.Sharpley: "Electric strength of transformer oil for large electrode areas" Proc. IEE, Vol. 118, n° 2, 1971, pp. 388-393

CHEMICAL RELAXATION OF THE NON-LINEAR DIELECTRIC EFFECT IN DIPOLE EQUILIBRIA

L. Hellemans*

At high field strength the non-linear terms in the description of the polarization become important. Changes of the dielectric permittivity with the electric field are usually studied by noting the frequency shift of an oscillator, whose resonance condition depends on the properties of the dielectric to be investigated (1). The system is subjected at the same time to a high static field (which may last for no longer than 10^{-3} s) and a rf voltage of low amplitude. In the presence of an electric field E the polar molecules are oriented, the average of the cosine of the angle θ between molecular dipole vector and field direction being expressed by the Langevin function $L(\mu E/kT)$. The higher terms of the series expansion of L account for the gradual decrease of the permittivity with field strength in the case of non-interacting systems. Thiébaut has derived an equation for the normal saturation in liquids, corresponding to the second term of the expansion of L (2).
It is the purpose of the present communication to analyse the origin of the different contributions to the total non-linear dielectric effect measured for interacting systems. However, the non-linearity due to orientation of the polarizability ellipsoid, to hyperpolarizability and to electrostriction will be neglected.
If a system in chemical equilibrium is subjected to the action of a high field, the chemical composition will be modified to reach a new equilibrium state, in which the expense of chemical free energy is compensated by the gain of electrostatic stability.
Bergmann, Eigen and De Maeyer (3) have shown on thermodynamic grounds that the change of equilibrium constant K with field strength is given by

$$d\ln K/dE = (1/RT)(\delta M/\delta \xi)_E \qquad (1)$$

M represents the total electric moment, which to first order depends on E. The generalized unit of reaction ξ (extent of reaction) is introduced, $d\xi$ being related to the concentration change dc_i of the reaction partners i by $d\xi = dc_i/\nu_i$, in which ν_i is the stoichiometric coefficient (positive for reaction products, negative for reactants).
Integration of Eq. (1) learns that the relative change

*Department of Chemistry, University of Leuven, Celestijnenlaan 200 D, 3030 Heverlee, Belgium

of the equilibrium constant is proportional to the square of the field strength applied. This fact implies that a small alternating voltage superposed on a high static field such as currently used to measure field dependent permittivity changes, perturbs the equilibrium composition if the molar moment of reaction $\delta M/\delta\xi$ is not zero. In the limit of zero field strength no perturbation is possible either.

The authors of ref. (3) have used this quadratic relationship to study the relaxation behavior of the chemical transformation involved. Excess losses $\Delta\tan\delta$ induced by the presence of the high field are measured as a function of the frequency of the alternating HF voltage. Non-linear field effects are determined experimentally as the difference between the permittivity measured at high field conditions and that quantity measured at zero field strength. Therefore, we shall define the permittivity by the derivative of the displacement. At high field strength E we have

$$\varepsilon_0\varepsilon(E) = dD/dE \qquad (2)$$

with ε_0 as the permittivity of free space.

It is advantageous to separate that part of the permittivity, which arises as the consequence of the perturbation of the chemical equilibrium by the alternating voltage in the presence of the high field.

$$\varepsilon_0\varepsilon(E) = (\delta D/\delta E)_\xi + (\delta D/\delta\xi)_E(\delta\xi/\delta E) \qquad (3)$$

Since this analysis is concerned with perturbation, the first term at the right hand side is to be taken at constant composition ξ, which corresponds to the equilibrium condition at field E.

We relate this quantity to the permittivity of the system, which would be recorded at high field strength under the condition that by some means the chemical composition is prevented from deviating from the equilibrium established at zero field.

$$(\delta D/\delta E)_\xi = (\delta D/\delta E)_{\xi(0)} + (\delta/\delta\xi)(\delta D/\delta E)_{\xi=\xi(0)}\cdot\Delta\xi \qquad (4)$$

wherein $\Delta\xi$ is defined by $\Delta\xi = \xi - \xi(0)$.

The last term may be approximated by $\varepsilon_0(\delta\varepsilon/\delta\xi)\cdot\Delta\xi$ without specifying exactly at what field strength and at which equilibrium composition the derivative is to be determined. Indeed, the field induced concentration changes are extremely small, as is the effect of the field strength on the permittivity.

By subtracting the zero field permittivity $\varepsilon_0\varepsilon(0) = \delta D/\delta E$ from Eq. (3) in which the result of Eq. (4) is substituted, one obtains the field induced permittivity increment $\Delta\varepsilon$

$$\varepsilon_0\Delta\varepsilon = L + \varepsilon_0(\delta\varepsilon/\delta\xi)\Delta\xi + (\delta D/\delta\xi)_E(\delta\xi/\delta E) \qquad (5)$$

The term L represents the decrease of permittivity by normal saturation at constant composition $\xi(0)$. The other terms at the right hand side are positive contributions to $\Delta\varepsilon$. The first of these terms accounts for the equilibrium shift produced by the high static field. This contribution will vanish at frequencies of the order of the inverse orientational relaxation time of the polar reaction partners. We calculate $\Delta\xi$ by

$$\Delta\xi = \int_0^E (\delta\xi/\delta E) \cdot dE \qquad (6)$$

The second term is the consequence of an increase in polarization, when the chemical composition is changed by the action of the HF voltage in the presence of a high field. The relaxation behavior of this contribution depends entirely on the chemical rate processes and can be expected inthe MHz region for diffusion controlled reactions. In the following we shall only have concern for the frequency dependence of this particular term. The partial derivatives in Eq. (5) and (6) have been evaluated (3). The final result for the complex chemical contribution to the field induced change of ε can be written as

$$\Delta\varepsilon = (\varepsilon_0\Gamma/RT)(\delta\varepsilon/\delta\xi)^2 E^2 \left[1/2 + 1/(1 + j\omega\tau)\right] \qquad (7)$$

wherein the frequency dependence of the chemical perturbation term is indicated explicitly. The factor $\Gamma = (\sum\nu_i^2/\bar{c}_i)^{-1}$ is the reaction capacity, depending on the equilibrium concentrations $\bar{c}_i$ of the reaction partners. The change of ε with reaction advancement is calculated from Onsager's equation

$$\delta\varepsilon/\delta\xi = \left[\varepsilon^2(n^2+2)^2/(n^4+2\varepsilon^2)\right] \cdot (N_0/9\varepsilon_0 kT)\sum\nu_i\mu_i^2 \qquad (8)$$

with n as the refractive index, N_0 Avogadro's number and μ_i the dipole moment of the reaction partner i. We have developed in our laboratory a differential measuring technique using alternating high fields (4). Both real and imaginary parts of the field induced $\Delta\varepsilon$ can be determined.

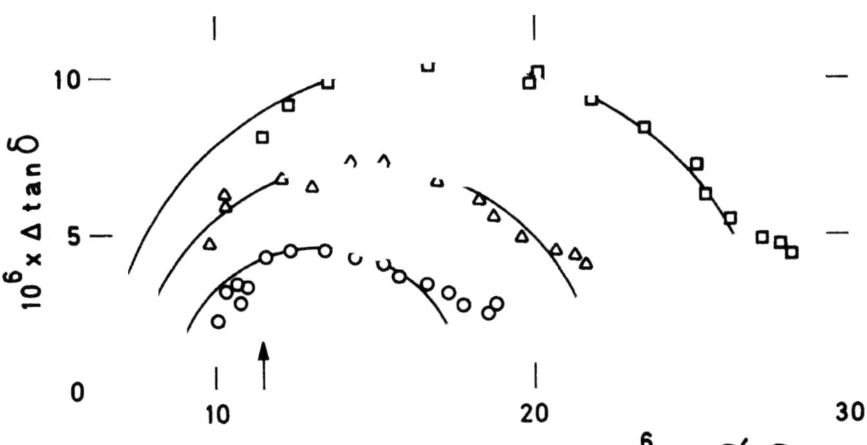

Fig. 1

Partial results for the dimerization of ε-caprolactam in CCl_4 at 22°C and E = 2.08 x 10^7 V/m are represented as Cole-Cole plots in Fig. 1. The concentrations of lactam are 0.0116, 0.0249 and 0.0420 M with increasing radius of the semicircle. The data are taken in the frequency range of 3 to 90 MHz. The relative permittivity increase noted for the pure solvent is indicated by an arrow.

Relaxation times and amplitudes agree with a dimerization mechanism with equilibrium constant of 200 M^{-1} and dipole moments of 4.0 and 0 D for monomer and dimer respectively. This is in good agreement with Hopmann's results obtained from the field induced incremental losses in this system (5).

The position of each semicircle with respect to the $\Delta\varepsilon'/\varepsilon$ value measured for the solvent is in accordance with the values of $\Delta\varepsilon$ from Eq. (7) and of L in Eq. (5), when they are calculated with the help of the above parameters K and μ_i.

(1) S. Kielich in "Dielectric and Related Molecular Processes", M. Davies ed., Vol.I, The Chemical Society, London (1972) p 325

(2) J.M. Thiébaut, J.L. Rivail and J. Barriol, J.C.S. Faraday II 68, 1253 (1972)

(3) K. Bergmann, M. Eigen and L. De Maeyer, Ber. Bunsenges. Phys. Chem. 67, 819 (1963)

(4) L. Hellemans and L. De Maeyer, to be published

(5) R. Hopmann, J. Phys. Chem. 78, 2341 (1974)

POWER FACTOR OF SILICONE LIQUIDS

A.Ohashi, T.Teranishi and M.Ueda *

Introduction

Silicone liquids are coming increasingly in use in dielectric and insulating systems because of their excellent thermal, chemical and electrical characteristics. In order to more fully characterize their dielectric properties, the measurement has been made in a wide range of viscosity (0.65 CS to 1000 CS), temperature (-180 °C to 120 °C), and frequency (30 Hz to 1 MHz) with two different dielectric cells (a) and (b); (a) is the ordinary coaxial cylindrical electrodes with a spacing gap of 1.5 mm, and (b) is such that the metal electrodes, which are evaporated on optical flats, are placed face to face with a narrow gap of several microns, and so the field strength in this cell becomes several ten KV/cm.

Experimental

The silicone liquids used for the investigation are poly-dimethyl-siloxanes designated by M_2D_n [1] having the structure $(CH_3)_3$-Si-[O-Si-$(CH_3)_2]_n$ -O-Si-$(CH_3)_3$. For convenience, we classify the liquid samples into 4 groups; A(M_2, M_2D, M_2D_2), B(M_2D_3, M_2D_4, 5 CS), C(10 CS, 20 CS), and D(50 CS to 1000 CS). These samples are poured into the dielectric cell housed within the temperature controlled chamber. The cooling or heating rate are 0.1 to 1 °C/min. In addition, the temperature controlled chamber has the plastic window to observe the change of appearance of sample according to the temperature change.

Results and discussion

In the first place, we depict the results obtained under weak field condition using the dielectric cell (a). Fig.1 and Fig.2 show the temperature dependence of dielectric constant ε' and power factor $\tan\delta$ for silicone liquids from 20 °C down to -160 °C at 1 KHz, in which the arrow marks indicate their freezing points. It is seen that, in liquid state in every specimen, ε' decreases lineary with decreasing temperature and no $\tan\delta$ peak appears.
In the case of liquids A, ε' decreases discontinuously corresponding to the sudden change of their appearance from water white to snow white at the freezing points, but small $\tan\delta$ peaks appear in solid state at about -150 °C with the order of magnitude of 10^{-4} as shown in Fig.3. This meens that liquids A crystalize at the freezing points with trace amounts of amorphous components.
On the other hand, liquids B stay in liquid state even under the melting points, and solidify with clear white at about -140 °C, where the dielectric dispersion and absorption are observed. It is seen in Fig.4 that the sharp and high $\tan\delta$ peak appears at glass

*Dept. of Electrical Engineering, Nagoya University, Nagoya, Japan

transition temperature and shifts to higher temperature range with increasing frequency. For summary, the dielectric behavior of low viscous liquids A and B are characterized by Debye's viscous rotation theory for polar liquids in amorphous region.
In liquids C, on cooling, their appearance changes continuously from water white through semitransparent white to snow white finally, and correspondingly it seems that the crystalization progresses gradually. The dielectric behavior of liquids C differs from other liquids A, B and D in that their dielectric constant increases rapidly according to the progress of crystalization, and tanδ peaks become sharper and higher with increasing frequency as shown in Fig.2 and Fig.5. This suggests that the dipole moment in amorphous region increases due to the crystalization of neighboring molecules.
In liquids D, the change of appearance on cooling is the same as that of liquids A. But, in contrast, no sharp transition is seen in the dielectric constant, instead, a small jump is seen at freezing point of about -70 °C, and the broad tanδ peak appears in solid state as shown in Fig.6. It seems that the dielectric loss in liquids D is caused by the various modes of segmental motion of long chain molecule in amorphous region. Note that liquids D have a value of 0.16 for Cole's parameter (introduced by Cole and Cole to account for a possible distribution of relaxation times), while liquids A and B have a Cole's parameter of 0.8 (corresponds to almost single relaxation). This agrees with the idea that the unit of flow for poly-dimethylsiloxane is a constant size of approximately 6 silicon atoms[1].

In the next place, we depict the results obtained under high field condition using the dielectric cell (b). Fig.7 shows the tanδ vs. frequency in log-log scale for M_2D_3 with temperature as a parameter. It is seen that tanδ decreases gradually with increasing frequency, and the slope increases with increasing temperature at lower frequency range. Replotting the data on semi-log scale, it is seen in Fig.8 that log tanδ is a linear function of temperature and the slope increases with decreasing frequency. These results are similar with those of glasses.[2] In our case, it seems that the dielectric loss is caused by the ionic conduction due to the presence of trace amounts of water. It is interesting to note the dielectric behavior of silicone liquid with a large amounts of impurity water under high field condition. Fig.9 shows the tanδ vs. frequency curves for such a sample including city water of about 400 ppm. with field strength as a parameter. It is seen that, under 30 KHz, tanδ depends on field strength, and at intermediate frequency range the slope of log tanδ vs. log frequency curves steepens with increasing field strength (approachs to 1), and at lower frequency range tanδ peak appears and shifts to higher frequency range with increasing field strength. The relation between tanδ and field strength is shown in Fig.10. It is seen that the field strength giving the maximum tanδ increases with increasing frequency. These behaviors suggest that the ionic space-charge loss predominates and the origin of ions can be attributed to the dissociation of impurity water molecules and/or the impurity ions dissolved from impurity water into silicone fluid. Note that the order of magnitude of 10^{-4} cm^2/v.sec for the mobility of

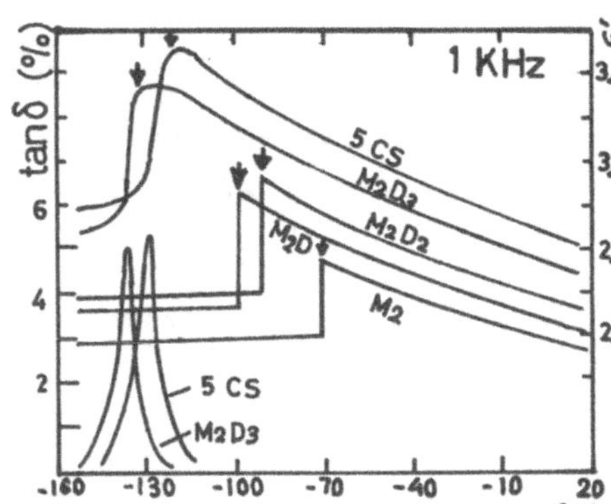

Fig.1 temperature (°C)

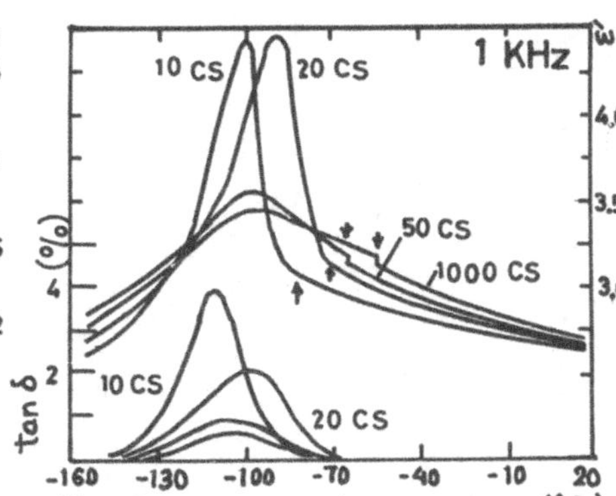

Fig.2 temperature(°C)

Fig.1, Fig.2 The dielectric constant and power factor as a function of temperature for silicone liquids at 1 KHz.

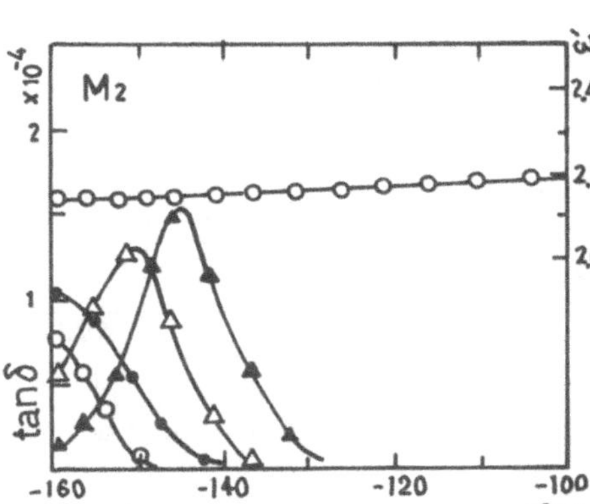

Fig.3

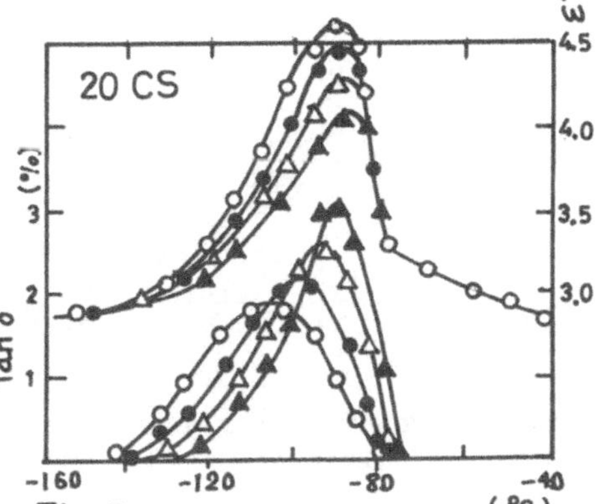

Fig.5

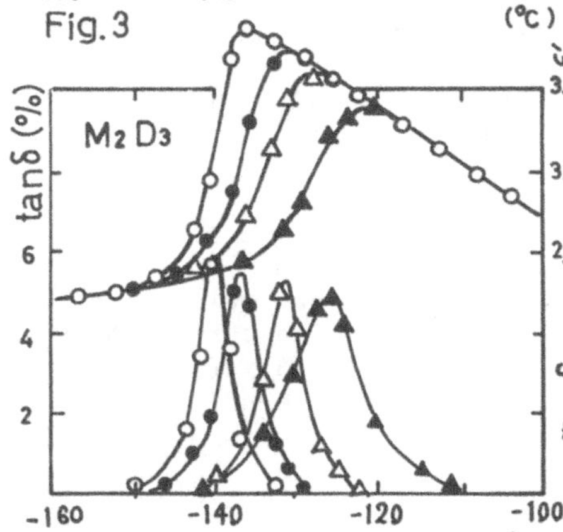

Fig.4 temperature (°C)

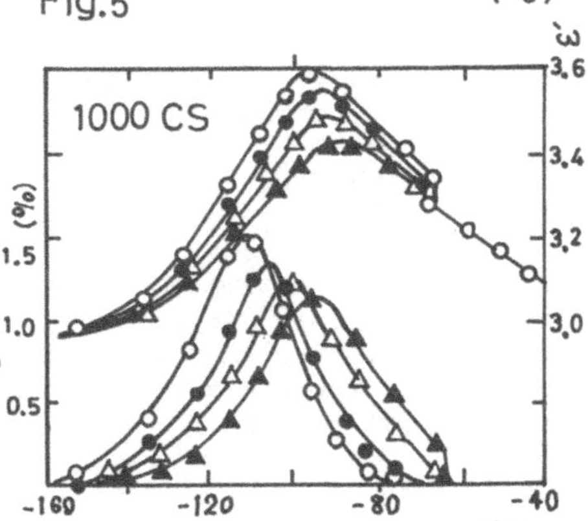

Fig.6 temperature (°C)

Fig.3 - Fig.6 The effect of frequency on the dielectric constant vs. temperature characteristic for M , M D , 20 CS, and 1000 CS silicone respectively at ○ 30 Hz; ● 1 KHz; △ 30 KHz; and ▲ 1 MHz.

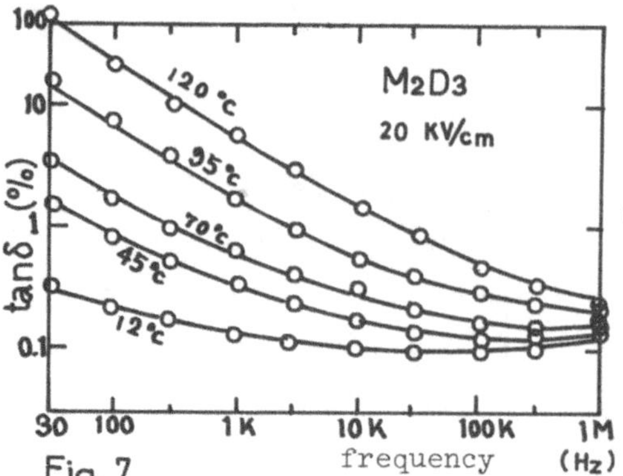

Fig. 7

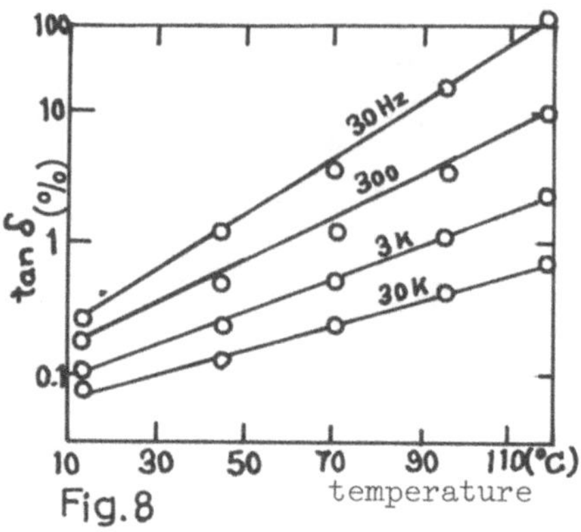

Fig. 8

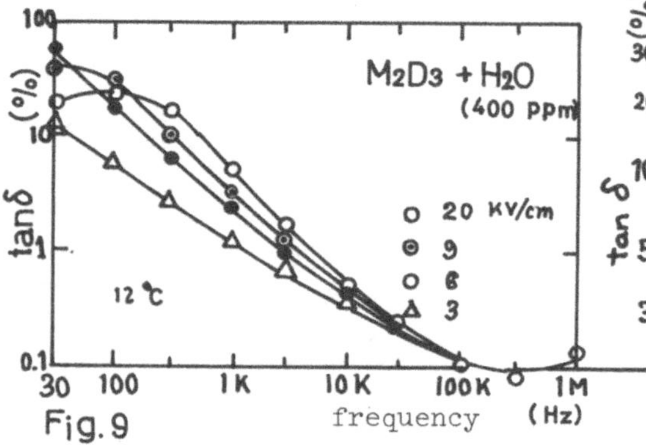

Fig. 9

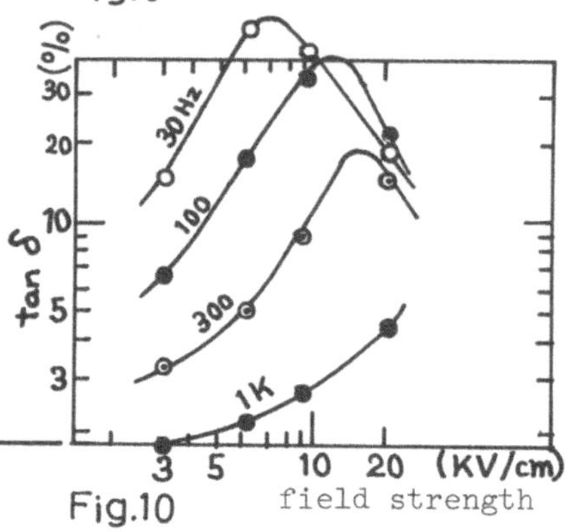

Fig.10

Fig.7 Power factor as a function of frequency for M D under the field strength of 20 KV/cm.

Fig.8 Power factor as a function of temperature.

Fig.9 Power factor as a function of frequency with field strength as a parameter for impure silicone liquid.

Fig.10 Power factor as a function of field strength.

carrier can be estimated by the assumption that the relaxation time is expressed simply as the time required for a carrier to move half of through the thickness of the liquid. A further analysis of the dielectric behavior in high field condition is under way at present.

References

1) D.W.Scott: J.Amer.Chem.Soc. 68 2294 (1946)
2) P.M.Sutton: The dielectric properties of glass,Progress in Dielectrics 2, London Heywood and Company, (1960)

ELECTRODISINTEGRATION OF DIELECTRIC LIQUIDS

P.G. Arnold and D. Michelson*

Introduction

The work is devoted to the study of the break up of the surface of dielectric liquids on the application of high electric fields. In view of the fact that at very high fields the break up of the liquid is very violent, the present study is restricted to the voltage range close to threshold where the discharge is stable and conditions are reproducible.

Experimental method

The dielectric liquid was supplied through a capillary tube which has a horizontal section and a vertical section drawn down to the required diameter and was a modified form of that used by Shorey and Michelson (Nucl. Inst & Meth 82(1970) 295-6). With zero field a small pendant drop is formed at the narrow tip of the capillary tube. On applying the voltage above the threshold value between a wire electrode inserted into the capillary tube and a horizontal plate placed 10^{-2} m from the tip, the pendant drop becomes conical and a fine spray of droplets is ejected. The base of this liquid cone is the same as that of the drawn down end of the capillary and had a diameter of 10^{-3} m in the experiments described below. The hydrostatic pressure P on the liquid cone could be controlled to within 5 Nm^{-2}. The height to base ratio k of the liquid cone was measured using a specially graduated eyepiece in the microscope. Photographs were also taken for different conditions. Measurements were made of the mass flow rates of the ejected liquid J, the applied voltage V, and the current carried by the liquid spray I. The variation of mass flow rates with pressure at constant voltage, the variation of the cone shape with voltage at constant hydrostatic pressure and the relation between the flow rate and the current were obtained for ethanol, methanol, n-propanol and n-butanol. Some experiments were also performed with formic acid acetic acid and water. The range of applied voltages was from 2.5 to 6 kV and the corresponding currents varied over the range 3nA to 1mA.

Results and Discussion

The variation of mass flow rate with the applied hydrostatic pressure for methanol is shown in figure 1. The line (a) corresponds to the flow characteristics of the tube under experimental conditions since this was obtained with V=0. On the application of the potential, the gradient of the line decreases and is independent of the voltage. This behaviour was typical for the alcohols. For the acids and water, however,

*Department of Physics, Brunel University, Kingston Lane, Uxbridge, Middx. UB8 3PH.

Figure 1

(a) V=0
(b) V=4.1 kV
(c) V=3.8 kV
(d) V=3.6 kV

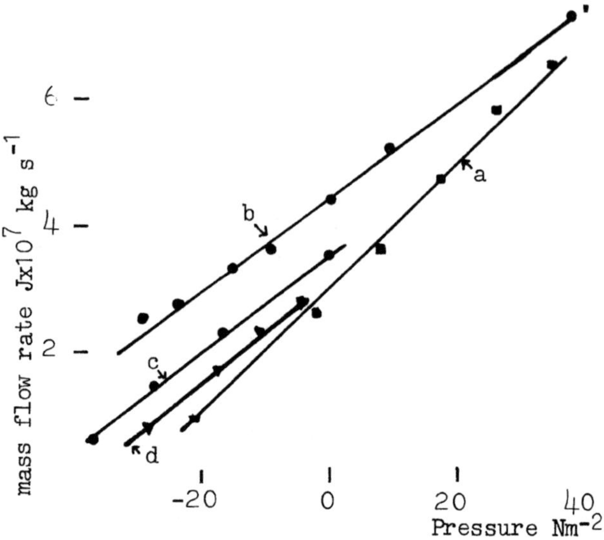

Figure 2

• n-butanol
o n-propanol
∇ ethanol
x methanol

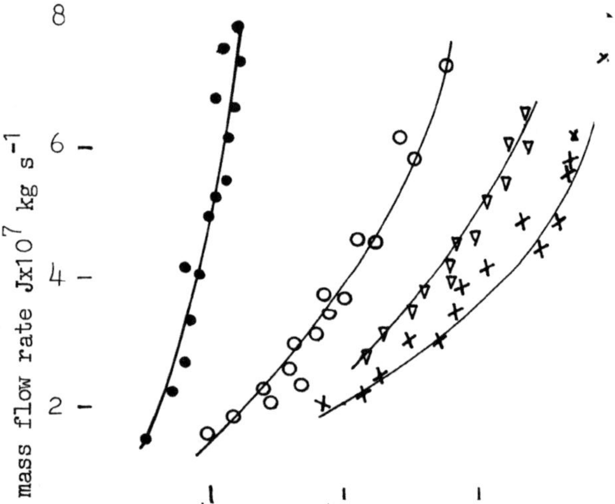

Figure 3

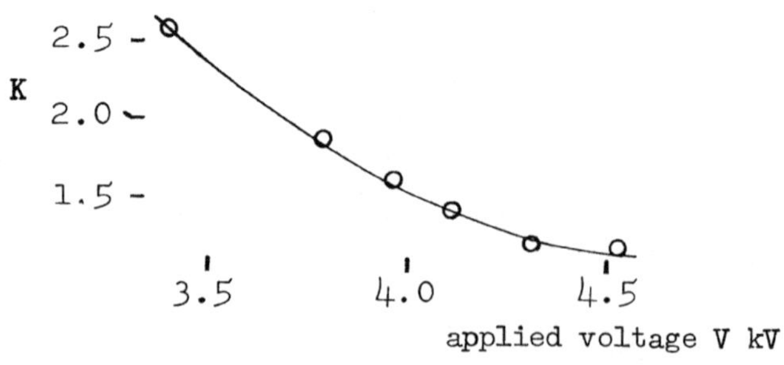

the gradient is greater and varies with the applied voltage. For the alcohols the gradient of the characteristics with applied voltage taken relative to the line (a) was found to vary linearly with the dielectric constant ε. For $\varepsilon < 12$ the difference in the gradients was less than the experimental error.
The variation of mass flow rate J with current for the alcohols is shown in figure 2. It should be stressed that these curves were obtained with the fluid issuing from the apex of the cone under steady conditions. At higher voltages when there was a tendency for the discharge to change to a sputtering mode the drop size increased and this was accompanied by a decrease in current by more than 50%. The analysis of the curves in figure 2 also shows a systematic variation with dielectric constant. The behaviour of acids and water is different from that of the alcohols. The flow rate is much more sensitive to the hydrostatic pressure. It is much smaller and the corresponding currents are much higher than in the alcohols. It is thus more difficult to maintain a stable cone.
Figure 3 shows the ratio k as a function of voltage at constant pressure. As the voltage increases the cone becomes more obtuse and the current increases.
A high degree of turbulence was observed in the fluid cone on switching on the field. This is consistent with the turbulence observed by Hopfinger and Gosse (Physics of Fluids 14(1971)1671).

Preliminary calculation of the forces on the fluid and also the difference in behaviour between the acids, water and alcohols indicate that the conditions at the surface cannot be neglected when considering the break up mechanism.

The authors are indebted to the Science Research Council for the support of the work.

THE FARADAY - SUMOTO EFFECT

J. S. Mirza*, M. S. Zafar* and C. W. Smith**

Introduction

The effect was discovered by Faraday[1] in 1836 when he found that if two high voltage leads were dipped into turpentine, the liquid climbed along the anode to form a sheath several inches in length.
In 1926, Gyemant[2] investigated the rise of oil into the gap of a concentric spheres capacitor and found that the rise of the liquid was proportional to the square of the applied voltage. The effect received further attention in 1955 when Sumoto[3] re-discovered it. Mirza and Smith[4,5] have given an expression for the rise of a dielectric liquid under such conditions, based on the mechanical stresses at an electrically stressed surface. The present work extends previous measurements with concentric cylinder electrodes to parallel plate and separated equal-radii parallel-cylinder electrodes. The results are compared with the theory.

Experimental

The electrodes were made by cutting a brass rod of diameter 12.0 mm and length 75 mm along a diameter and parallel to its axis to give two equal segments, all the corners were radiused and the surfaces well polished. The segments were thoroughly cleaned, rinsed in water, dried and finally rinsed in the liquid to be used for the experiments.
The two segments were mounted with their axes parallel and the electrode surfaces separated by glass spacers of known thickness. Each end of the assembly was held in an ebonite clamp and the whole was suspended from metal rods screwed into the end of each electrode so as to stand them vertically in a 100 ml glass beaker which was then half filled with the liquid for the experiment.
A travelling microscope was used to determine the level of the liquid at the electrodes. The zero level of reference for these measurements was taken to be the reading on the travelling microscope set at the liquid level determined by surface tension in the absence of an applied electric field.
The concentric cylinder electrodes comprised an outer cylinder of sheet metal with a narrow slit parallel to its axis for observing the rise of the liquid up the central rod which formed the other electrode.
A calibrated high voltage supply was connected between the electrodes and the voltage slowly raised in a stepwise manner

*Department of Physics, New Campus, University of the Punjab, Lahore, Pakistan.
**Department of Electrical Engineering, University of Salford, Salford M54WT, Lancs, England.

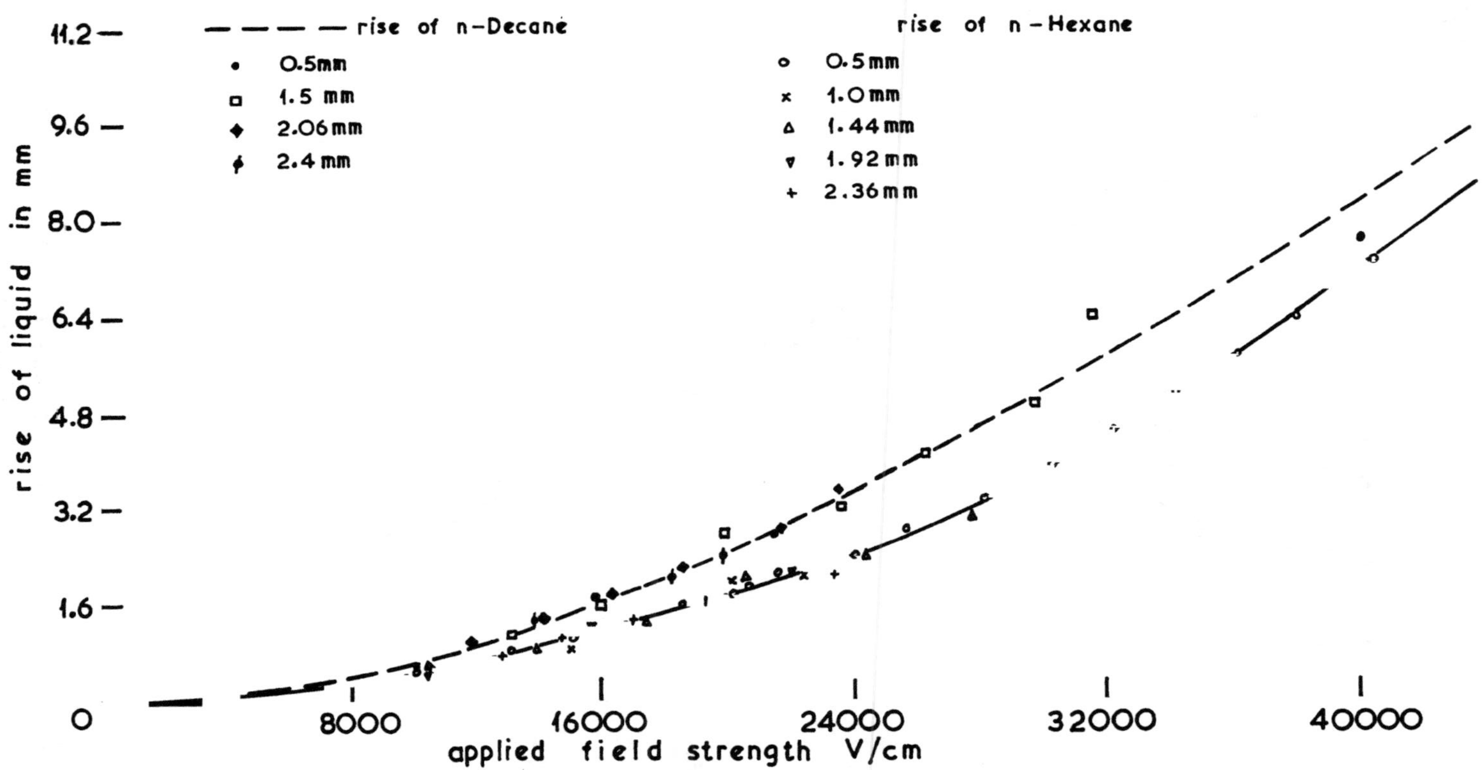

rise of n-Decane
0.5mm
1.5 mm
2.06mm
2.4 mm
rise of n-Hexane
0.5mm
1.0mm
1.44mm
1.92mm
2.36mm
11.2
9.6
8.0
6.4
4.8
3.2
1.6
0
8000
16000
24000
32000
40000
applied field strength V/cm
rise of liquid in mm

Figure 1

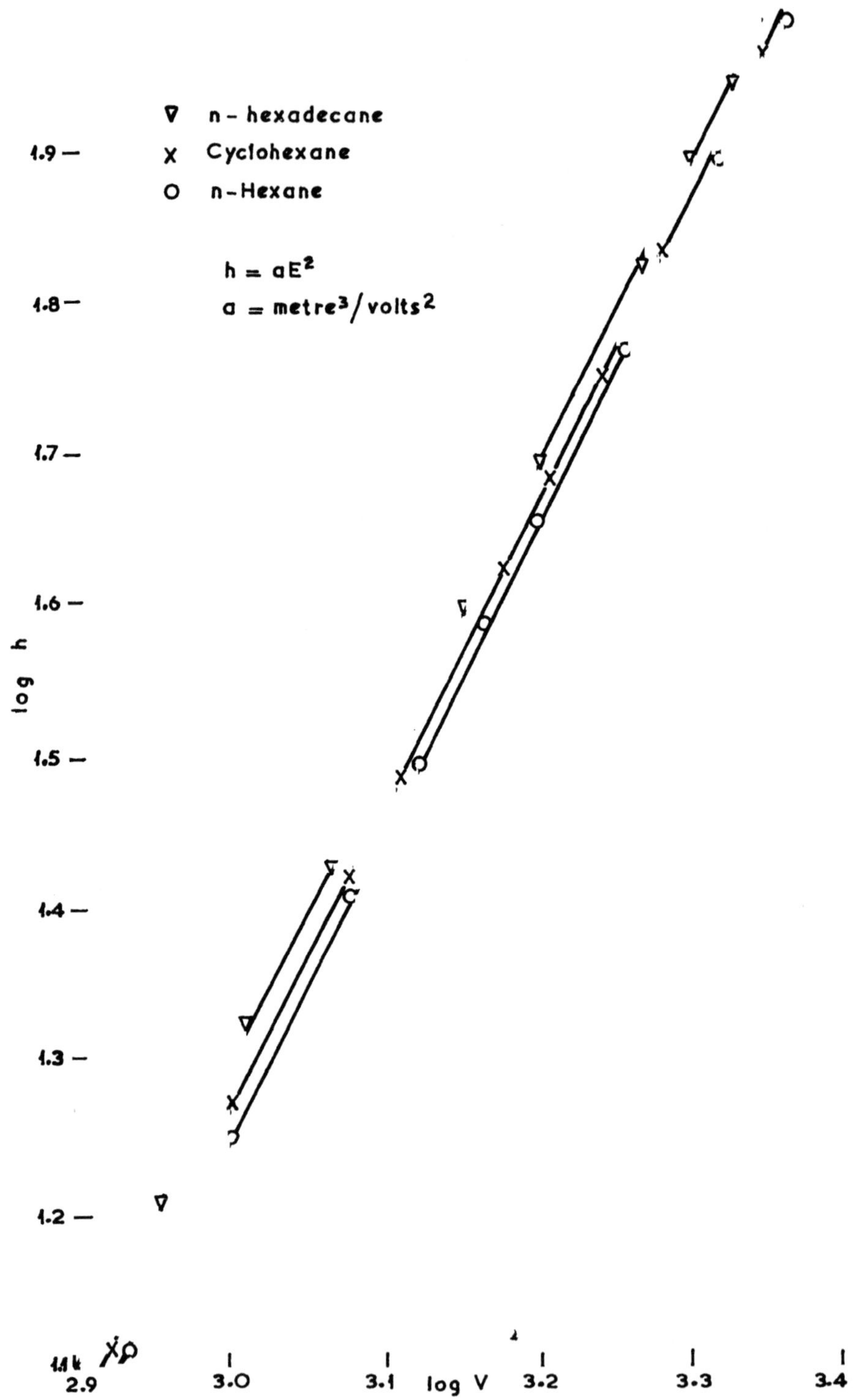

n - hexadecane
Cyclohexane
n-Hexane
h = aE²
a = metre³/volts²
1.9
1.8
1.7
1.6
1.5
1.4
1.3
1.2
log h
2.9
3.0
3.1
3.2
3.3
3.4
log V

Figure 2

with these readings taken at each setting. Five different thicknesses of spacer were used having the values shown in figure 1 for those particular measurements with n-hexane and n-decane. Enhanced evaporation of the liquid was observed when the high voltages were applied[5], this resulted in a slight drift in the zero reference level for which correction was made.

Readings were taken as the voltage was both raised and lowered. The latter readings were slightly higher in value due to a thin film of the liquid left "tailing" on the electrodes until it evaporated.

Results

Figure 2 shows the results of measurements on n-hexane, n-hexadecane and cyclohexane plotted logarithmically. They can be expressed in the empirical form

$$h = a\,E^{2 \pm 0.1}$$

In figure 2, h is the rise in liquid level (in millimetres x 50) and V is the applied voltage. The field $E = V \times 10^3/1.26$ volts/metre. The constant a has the following values:
n-hexane $a = 5.1 \times 10^{-16}\ m^3/V^2$
n-hexadecane $a = 5.7 \times 10^{-16}\ m^3/V^2$
cyclohexane $a = 5.6 \times 10^{-16}\ m^3/V^2$.

The theoretical equation[4,5] for the rise of a liquid dielectric due to mechanical stresses produced by the electric field gives the constant

$$a = \frac{1}{2\rho g}(\varepsilon - \varepsilon_o)(\mathrm{Sin}^2\theta + \frac{\varepsilon_o}{\varepsilon}\mathrm{Cos}^2\theta)\,\mathrm{Sin}\,\theta$$

where ε is the permittivity of the dielectric liquid and ρ is its density, ε_o is the permittivity of the medium above the liquid (assumed to be free space), $g = 9.81\ m/sec^2$ and θ is the angle of contact between the liquid surface and the electrode. Although separate experiments to determine the angle of contact on the material of the electrodes gave $\theta = 3^o$, the above values of 'a' seem to require a value of θ which is close to 90°. Further work will be necessary to resolve this discrepancy.

References

1. Faraday, M., 1838, Phil. Trans. 128, 125.
2. Gyemant, A., 1926, Wise. Veroff. Siem. Konz. 5, No. 2, 55.
3. Sumoto, I., 1956, Oyo Butsuri 25, 264.
4. Mirza, J.S. and Smith, C.W., 1970, Physics Exhibition Handbook, (London: Institute of Physics), 69.
5. Mirza, J.S., Smith, C.W. and Calderwood, J.H., 1974, Proc. 43rd, Conf. Elect. Insul. and Dielect. Phen. (Washington DC: N.A.S./N.R.C.), (to be published).

PULSE GENERATORS WITH LIQUID DIELECTRIC GAPS

N.Shammas* C.W. Smith* and J.H.Calderwood*

Introduction

Previous work on electrical breakdown in insulating liquids had shown that breakdown in n-hexane could occur in less than nano-second (Smith and Calderwood, 1968). The present work was carried out in order to provide a simple pulse generator which could be charged from the high voltage, low current, power supplies already available and which would give high voltage, nanosecond pulse suitable for investigating the extent to which the various prebreakdown phenomena could develop in this time interval. Two methods were tried, the three electrode "trig-atron" arrangement with the triggered breakdown occuring in n-hexane; also the use of a metal sphere moving in n-hexane between two electrodes one of which was kept at a high voltage.

The 'Trigatron'

Figure 1 shows the experimental arrangement of the electrodes used to determine the conditions under which the triggering gap d_2 and the main gap d_1 would breakdown. The apparatus consisted of a glass tube approximately 40 mm diameter and 120mm long, with two side tubes 10mm diameter and 70mm long. The electrodes were brazed to brass rods and subsequently the tungsten electrodes were etched to a point of the order of a micron radius; the positions of the electrodes could be adjusted by screwing the rods in or out. The gap lengths were measured to 0.1mm with a travelling microscope.
Figure 2 shows the triggered breakdown characteristics for n-hexane, for example if the main gap length is set to 5 mm and $V_1 \geqslant 34$ kV and the trigger gap is set to 0.5mm then triggered breakdown will occur if $V_2 \geqslant 15$ kV. Figure 2 gives the characteristics for negative V_1. With the spherical electrode positive they are the same for d_1 less than 3mm, for d_1 greater than 3mm the values of V_1 are approximately 20% higher.
The n-hexane could be used repeatedly until discoloured without significant change in the characteristics. The main gap was in the region of the electrohydrodynamic motion and "spray" (Singh et al, 1972) with the electrodes at the angles shown in figure 1. Moving the lower electrode above the trigger gap and clear of this region, only raised V_1 by 4%.
Capacitors were charged to the value of V_2 appropriate to the other quantities in figure 2 and then connected across the trigger electrodes; energies from 640 mJ to 1 mJ would trigger the main gap provided the voltage was correct. The energy stored in the capacitance between the trigger electrodes at their breakdown voltage was estimated to be 0.1 mJ.

*Department of Electrical Engineering, University of Salford, Salford M5 4WT England

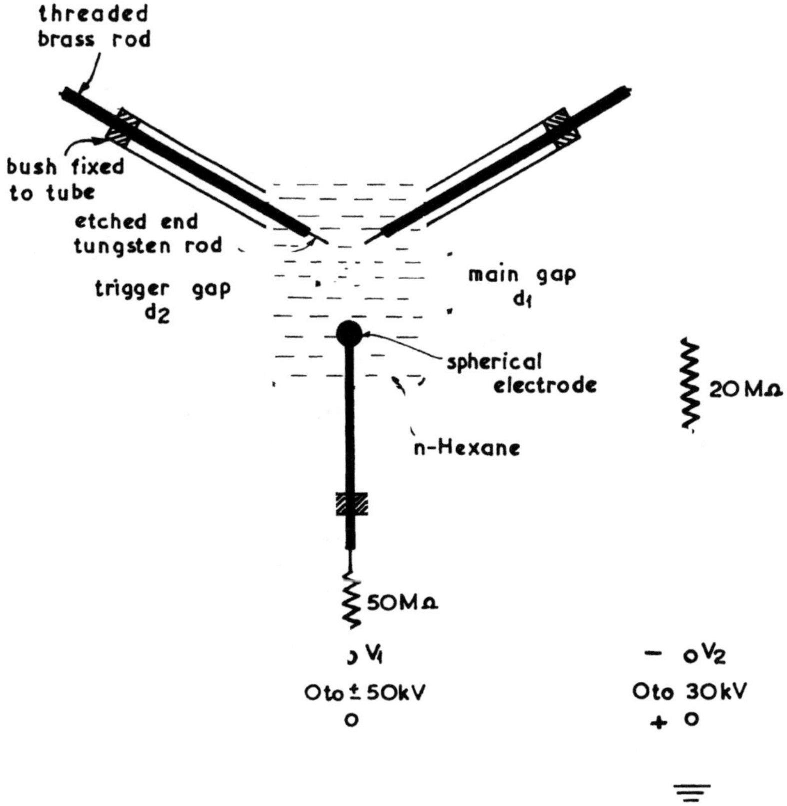

Electrode assembly and test cell showing circuit used.

Figure 1

Moving Sphere Pulse Generator

Figure 3 shows this pulse generator and a test cell used to investigate corona discharges. The high pressure nitrogen used by Gunn (1966) is replaced by a liquid insulator (n-hexane). The outer conductors were 1" (copper) water piping and fittings, the inner conductors were UR 70 cable with the braiding removed and the two fixed spheres brazed in position. The three spheres were kept in alignment by a cage of PTFE. When the high voltage was applied, the moving central sphere oscillated between the fixed spheres generating a nanosecond pulse each time it discharged. Single pulses could be obtained using a steel moving ball which was controlled by an external electromagnet.

References

Smith, C W, and Calderwood, J H, 1968, Proc.Int.Conf.Phén.Cond. Isol, 161-189, (CNRS:Paris).
Singh, B, Chadband, W G, Smith, C W, and Calderwood, J H, 1972, J.Phys.D:Appl.Phys, 5, 1457-1464.
Gunn, J B, Electronics Letters, 1966, 2, 172.

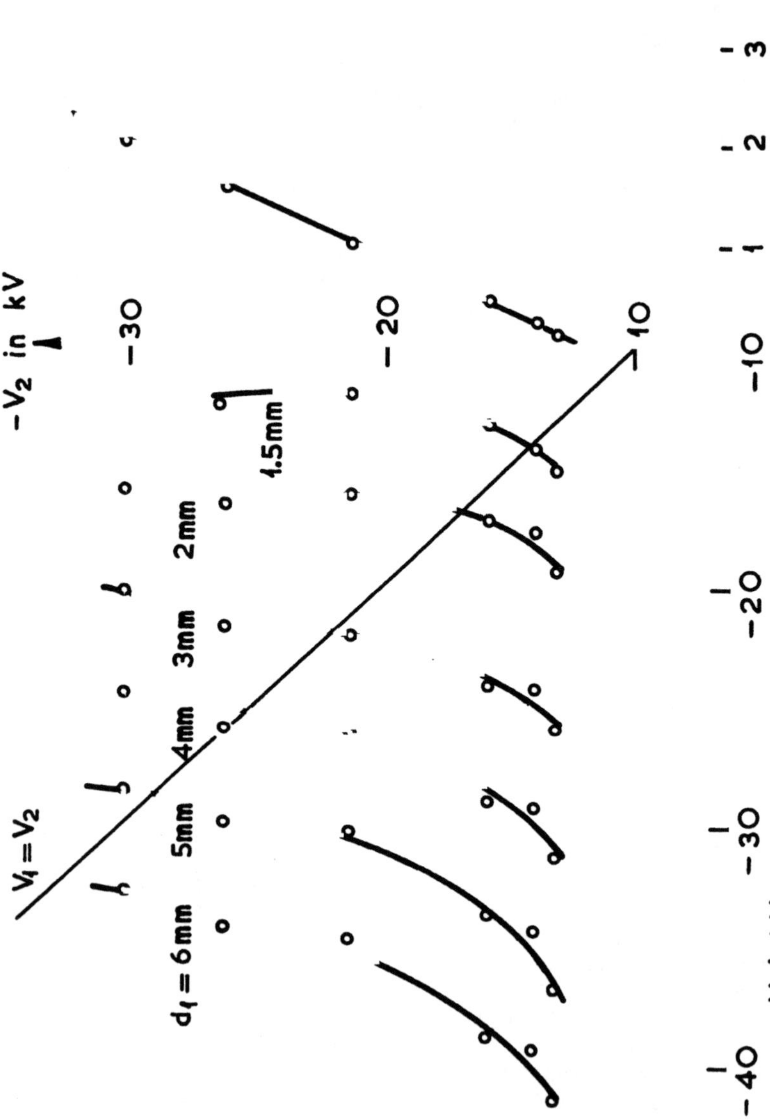

$V_1 = V_2$
$d_1 = 6mm$
5mm
4mm
3mm
2mm
4.5mm
$-V_2$ in kV
−30
−20
10
V_1 in kV
−40
−30
−20
−10
d_2 in mm
1
2
3

Figure 2

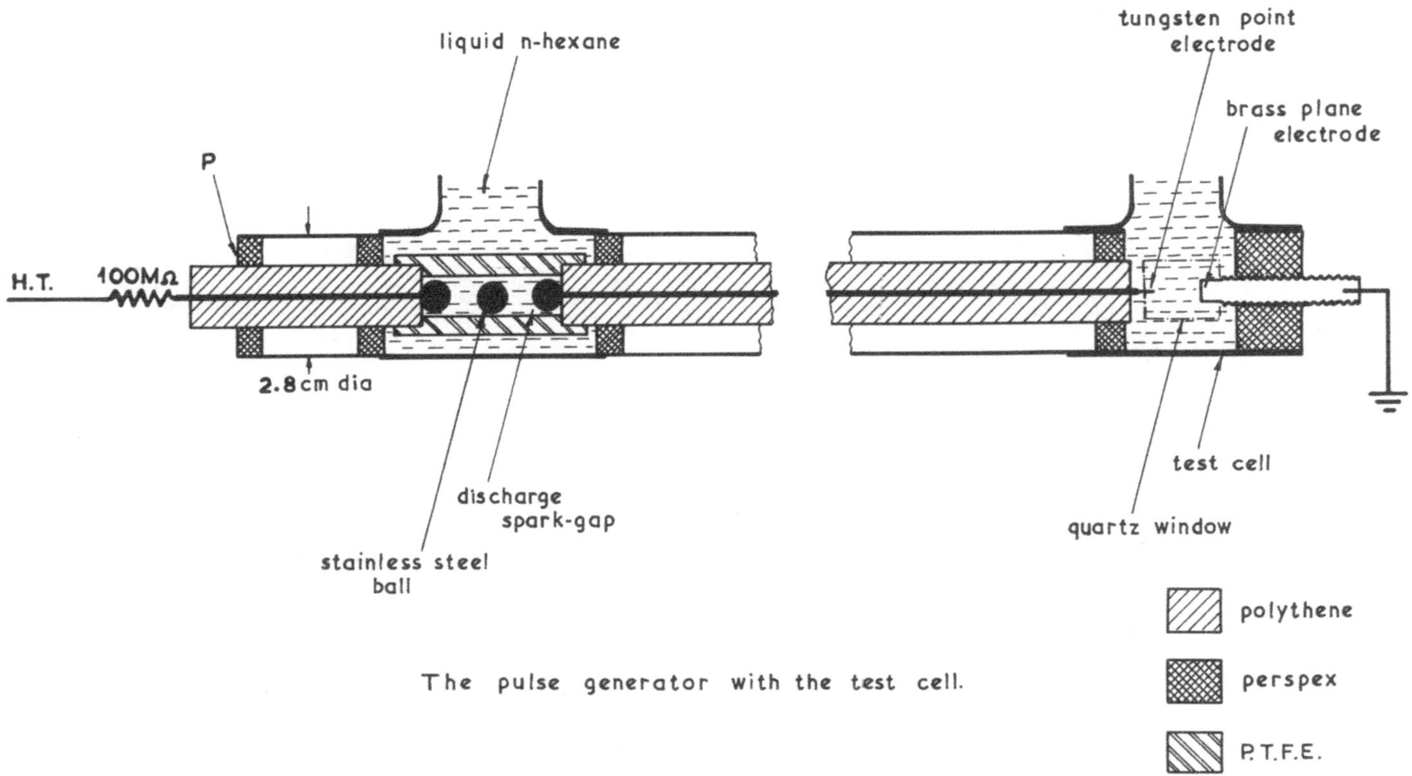

The pulse generator with the test cell.

GPSR Compliance

The European Union's (EU) General Product Safety Regulation (GPSR) is a set of rules that requires consumer products to be safe and our obligations to ensure this.

If you have any concerns about our products, you can contact us on ProductSafety@springernature.com

In case Publisher is established outside the EU, the EU authorized representative is:

Springer Nature Customer Service Center GmbH
Europaplatz 3
69115 Heidelberg, Germany

Batch number: 10620141

Printed by Printforce, the Netherlands